The Optical Constants of Bulk Materials and Films

Second Edition

The IOP series on Optics and Electronics

Series Editors: **E R Pike FRS, B E A Saleh, S Lowenthal** and **W T Welford FRS** (deceased)

Other books in the series

Aberrations of Optical Systems
W T Welford

Cutting and Polishing Optical and Electronic Materials
G W Fynn and W J A Powell

Ferroelectric Crystals for Laser Radiation Control
A M Prokhorov and Yu S Kuz'minov

The Fabry-Perot Interferometer
M Vaughan

Infrared Optical Fibres
T Katsuyama and H Matsumura

Interferometry of Fibrous Materials
N Barakat and A A Hamza

KDP – Family Single Crystals
L N Rashkovich

Laser Analytical Spectrochemistry
edited by V S Letokhov

Laser Damage in Optical Materials
R Wood

Laser Heating of Metals
A M Prokhorov, V I Konov, I Ursi and I N Mihailescu

Laser Picosecond Spectroscopy and Photochemistry of Biomolecules
V S Letokhov

Laser Resonators and the Beam Divergence Problem
Y A Anan'ev

The Optical Constants of Bulk Materials and Films – 2nd Edition
L Ward

Physics and Chemistry of Crystalline Lithium Niobate
A M Prokhorov and Yu S Kuz'minov

Prism and Lens Making
F Twyman

Waves in Focal Regions
J J Stamnes

Optics and Optoelectronics Series

The Optical Constants of Bulk Materials and Films

Second Edition

L Ward

Honorary Research Fellow
Coventry University

Institute of Physics Publishing,
Bristol and Philadelphia

British Library Cataloguing in Publication Data

A catalogue record for this book is available from the British Library

ISBN 0 7503 0324 7

Library of Congress Cataloging-in-Publication Data

Ward, L. (Leslie)
The optical constants of bulk materials and films / L. Ward. —
2nd ed.
p. cm. — (Optics and optoelectronics series)
Includes bibliographical references and index.
ISBN 0-7503-0324-7 : $140.00
1. Thin films—Optical properties. 2. Materials—Optical properties. 3. Optical constants. I. Title. II. Series.
QC176.84.O7W37 1994
620.1′1295—dc20 94-13114
CIP

First edition published 1988 by Adam Hilger Ltd
Second edition published 1994 by IOP Publishing Ltd

Published by IOP Publishing Ltd, a company wholly owned by the Institute of Physics, London
Techno House, Redcliffe Way, Bristol BS1 6NX, England
US Editorial Office: The Public Ledger Building, Suite 1035, Independence Square, Philadelphia, PA 19106, USA

Typeset by Mathematical Composition Setters Ltd, Salisbury
Printed in Great Britain by J W Arrowsmith Ltd, Bristol

Contents

Series Editors' Preface

Optics has been a major field of pure and applied physics since the mid 1960s. Lasers have transformed the work of, for example, spectroscopists, metrologists, communication engineers and instrument designers in addition to leading to many detailed developments in the quantum theory of light. Computers have revolutionised the subject of optical design and at the same time new requirements such as laser scanners, very large telescopes and diffractive optical systems have stimulated developments in aberration theory. The increasing use of what were previously not very familiar regions of the spectrum, e.g. the thermal infrared band, has led to the developments of new optical materials as well as new optical designs. New detectors have led to better methods of extracting the information from the available signals. These are only some of the reasons for having an *Adam Hilger Series on Optics and Optoelectronics.*

The name Adam Hilger, in fact, is that of one of the most famous precision optical instrument companies in the UK; the company existed as a separate entity until the mid 1940s. As an optical instrument firm Adam Hilger had always published books on optics, perhaps the most notable being Frank Twyman's *Prism and Lens Making.*

Since the purchase of the book publishing company by The Institute of Physics in 1976 their list has been expanded into all areas of physics and related subjects. Books on optics and quantum optics have continued to comprise a significant part of Adam Hilger's output, however, and the present series has some twenty titles in print or to be published shortly. These constitute an essential library for all who work in the optical field.

Preface to the Second Edition

The opportunity has been offered to me to produce a new edition of *The Optical Constants of Bulk Materials and Films* and this has enabled me to make several changes from the first edition.

In the first place, the chance has been taken to amend at least some of the errors, mostly algebraic, which had crept into the text of the first edition.

Since 1988, the spate of publications in this field of optics has not slackened and an attempt has been made to incorporate into the new edition some of the relevant information which has been published in the last five years. In particular, sections have been added on the optical properties of the sulphides, selenides and tellurides of cadmium, zinc and lead, all of which have important uses in optics; two other new sections deal with (*a*) low dimensional structures and (*b*) the sensitivities of different methods for measuring n, k and d/λ.

At the Editor's suggestion, a completely new chapter has been added dealing with the applications of thin films in the construction of filters, beam splitters and antireflection coatings.

Finally, a table has been included in the form of an appendix giving values of n, k, e_1 and e_2 for a large number of elements, insulators and semiconductors at two common laser wavelengths (488 and 633 μm).

I would like to express my thanks to the following organisations for permission to reproduce certain additional figures in this second edition: Academic Press Inc. (San Diego) (figures 4.10(*b*) and 4.19(*a*, *b*)); American Institute of Physics (figure 4.22); Butterworth Heinemann (Oxford) (figure 1.2(*a*, *b*)); Pergamon Press (Oxford) (figure 8.9).

Preface to the First Edition

The study of the optical constants of materials is interesting for a number of reasons, of which two are particularly relevant. Firstly, the use of many kinds of materials in optical applications such as interference filters, optical fibres and reflective coatings requires accurate knowledge to be available of their optical constants over wide ranges of wavelengths. Secondly, the optical properties of all materials may be related to their atomic structure and especially to their electronic band structure and electrical properties. This aspect of the study of optical constants thus relates to fundamental theory, and the values of the optical constants, together with their variation with frequency, enable correlation and corroboration to be made with the band structures derived by other methods.

Although the phenomenon of refraction has been known for a very long time, the optical study of materials having complex refractive indices dates back only about a century. However, during that period a considerable amount of information has been accumulated on the experimental techniques, the values of the optical properties themselves and the theories which have evolved to attempt to explain them.

The aim of this book is to present a general account of this work and to cover a wide range of materials. The format is to survey the theories, experimental techniques and results of studies of the optical properties of materials over a wide range of the electromagnetic spectrum between soft x-rays and microwaves. It is convenient to divide the materials into two main groups: (i) insulators (dielectrics), where, over most of the radiation spectrum, the absorption of electromagnetic radiation is small and a wide range of optical properties can be defined in terms of a real refractive index, n; and (ii) metals, where a complex refractive index, $n-\mathrm{i}k$, is needed because of the strong absorption of electromagnetic waves. Another useful division, which will be followed in the layout of the book, is between bulk materials, continuous thin films and discontinuous films; the boundaries between these categories, however, are quite arbitrary and the properties of one of these groups may well be present in the others.

The optical properties of a solid material are described in terms of electromagnetic theory and solid state physics, both of which topics can, and do, fill volumes much larger than this one. It is therefore the intention to discuss such topics only to a point sufficient for the understanding of the main items in the remainder of the book and to refer the reader to the standard works if a fuller treatment is required.

The level of treatment in this book is aimed at postgraduate students and postdoctoral workers in this field.

A large number of other books cover at least parts of my chosen field. In particular, I would like to mention Sokolov, *Optical Properties of Metals* (Blackie, 1967), Palik, *Handbook of Optical Constants of Solids* I and II (Academic, 1985, 1991), the American Institute of Physics Handbook (1976), Heavens, *Optical Properties of Thin Solid Films* (Dover, 1965), Ditchburn, *Light* (Academic, 1976) and Born and Wolf, *Principles of Optics* (Pergamon, 1975) which are all good source books.

In addition, I wish to express my thanks to the following who have kindly given their permission for me to use illustrations in this book: Academic Press, Orlando, Florida (figures 1.4, 1.6, 4.1(*a*), 4.5, 4.13, 4.15, 4.16, 4.18 and 4.21); Butterworth Scientific Ltd, Guildford (figures 5.1, 6.6, 6.8, 6.9 and 6.10); IOP Publishing Ltd, Bristol (figures 2.11, 4.2, 6.1, 6.2, 6.4, 6.5, 6.11, 6.12 and 8.3); The American Institute of Physics (figures 1.5, 4.6, 4.9, 4.10(*a*), 4.18, 7.1(*b*), 7.2. 7.3, 7.4, 7.15, 7.16, 8.5, 8.10, 8.11, 8.14 and 8.15); Journal of the Optical Society of America (figures 2.12, 4.4(*a*, *b*), 7.1(*a*), 7.5, 7.6, 7.7, 7.8, 7.11, 7.12, 7.13, 7.14, 7.17, 8.12 and 8.13); Springer Verlag, Heidelberg (figure 7.9); Taylor and Francis Ltd, London (figure 2.9); Chartwell-Bratt Ltd, Bromley and Studentlitteratur, Lund (figure 1.9); North-Holland Publishing Co., Amsterdam (figures 4.3(*a*, *b*) and 4.14); Fachinformationszentrum, Karlsruhe (figures 4.11 and 4.12); Elsevier Sequioa (figure 7.10) Akademie Verlag, Berlin (figures 8.2 and 8.8); Pergamon Press (figure 8.7); Cambridge University Press (figure 3.1); Physical Society of Japan (figure 5.2).

Finally, I would like to acknowledge the extensive help which I have received in computer and library facilities from Coventry Polytechnic.

L Ward

List of Symbols

a	mean diameter of film island
b	height of film island
$\boldsymbol{B}$	magnetic flux density vector
c	speed of electromagnetic radiation in free space
$\boldsymbol{D}$	electric displacement vector
d	film thickness
$\boldsymbol{E}$	electric field vector
e	charge on electron
$\boldsymbol{F}$	island shape factor, including interactions
f_j	oscillator strength
f	island shape factor, excluding interactions
$\boldsymbol{G}$	Poynting's vector
$\boldsymbol{H}$	magnetic field vector
h	Planck's constant
j	quantum number
K	wavenumber $= 2\pi/\lambda$
k	imaginary part of complex refractive index
l	mean free path
m	mass of electron
m^*	effective mass of electron
m_0	optical mass of electron
N	number of electrons per unit volume
n	real part of complex refractive index
$\tilde{n}$	complex refractive index
n_0	refractive index of free space
n_1, n_2	refractive index of medium
n_s	refractive index of substrate
n_e	refractive index, extra-ordinary ray
n_o	refractive index, ordinary ray
$\boldsymbol{P}$	electrical polarisation
p	wave with electric vector in plane of incidence

q	volume fraction of film
R_s	reflectance of s-polarised radiation
R_p	reflectance of p-polarised radiation
r	amplitude reflection coefficient
s	wave with electric vector at right angles to plane of incidence
T_s	transmittance of s-polarised radiation
T_p	transmittance of p-polarised radiation
U	photon energy
U_F	Fermi energy
u	real part of $\tilde{n}\cos\theta$
v	imaginary part of $\tilde{n}\cos\theta$
x	optical thickness $= 2\pi d/\lambda$
Y_0	admittance of free space
Y_1	admittance of medium
Y_s	admittance of substrate
Z_0	impedance of free space
Z_1	impedance of medium
Z_s	impedance of substrate
α	polarisability
β	restoring force per unit displacement for electron
γ	electron relaxation frequency
Δ	phase difference between s and p components of light
δ	optical path length $= 2\pi d \cos\theta/\lambda$
ε_0	permittivity of free space
ε_r	relative permittivity or dielectric constant
ε_1	real part of complex dielectric constant, ε
ε_2	imaginary part of complex dielectric constant, ε
ε_3	$(n \sin\theta)^2$
θ_i, θ_1	angle of incidence
θ_r, θ_2	angle of refraction
θ_{pr}	principal angle of incidence
λ	wavelength
μ_0	permeability of free space
μ_r	relative permeability
ρ	density of free charges
ρ_1, ρ_2	amplitude of reflected wave
σ	electrical conductivity
τ	amplitude transmission coefficient
τ_c	collision relaxation time of electron
ϕ	phase change on reflection
χ	phase change on transmission
Ψ	azimuth
Ψ_{pr}	principal azimuth
ω	angular frequency

ω_0	angular frequency of bound electron
ω_p	plasma frequency
ω_s	angular frequency of surface plasma
μ	absorption coefficient

Values of constants

ε_0	$8.8542 \times 10^{-12}\ \mathrm{F\,m^{-1}}$
μ_0	$4\pi \times 10^{-7}\ \mathrm{H\,m^{-1}}$
h	$6.6262 \times 10^{-34}\ \mathrm{J\,s} = 4.1357 \times 10^{-15}\ \mathrm{eV\,s}$
e	$1.6022 \times 10^{-19}\ \mathrm{C}$
m	$9.1095 \times 10^{-31}\ \mathrm{kg}$
c	$2.9979 \times 10^{8}\ \mathrm{m\,s^{-1}}$
Y_0	$0.00266\ \Omega^{-1}$
Z_0	$377\ \Omega$
1 eV	$1.6022 \times 10^{-19}\ \mathrm{J} \equiv 1.2399\ \mu\mathrm{m}$

1 Theoretical Background

1.1 Introduction

This first chapter is used to develop expressions for the optical functions of reflectance, transmittance and the phase changes on reflection and transmission at the junction between two media, and also to give the background to the electron theories of materials such as the free and bound electron theories and the band theory of solids.

A basic knowledge of Maxwell's electromagnetic theory is needed in order to derive the expressions for the optical functions. However, only an outline account will be given here, sufficient to enable the remaining arguments in this book to be followed. Many books (e.g. Ditchburn 1976, Heavens 1965, Born and Wolf 1975) give much more detailed derivations of this theory and the reader is referred to them if a treatment in greater depth is needed.

The four laws of electromagnetism can be stated as follows:

$$\text{div } \boldsymbol{D} = \varepsilon_0\varepsilon_\mathrm{r} \text{ div } \boldsymbol{E} = \rho \tag{1.1}$$

$$\text{div } \boldsymbol{B} = \mu_0\mu_\mathrm{r} \text{ div } \boldsymbol{H} = 0 \tag{1.2}$$

$$\text{curl } \boldsymbol{E} = -\mu_0\mu_\mathrm{r}\frac{\partial \boldsymbol{H}}{\partial t} = -\frac{\partial \boldsymbol{B}}{\partial t} \tag{1.3}$$

$$\text{curl } \boldsymbol{H} = \sigma\boldsymbol{E} + \frac{\partial \boldsymbol{D}}{\partial t} = \varepsilon_0\varepsilon_\mathrm{r}\frac{\partial \boldsymbol{E}}{\partial t} + \sigma\boldsymbol{E} \tag{1.4}$$

where $\boldsymbol{D}$ is the electric flux density, or displacement due to the electric field, $\boldsymbol{E}$, $\boldsymbol{B}$ is the magnetic flux density resulting from magnetic field, $\boldsymbol{H}$, ε_0 is the permittivity of free space, μ_0 is the permeability of free space, ε_r is the relative permittivity of the medium (dielectric constant), μ_r is the relative permeability of the medium, ρ is the charge density other than that due to polarisation, σ is the electrical conductivity and t is time. $\boldsymbol{D}$ and $\boldsymbol{B}$ can be eliminated in turn from these equations and, for the situation when $\rho = 0$, this leads to the following two expressions:

$$\nabla^2 \boldsymbol{E} = \mu_0\mu_r\varepsilon_0\varepsilon_r \frac{\partial^2 \boldsymbol{E}}{\partial t^2} + \mu_0\mu_r\sigma \frac{\partial \boldsymbol{E}}{\partial t} \tag{1.5}$$

$$\nabla^2 \boldsymbol{H} = \mu_0\mu_r\varepsilon_0\varepsilon_r \frac{\partial^2 \boldsymbol{H}}{\partial t^2} + \mu_0\mu_r\sigma \frac{\partial \boldsymbol{H}}{\partial t}\,. \tag{1.6}$$

Both these equations are forms of the general equation for a damped wave; $\boldsymbol{E}$ and $\boldsymbol{H}$ are thus propagated through the medium as waves. It may be shown that the vibrations of $\boldsymbol{E}$ and $\boldsymbol{H}$ are transverse to the direction of propagation and are also in directions at right angles to each other.

1.2 Behaviour of insulators

The easiest case to consider is that when $\sigma = 0$, i.e. for a non-conducting or dielectric medium. Equations (1.5) and (1.6) then reduce to

$$\nabla^2 \boldsymbol{E} = \mu_0\mu_r\varepsilon_0\varepsilon_r \frac{\partial^2 \boldsymbol{E}}{\partial t^2} \tag{1.7}$$

$$\nabla^2 \boldsymbol{H} = \mu_0\mu_r\varepsilon_0\varepsilon_r \frac{\partial^2 \boldsymbol{H}}{\partial t^2}\,. \tag{1.8}$$

By comparison, the general wave equation for a disturbance, Q, in a non-absorbing medium may be written in the form

$$\frac{\partial^2 Q}{\partial z^2} = \frac{1}{v^2}\frac{\partial^2 Q}{\partial t^2}\,. \tag{1.9}$$

Here, v is the speed of the wave in the medium and is equal to the product of the wavelength, λ, and the frequency, f. Thus, from equation (1.7), the variation of the electric field vector in the x-direction, E_x, for a wave propagated in the z-direction can be written in the form

$$E_x = E_0 \exp[\mathrm{i}(\omega t - Kz)] = E_0 \exp[2\pi \mathrm{i}/\lambda(vt - z)] \tag{1.10}$$

where ω ($= 2\pi f$) is the angular velocity or angular frequency, K ($= 2\pi/\lambda$) is the wavenumber and E_0 is the amplitude or maximum displacement of the wave. This solution may be readily confirmed by appropriate differentiation of equation (1.10) and substitution in equation (1.7).

In the same way, from equation (1.8), the variation of the magnetic vector in the x-direction is given by the equation

$$H_x = H_0 \exp[\mathrm{i}(\omega t - Kz)]\,. \tag{1.11}$$

Similar equations may be obtained for the electric and magnetic fields in the y-direction, E_y and H_y.

The speed of propagation of the waves is obtained by comparing equations (1.7) and (1.8) with equation (1.9) so that

$$v = 1/(\mu_0\varepsilon_0\mu_r\varepsilon_r)^{1/2}. \tag{1.12}$$

In free space μ_r and ε_r are both equal to unity and hence the speed of propagation of electromagnetic radiation in free space, c, is given by

$$c = 1/(\mu_0\varepsilon_0)^{1/2}. \tag{1.13}$$

It thus follows that

$$v = c/(\mu_r\varepsilon_r)^{1/2}. \tag{1.14}$$

For all materials except the ferromagnetics $\mu_r \simeq 1$, leading to

$$v = c/\varepsilon_r^{1/2}. \tag{1.15}$$

The refractive index, n, is defined as the ratio of the speeds of propagation of the waves in free space and in the medium; hence, for a non-absorbing medium, $n = c/v = \varepsilon_r^{1/2}$.

1.2.1 Poynting's vector

It was shown by Poynting that the rate of flow of energy per unit area associated with electromagnetic waves is given by the vector $\boldsymbol{G} = (\boldsymbol{E}\cdot\boldsymbol{H})$. Considering the real part of $\boldsymbol{G}$ it may be shown (e.g. Lipson and Lipson 1969) that the mean value of $\boldsymbol{G}$ is equal to $\boldsymbol{E}_0^2(\varepsilon_r\mu_r)^{1/2}$. As $\mu_r \simeq 1$ for the majority of materials, it follows that

$$\boldsymbol{G} = \boldsymbol{E}_0^2\varepsilon_r^{1/2} = \boldsymbol{E}_0^2 n. \tag{1.16}$$

When a wave passes from a medium of refractive index n_1 into another of refractive index n_2, Poynting's vector requires that the ratio of the rates of flow of energy per unit area in the two media is given by

$$G_1/G_2 = n_1 \cos\theta_1/n_2 \cos\theta_2 \tag{1.17a}$$

where θ_1 and θ_2 are the angles of incidence and refraction respectively.

Also, by Snell's law,

$$n_1 \sin\theta_1 = n_2 \sin\theta_2. \tag{1.17b}$$

1.2.2 Nomenclature of planes of polarisation

As noted above, the solutions to the electromagnetic equations indicate two transverse waves vibrating at right angles to each other. In discussing the passage of the waves through a medium it is important to distinguish between these two waves, and this is done by defining planes of polarisation for the waves. The nomenclature of the planes of polarisation of an electromagnetic wave is based on the direction of the $\boldsymbol{E}$ vector. Only two planes of polarisation need to be defined: (1) where the $\boldsymbol{E}$ vector is parallel to the plane of incidence—this is known as p-polarisation, sometimes referred to as $E_{\parallel}$, and (2) where the $\boldsymbol{E}$ vector is at right angles to the plane of incidence—known as s-polarisation or $E_{\perp}$.

1.2.3 Birefringence
Some anisotropic transparent materials exhibit the phenomenon of birefringence in which two refractive indices are needed to explain their optical behaviour. The so-called ordinary ray obeys Snell's law with a refractive index n_o, whilst the extra-ordinary ray has a refractive index n_e which depends on the orientation of the ray to the optic axis; when the ray is along the optic axis, $n_o = n_e$. In the case of calcite, $n_o > n_e$ whilst for quartz, $n_o < n_e$.

1.3 Theory of reflection and refraction of electromagnetic waves by insulators

When electromagnetic waves pass from one medium into another the values of ε_r and μ_r involved in equations (1.1) to (1.10) must be altered accordingly. In addition, a set of boundary conditions is needed to ensure that the waves in the two media match at the boundary. From Maxwell's equations it may be shown that

(1) the tangential components of E and H must be continuous across a boundary, and

(2) the normal components of D and B must also be continuous across a boundary.

Thus, for a wave parallel to the OXY plane in a Cartesian system

$$E_{1x} = E_{2x} \qquad E_{1y} = E_{2y} \qquad H_{1x} = H_{2x} \qquad H_{1y} = H_{2y}$$

and

$$\varepsilon_0\varepsilon_1 E_{1z} = \varepsilon_0\varepsilon_2 E_{2z} \qquad \mu_0\mu_1 H_{1z} = \mu_0\mu_2 H_{2z}. \tag{1.18}$$

The final two conditions are not independent and it is usually only necessary to consider the first four boundary conditions.

1.3.1 Optical impedance
The concept of optical impedance is extremely useful in considering the reflection and transmission of electromagnetic waves across a boundary. For an isotropic medium, optical impedance is defined as

$$Z = E_x/H_y = E_y/H_x. \tag{1.19}$$

Now in a medium the following equations obtain

$$E_x = E_{0x} \exp[\mathrm{i}(\omega t - Kz)] \qquad H_y = H_{0y} \exp[\mathrm{i}(\omega t - Kz)]$$

as in equations (1.10) and (1.11).

These equations may be substituted into Maxwell's third equation (equation (1.3)). leading to

$$-\mu_0\mu_r \frac{\mathrm{d}H_y}{\mathrm{d}t} = \frac{\mathrm{d}E_x}{\mathrm{d}z} - \frac{\mathrm{d}E_z}{\mathrm{d}x}. \tag{1.20}$$

This gives

$$-\mu_0\mu_r H_y \mathrm{i}\omega = -E_x \mathrm{i}K.$$

Hence,

$$Z = E_x/H_y = \mu_0\mu_r\omega/K = \mu_0\mu_r v$$
$$= (\mu_0\mu_r/\varepsilon_0\varepsilon_r)^{1/2}. \quad (1.21)$$

On substituting values for ε_0 and μ_0, the impedance of free space $Z_0 = (\mu_0/\varepsilon_0)^{1/2} = 377\ \Omega$. The admittance of free space, Y_0, is given by

$$Y_0 = 1/Z_0 = (\varepsilon_0/\mu_0)^{1/2} = 0.00266\ \Omega^{-1}. \quad (1.22)$$

In a dielectric of relative permittivity ε_r, and when the relative permeability, μ_r, is equal to unity, the admittance is given by

$$Y = (\varepsilon_r\varepsilon_0/\mu_0)^{1/2} = Y_0\varepsilon_r^{1/2} = Y_0 n. \quad (1.23)$$

1.3.2 Reflection and transmission coefficients

Optical impedance may be used to consider the reflection and transmission of waves at the junction of two media in either s-polarisation or p-polarisation. The situation for p-polarisation is shown in figure 1.1. A wave with its ***E*** vector vibrating in the plane of incidence is incident in medium 1 at an angle θ_1 and is partially reflected at the same angle. At the same time, part of the wave is refracted into the second medium of refractive

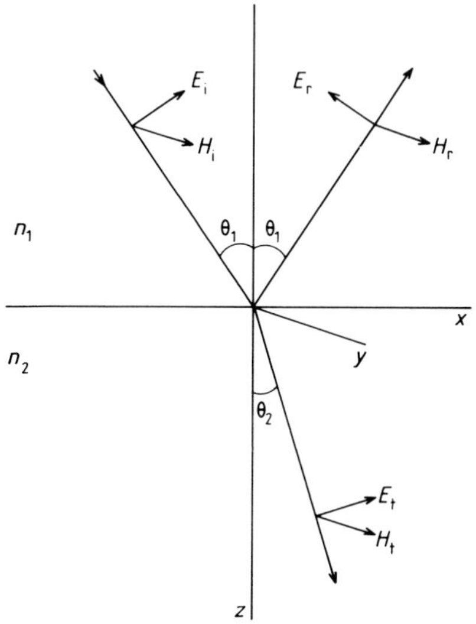

Figure 1.1 *E* and *H* vectors at the junction of two media (p-polarisation).

index n_2 at an angle θ_2; the $\boldsymbol{H}$ vectors are parallel to the y-axis which is at right angles to the plane of the paper. The tangential components of both $\boldsymbol{E}$ and $\boldsymbol{H}$ each balance on both sides of the boundary. Thus

$$H_i + H_r = H_t \tag{1.24}$$

and

$$(E_i - E_r)\cos\theta_1 = E_r\cos\theta_2. \tag{1.25a}$$

In addition, $Z_1 = E_i/H_i = E_r/H_r$ and $Z_2 = E_t/H_t$. Consequently,

$$(E_i + E_r)Z_2 = E_t Z_1 \tag{1.25b}$$

which leads to an expression for r_p, the amplitude reflection coefficient defined as E_r/E_i, as follows

$$r_p = \frac{Z_1\cos\theta_1 - Z_2\cos\theta_2}{Z_1\cos\theta_1 + Z_2\cos\theta_2} \tag{1.26a}$$

$$= \frac{\cos\theta_1/Y_1 - \cos\theta_2/Y_2}{\cos\theta_1/Y_1 + \cos\theta_2/Y_2} \tag{1.26b}$$

$$= \frac{Y_2/\cos\theta_2 - Y_1/\cos\theta_1}{Y_2/\cos\theta_2 + Y_1/\cos\theta_1}. \tag{1.26c}$$

Now, $Y_1 = Y_0 n_1$ and $Y_2 = Y_0 n_2$ so that equation (1.26c) may be written

$$r_p = \frac{n_2/\cos\theta_2 - n_1/\cos\theta_1}{n_2/\cos\theta_2 + n_1/\cos\theta_1}. \tag{1.26d}$$

These equations may be transformed to give

$$r_p = \frac{\tan(\theta_1 - \theta_2)}{\tan(\theta_1 + \theta_2)}. \tag{1.27}$$

It is easily shown that $r_p = 0$ when $\theta_1 + \theta_2 = \pi$; this condition applies when $\theta_1 = \arctan(n_2/n_1)$, which is known as Brewster's angle. For angles of incidence less than Brewster's angle and for $n_2 > n_1$, r_p is positive and no change in phase occurs on reflection; at angles of incidence greater than Brewster's angle, r_p is negative and a phase change of π takes place on reflection. The situations are reversed for internal reflection when $n_2 < n_1$.

As the reflected and incident waves are in the same medium, the reflectance, R_p, is given by Poynting's vector as r_p^2 and is thus equal to

$$R_p = \frac{(Y_2/\cos\theta_2 - Y_1/\cos\theta_1)^2}{(Y_2/\cos\theta_2 + Y_1/\cos\theta_1)^2}. \tag{1.28}$$

In the case of transmission, the amplitude transmission coefficient is given by

$$t_p = E_t/E_i = Y_1/Y_2(1 + r_p). \tag{1.29}$$

Thus

$$t_p = \frac{Y_1/Y_2(2Y_2/\cos\theta_2)}{Y_2/\cos\theta_2 + Y_1/\cos\theta_1} \tag{1.30a}$$

$$= \frac{2Y_1/\cos\theta_2}{Y_2/\cos\theta_2 + Y_1/\cos\theta_1} \tag{1.30b}$$

$$= \frac{2n_1/\cos\theta_2}{n_2/\cos\theta_2 + n_1/\cos\theta_1}. \tag{1.30c}$$

Equations (1.30a) to (1.30c) may be expressed in the form

$$t_p = 2\sin\theta_2\cos\theta_1[\sin(\theta_1+\theta_2)\cos(\theta_1-\theta_2)]^{-1}. \tag{1.31}$$

In order to comply with the requirements of the Poynting vector (equation (1.17)), the transmittance, T, is given by

$$T_p = t_p^2 Y_2\cos\theta_2/Y_1\cos\theta_1. \tag{1.32}$$

Thus

$$T_p = 4\frac{n_1}{\cos\theta_1}\frac{n_2}{\cos\theta_2}\left(\frac{n_2}{\cos\theta_2} + \frac{n_1}{\cos\theta_1}\right)^{-2}. \tag{1.33}$$

Figure 1.1 can also be used to illustrate the case for s-polarisation with the ***E*** and ***H*** vectors interchanged. The tangential components of E and H are now as follows:

$$E_i + E_r = E_t \qquad (H_i - H_r)\cos\theta_1 = H_t\cos\theta_2. \tag{1.34}$$

These equations lead to the following expression for the amplitude reflection coefficient, r_s:

$$r_s = \frac{E_r}{E_i} = \frac{Z_2\cos\theta_1 - Z_1\cos\theta_2}{Z_2\cos\theta_1 + Z_1\cos\theta_1} \tag{1.35a}$$

$$= \frac{Y_1\cos\theta_1 - Y_2\cos\theta_2}{Y_1\cos\theta_1 + Y_2\cos\theta_2} \tag{1.35b}$$

$$= \frac{n_1\cos\theta_1 - n_2\cos\theta_2}{n_1\cos\theta_1 + n_2\cos\theta_2}. \tag{1.35c}$$

These equations may be transformed to give

$$r_2 = -\frac{\sin(\theta_1-\theta_2)}{\sin(\theta_1+\theta_2)}. \tag{1.36}$$

If $n_2 > n_1$, $Y_2\cos\theta_2$ is always greater than $Y_1\cos\theta_1$ and hence r_s will be negative indicating a phase change of π at an external reflection. On the other hand, for an internal reflection, $Y_1\cos\theta_1 > Y_2\cos\theta_2$ and no phase change will occur.

The reflectance, R_s, is equal to r_s^2 and is thus given by

$$R_s = \frac{(Y_1 \cos\theta_1 - Y_2 \cos\theta_2)^2}{(Y_1 \cos\theta_1 + Y_2 \cos\theta_2)^2}. \tag{1.37}$$

The amplitude transmission coefficient, t_s, is given by

$$t_s = E_t/E_i = 1 + r_s$$

$$= \frac{2Y_1 \cos\theta_1}{Y_1 \cos\theta_1 + Y_2 \cos\theta_2}. \tag{1.38a}$$

Transformation of this equation leads to

$$t_s = \frac{2 \sin\theta_2 \cos\theta_1}{\sin(\theta_1 + \theta_2)}. \tag{1.38b}$$

These equations show that there is no phase change on transmission across the boundary.

The transmittance, T_s, is given by $t_s^2 Y_2 \cos\theta_2 / Y_1 \cos\theta_1$ in accordance with Poynting's vector. Hence

$$T_s = \frac{4Y_1 \cos\theta_1 Y_2 \cos\theta_2}{(Y_1 \cos\theta_1 + Y_2 \cos\theta_2)^2}. \tag{1.39}$$

Equations (1.27), (1.31), (1.36) and (1.38b) are collectively known as Fresnel's equations.

1.4 Behaviour of absorbing materials

In the case of an absorbing medium, the electrical conductivity term σ must be included in the electromagnetic field equations (1.5)–(1.8). This leads to equations of the form of the general wave equation for a damped oscillation, and the solutions must therefore contain an imaginary term in the refractive index in order to balance that in E; let the solution be of the form

$$E_x = E_0 \exp\{i\omega[t - z/c(n - ik)]\}. \tag{1.40}$$

Substitution of the appropriate differentials of equation (1.40) into equation (1.5) leads to

$$(n^2 - k^2 - 2ink)(\omega/c)^2 = -\mu_0\varepsilon_0\mu_r\varepsilon_r\omega^2 - \mu_0\mu_r\sigma i\omega. \tag{1.41}$$

Now, equating separately the real and the imaginary terms in equation (1.41) gives

$$n^2 - k^2 = c^2\mu_r\varepsilon_r\varepsilon_0\mu_0 \tag{1.42}$$

for the real terms and

$$2nk(\omega/c)^2 = \mu_0\sigma\omega \tag{1.43}$$

for the imaginary terms. Hence,

$$2nk = \sigma/\omega\varepsilon_0. \tag{1.44}$$

Thus Maxwell's equations for an absorbing medium are satisfied by using a complex form of the refractive index. The relative permittivity (dielectric constant) was previously equated to n^2 and this convention may also be used with an absorbing medium. Thus

$$\varepsilon = \varepsilon_1 - \mathrm{i}\varepsilon_2 = (n - \mathrm{i}k)^2 = \tilde{n}^2 \tag{1.45}$$

giving $\varepsilon_1 = n^2 - k^2$ and $\varepsilon_2 = 2nk$ as in equations (1.42) and (1.44).

A complex dielectric constant is interpreted as implying that the electric field and the electric polarisation vectors are out of phase by an angle given by $\arctan(\varepsilon_2/\varepsilon_1)$. The complex refractive index is given by the term $n - \mathrm{i}k$, although, particularly in older treatments, the expression $n(1 - \mathrm{i}k)$ may be found, or even $n(1 + \mathrm{i}k)$. Some more recent texts use $n + \mathrm{i}k$. It is the intention to use the form $n - \mathrm{i}k$ throughout this book. n is still generally called, confusingly, the refractive index, although it cannot be used directly as such in Snell's law for absorbing materials (see §1.4.2) whilst k is frequently referred to as the extinction coefficient; n and k are called the optical constants of the medium.

Now, the intensity of the wave

$$I = E_x \cdot E_x^* \tag{1.46a}$$

$$= E_0^2 \exp(-2\omega zk/c) \tag{1.46b}$$

$$= E_0^2 \exp(-4\pi zk/\lambda) \tag{1.46c}$$

$$= E_0^2 \exp(-\mu z) \tag{1.46d}$$

where μ = absorption coefficient $= 4\pi k/\lambda$.

1.4.1 Optical functions for absorbing materials

Expressions for the optical functions in both the s and p states of polarisation for absorbing materials may be obtained from the Fresnel's equations derived in §1.3. Consider a beam of radiation incident in a medium of refractive index n_1 on an absorbing medium of complex refractive index $\tilde{n}_2 = n_2 - \mathrm{i}k_2$, the angle of incidence being θ_1. The quantity $\tilde{n}_2 \cos\theta_2$ for the second medium is now also complex; let it be represented by $u_2 - \mathrm{i}v_2$.

Snell's law in complex form is

$$n_1 \sin\theta_1 = \tilde{n}_2 \sin\theta_2 = (n_2 - \mathrm{i}k_2)\sin\theta_2 \tag{1.47}$$

also, as above,

$$\tilde{n}_2 \cos\theta_2 = u_2 - \mathrm{i}v_2. \tag{1.48}$$

Squaring and adding equations (1.47) and (1.48) gives

$$u_2^2 - v_2^2 = n_2^2 - k_2^2 - n_1^2 \sin \theta_1^2 \qquad \text{(real parts)} \qquad (1.49a)$$

$$2n_2k_2 = 2u_2v_2 \qquad \text{(imaginary parts)} \qquad (1.49b)$$

and combining equations (1.49a) and (1.49b) leads to the following expressions for u_2 and v_2:

$$2u_2^2 = \varepsilon_1 - \varepsilon_3 + [(\varepsilon_1 - \varepsilon_3)^2 + \varepsilon_2^2]^{1/2} \qquad (1.50)$$

$$2v_2^2 = \varepsilon_3 - \varepsilon_1 + [(\varepsilon_1 - \varepsilon_3)^2 + \varepsilon_2^2]^{1/2} \qquad (1.51)$$

where $\varepsilon_3 = (n_1 \sin \theta_1)^2$.

Using this concept, the equations for the optical functions evolve as follows:

$$r_s = \frac{n_1 \cos \theta_1 - (u_2 - iv_2)}{n_1 \cos \theta_1 + (u_2 - iv_2)}. \qquad (1.52a)$$

r_s may also be expressed in terms of an amplitude, ρ_s, and a phase angle, ϕ_s:

$$r_s = \rho_s \exp(i\phi_2). \qquad (1.52b)$$

The reflectance, R_s, is then obtained by multiplying r_s by its complex conjugate, r_s^*; hence

$$R_s = r_s r_s^* = \frac{(n_1 \cos \theta_1 - u_2)^2 + v_2^2}{(n_1 \cos \theta_1 + u_2)^2 + v_2^2}. \qquad (1.53)$$

The phase change on reflection, ϕ_s, is given by

$$\tan \phi_s = \text{Im } r_s / \text{Re } r_s$$

$$= \frac{2n_1 \cos \theta_1 v_2}{(n_1 \cos \theta_1)^2 - u_2^2 - v_2^2}. \qquad (1.54)$$

Similarly, the amplitude of the transmitted wave at the boundary is

$$t_s = \frac{2n_1 \cos \theta_1}{n_1 \cos \theta_1 + u_2 - iv_2} \qquad (1.55a)$$

$$= \tau_s \exp(i\chi_s). \qquad (1.55b)$$

The phase change on transmission across the boundary, χ_s, is then given by

$$\tan \chi_s = v_2 / (n_1 \cos \theta_1 + u_2). \qquad (1.56)$$

In the case of p-polarised radiation, $n_1 \cos \theta_1$ must be replaced by

$\cos\theta_1/n_1$ and $\tilde{n}_2\cos\theta_2$ by $\cos\theta_2/\tilde{n}_2$; then

$$r_\mathrm{p} = \frac{(\cos\theta_1/n_1) - (\cos\theta_2/\tilde{n}_2)}{(\cos\theta_1/n_1) + (\cos\theta_2/\tilde{n}_2)} \tag{1.57}$$

$$= \frac{\tilde{n}_2^2\cos\theta_1 - n_1\tilde{n}_2\cos\theta_2}{\tilde{n}_2^2\cos\theta_1 + n_1\tilde{n}_2\cos\theta_2} \tag{1.58}$$

$$= \frac{(n_2^2 - k_2^2)\cos\theta_1 - n_1u_2 - \mathrm{i}(2n_2k_2\cos\theta_1 - n_1v_2)}{(n_2^2 - k_2^2)\cos\theta_1 + n_1u_2 - \mathrm{i}(2n_2k_2\cos\theta_1 + n_1v_2)}. \tag{1.59}$$

As before, $r_\mathrm{p} = \rho_\mathrm{p}\exp(\mathrm{i}\phi_\mathrm{p})$ and $R_\mathrm{p} = r_\mathrm{p}r_\mathrm{p}^*$. Thus

$$R_\mathrm{p} = \frac{(n_2^2 - k_2^2 - u_2n_1/\cos\theta_1)^2 + (2n_2k_2 - v_2n_1/\cos\theta_1)^2}{(n_2^2 - k_2^2 + u_2n_1/\cos\theta_1)^2 + (2n_2k_2 + v_2n_1/\cos\theta_1)^2}. \tag{1.60}$$

Considering the real and imaginary parts of r_p leads to the following expression for the phase change on reflection

$$\tan\phi_\mathrm{p} = \frac{2(n_1/\cos\theta_1)[(n_2^2 - k_2^2)v_2 - 2n_2k_2u_2]}{(n_2^2 - k_2^2)^2 + (2n_2k_2)^2 - (n_1/\cos\theta_1)^2(u_2^2 + v_2^2)}. \tag{1.61}$$

The corresponding equations for the transmitted radiation are

$$t_\mathrm{p} = \frac{2n_1(n_2 - \mathrm{i}k_2)}{(u_2 - \mathrm{i}v_2)n_1/\cos\theta_1 + (n_2 - \mathrm{i}k_2)^2} \tag{1.62a}$$

$$= \tau_\mathrm{p}\exp(\mathrm{i}\chi_\mathrm{p}) \tag{1.62b}$$

and

$$\tan\chi_\mathrm{p} = \frac{(n_1/\cos\theta_1)(n_2v_2 - k_2u_2) + k_2(n_2^2 + k_2^2)}{(n_1/\cos\theta_1)(n_2u_2 + k_2v_2) + n_2(n_2^2 + k_2^2)}. \tag{1.63}$$

1.4.2 Case of two absorbing media

In the case of radiation incident on the boundary between two absorbing media of characteristics $u_1 - \mathrm{i}v_1$ and $u_2 - \mathrm{i}v_2$ respectively, the above equations become

$$r_\mathrm{s} = \frac{(u_1 - \mathrm{i}v_1) - (u_2 - \mathrm{i}v_2)}{(u_1 - \mathrm{i}v_1) + (u_2 - \mathrm{i}v_2)} \tag{1.64}$$

$$t_\mathrm{s} = \frac{2(u_1 - \mathrm{i}v_1)}{(u_1 - \mathrm{i}v_1) + (u_2 - \mathrm{i}v_2)} \tag{1.65}$$

$$r_\mathrm{p} = \frac{(u_1 - \mathrm{i}v_1)(n_2 - \mathrm{i}k_2)^2 - (u_2 - \mathrm{i}v_2)(n_2 - \mathrm{i}k_2)^2}{(u_1 - \mathrm{i}v_1)(n_2 - \mathrm{i}k_2)^2 + (u_2 - \mathrm{i}v_2)(n_2 - \mathrm{i}k_2)^2} \tag{1.66}$$

and

$$t_p = \frac{2(n_1 - ik_1)(n_2 - ik_2)}{(n_2 - ik_2)^2 + [(n_1 - ik_1)(u_2 - iv_2)/(u_1 - iv_1)]}. \qquad (1.67)$$

It is instructive to display the values of R and ϕ for solid materials as contours in the n_2, k_2 plane. Figure 1.2(a) shows the contours of constant R_s for radiation *in vacuo* incident on the medium at an angle of incidence of 75°; these are a set of circles whose centres lie on the positive n_2-axis. Using equation (1.53) and noting that $u_2 \doteqdot n_2$ and $v_2 \doteqdot k_2$, it may be shown (Ward 1993) that the circles have radii given by $2R_s n_1 \cos\theta_1/(1 - R_s)$ and their centres are at values of n_2 equal to $n_1 \cos\theta_1 (1 + R_s)/(1 - R_s)$. Thus, as R_s increases, the radius of the contour line will increase and the centre will move to larger values of n_2. As the angle of incidence increases, the values of R_s for given values of n_2 and k_2 will also increase and the pattern appears to contract towards the origin. Figure 1.2(a) also shows the contours of constant ϕ_s, again for $\theta_1 = 75°$; from equation (1.54), these, too, can be shown to form a set of circles but now having their centres on the positive k_2-axis. The radii are given by $n_1 \cos\theta_1/\sin\phi_s$ and the centres by $n_1 \cos\theta_1/\tan\phi_s$. As the angle of incidence increases, the radii will decrease whilst the centres will move closer to the origin; the same effect occurs as ϕ_s increases.

In the case of R_p, the contours are again in the form of circles centred on the n_2-axis; the formulae for the radii and centres can be obtained by substituting R_p for R_s in the expressions in the preceding paragraph and also $n_1/\cos\theta_1$ for $n_1 \cos\theta_1$. The contours of R_p for $\theta_1 = 75°$ are shown in figure 1.2(b); as R_p increases the circles have larger radii but, as θ_1 increases, the pattern of circles will appear to expand. ϕ_p may be obtained by the same substitutions and the contours are again circles with centres on the k_2-axis. In this case, however, the circles do not pass through the origin but, for a given value of θ_1, through a common point on the n_2-axis; see figure 1.2(b) which is for $\theta_1 = 75°$. This convergence point is determined when the denominator in equation (1.61) becomes zero, i.e. when $n_2 = n_1/\cos\theta_1$; the values n_2 for this point for different θ_1 are as follows: $\theta_1 = 0$, $n_2 = 1$; $\theta_1 = 60°$, $n_2 = 2$; $\theta_1 = 75°$, $n_2 = 3.86$; $\theta_1 = 80°$, $n_2 = 5.75$ and $\theta_1 = 85°$, $n_2 = 11.47$.

No contour diagrams for T or χ are given at this point as, for a bulk absorbing material these optical functions are generally insignificant. However, when dealing with thin films, both T and χ are measurable and their contour diagrams are considered in Chapter 5.

1.4.3 Angle of refraction in an absorbing medium

It is useful to calculate the real angle of refraction in an absorbing medium arising from radiation incident from a non-absorbing medium.

Consider a plane wave incident from a medium of refractive index n_1 on

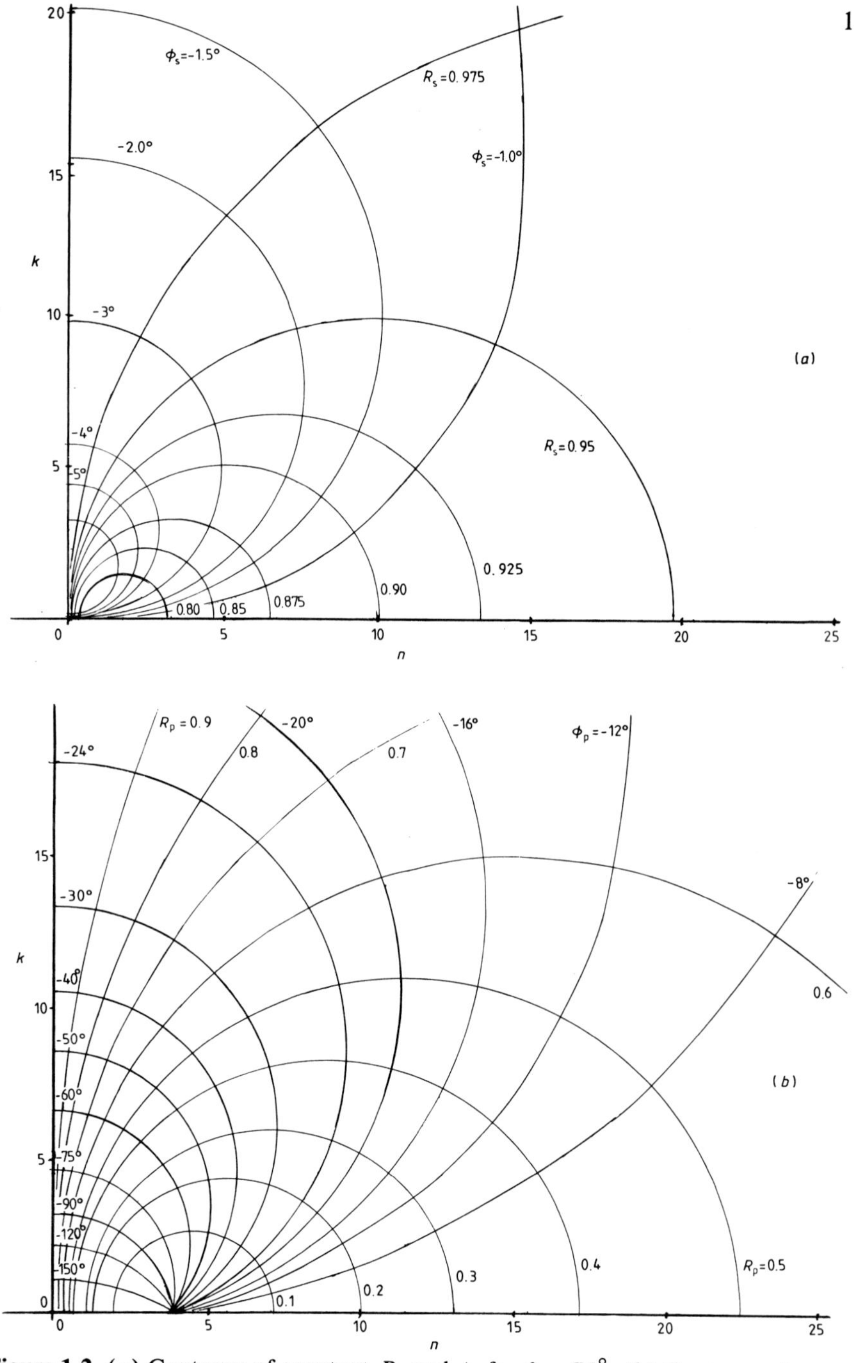

Figure 1.2 (*a*) Contours of constant R_s and ϕ_s for $\theta_1 = 75^\circ$. (*b*) Contours of constant R_p and ϕ_p for $\theta_1 = 75^\circ$. (From Ward (1993), by permission of the publishers, Butterworth Heinemann Ltd ©.)

to the surface of the absorbing medium which has optical constants, $n_2 - \mathrm{i}k_2$ (figure 1.3). Application of a Huygens type construction to the incident and refracted beams gives

$$\sin\theta_1/\sin\theta_2' = n'/n_1 = c/p \quad (1.68)$$

where n' is the effective (real) refractive index of the medium (not equal to either n_2 or $\tilde{n}_2$) and p is the phase velocity in the medium. The wave in the absorbing medium is attenuated proportionally to its pathlength; thus the surfaces of constant amplitude will be parallel to the interface and the amplitude will be proportional to exp($-k'z$) where k' is an extinction coefficient, not necessarily equal to k_2.

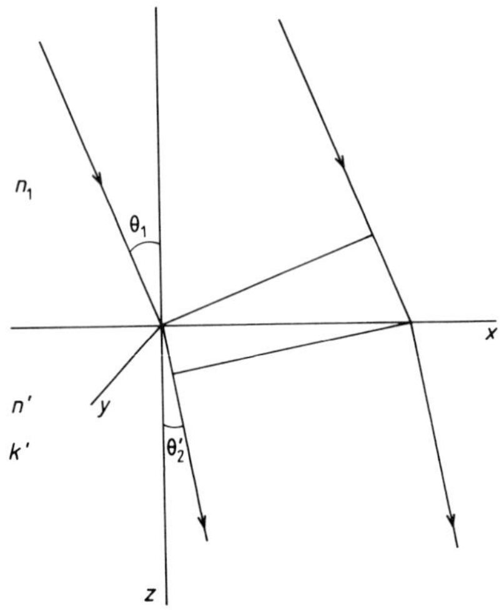

Figure 1.3 Refraction by an absorbing medium.

Taking into account both the velocity in the medium and the attenuation, the electrical vector of the wave in the medium may be written

$$E_{2y} = E_{02y}\exp(-\omega k'z/c)\exp\{\mathrm{i}\omega[(t-n')/c(x\sin\theta_2' + z\cos\theta_2')]\}. \quad (1.69)$$

This equation may also be expressed in the more usual form of equation (1.10)

$$E_{2y} = E_{02y}\exp\{\mathrm{i}\omega[t-(n_2-\mathrm{i}k_2)(x\sin\theta_2' + z\cos\theta_2')/c]\}. \quad (1.70)$$

Both these equations must satisfy the general wave equation:

$$\frac{\mathrm{d}^2E}{\mathrm{d}x^2} + \frac{\mathrm{d}^2E}{\mathrm{d}z^2} = \frac{n^2}{c^2}\frac{\mathrm{d}^2E}{\mathrm{d}t^2} \quad (1.71)$$

and thus

$$n'^2 \sin^2 \theta_2' + (n' \cos \theta_2' - \mathrm{i}k')^2 = (n_2 - \mathrm{i}k_2)^2. \tag{1.72}$$

Equating the real and the imaginary parts of this equation gives

$$n'^2 - k'^2 = n_2^2 - k_2^2 = \varepsilon_1 \tag{1.73}$$

$$n'k' \cos \theta_2' = n_2 k_2 = \varepsilon_2/2 \tag{1.74}$$

and from equation (1.68)

$$n' \sin \theta_2' = n_1 \sin \theta_1 = \sqrt{\varepsilon_3}. \tag{1.75}$$

These equations may be readily solved to give the expressions for n', k' and $\sin \theta_2'$ below.

$$2n'^2 = \varepsilon_3 + \varepsilon_1 + [(\varepsilon_1 - \varepsilon_3)^2 + \varepsilon_2^2]^{1/2} \tag{1.76}$$

$$2k'^2 = \varepsilon_3 - \varepsilon_1 + [(\varepsilon_1 - \varepsilon_3)^2 + \varepsilon_2^2]^{1/2} \tag{1.77}$$

$$\sin^2 \theta_2' = \varepsilon_3/n'^2$$

$$= 2\varepsilon_3/\{\varepsilon_3 + \varepsilon_1 + [(\varepsilon_1 - \varepsilon_3)^2 + \varepsilon_2^2]^{1/2}\}. \tag{1.78}$$

Table 1.1 Angles of refraction (in degrees) as functions of n_2 and k_2 for different angles of incidence. The Snell's law case is the column for $k_2 = 0$.

$\theta_i = 15°$

	$k_2 =$ 0	1	2	3
$n_2 = 1$	15.0	14.7	14.6	14.5
2	7.4	7.4	7.4	7.4
3	5.0	5.0	4.9	4.9
4	3.7	3.7	3.7	3.7

$\theta_i = 30°$

	$k_2 =$ 0	1	2	3
$n_2 = 1$	30.0	28.0	27.1	26.8
2	14.5	14.4	14.3	14.2
3	9.6	9.6	9.6	9.5
4	7.2	7.2	7.2	7.2

$\theta_i = 45°$

	$k_2 =$ 0	1	2	3
$n_2 = 1$	45.0	38.6	36.6	35.9
2	20.7	20.4	20.0	19.8
3	13.6	13.6	13.5	13.4
4	10.2	10.2	10.2	10.1

$\theta_i = 60°$

	$k_2 =$ 0	1	2	3
$n_2 = 1$	60.0	46.1	42.9	41.9
2	25.6	25.1	24.4	23.8
3	16.8	16.7	16.6	16.4
4	12.5	12.5	12.4	12.4

$\theta_i = 75°$

	$k_2 =$ 0	1	2	3
$n_2 = 1$	75.0	50.4	46.5	45.3
2	28.9	28.0	27.1	26.6
3	18.8	18.7	18.5	18.3
4	14.0	13.9	13.9	13.8

$\theta_i = 85°$

	$k_2 =$ 0	1	2	3
$n_2 = 1$	85.0	51.6	47.5	46.2
2	29.9	28.9	27.9	27.3
3	19.4	19.3	19.0	18.8
4	14.4	14.4	14.3	14.3

If $k = 0$, it is seen that $n' = n_2$, $k' = 0$ and $\sin\theta_2 = n_1 \sin\theta_1/n_2$ which is of the same form as Snell's law for a non-absorbing medium. In addition, at normal incidence, $\varepsilon_3 = 0$ again giving $n' = n_2$ and $k' = k_2$.

Table 1.1 shows values of the real angles of refraction for materials of various optical constants using angles of incidence of 15°, 30°, 45°, 60°, 75° and 85°. The greatest difference between the actual values of θ_2' and those predicted by a simple Snell's law using n_2 occurs at large angles of incidence, the difference becoming smaller as both n_2 and k_2 increase.

1.5 Variation with frequency—dispersion theory

Up to now the treatment of Maxwell's equations has assumed, in the cases of both absorbing and non-absorbing materials, that the medium may be treated as a continuum. However, this model is too crude to explain fully the variation of both n and k with frequency and it becomes necessary to consider the reaction between the electrical component of the incident electromagnetic wave and the electrons in the material. In dielectrics the electrons are assumed to be strongly bound to the constituent atoms but capable of displacement by an electric field, whilst in metals a major effect is produced by the reaction of the wave with the free electrons.

The general equation of motion of an electron, of mass m and charge e in an electromagnetic electric field of the form $E_0 \exp(\mathrm{i}\omega t)$ applied in the x-direction is

$$m\frac{\mathrm{d}^2x}{\mathrm{d}t^2} + m\gamma\frac{\mathrm{d}x}{\mathrm{d}t} + \beta x = -eE_0 \exp(\mathrm{i}\omega t) \tag{1.79}$$

where β is the restoring force per unit displacement and γ is the damping force per unit mass per unit velocity; γ thus has the dimensions of 1/time and is often treated as an angular frequency.

The first term in equation (1.79) is the inertial force, the third is the restoring force acting on the electron following its displacement and the second is the damping force which will be assumed to be proportional to the velocity of the electron. The natural angular frequency of vibration of the electron is given by $\omega_0 = (\beta/m)^{1/2}$. ω_0 is thus determined by the restoring forces acting on the displaced electron. Solution of equation (1.79) leads to the following expression for the displacement, x:

$$x = -eE_0/m[(\omega_0^2 - \omega^2) + \mathrm{i}\gamma\omega]. \tag{1.80}$$

The electrical polarisation, P, can be expressed by two relationships:

$$P = -Nex \tag{1.81}$$

where N is the number of electrons per unit volume, and

$$P = \varepsilon_0 E_0(\varepsilon_r - 1). \tag{1.82}$$

Thus

$$\begin{aligned}\varepsilon_r &= 1 + \frac{Ne^2}{\varepsilon_0 m\,[(\omega_0^2 - \omega^2) + i\gamma\omega]} \\ &= \varepsilon_1 - i\varepsilon_2\end{aligned} \tag{1.83}$$

giving the Lorentz type equations (1906)

$$\begin{aligned}\varepsilon_1 &= 1 + \frac{Ne^2(\omega_0^2 - \omega^2)}{\varepsilon_0 m\,[(\omega_0^2 - \omega^2)^2 + \gamma^2\omega^2]} \\ &= n^2 - k^2\end{aligned} \tag{1.84}$$

$$\begin{aligned}\varepsilon_2 &= \frac{Ne^2\gamma\omega}{\varepsilon_0 m\,[(\omega_0^2 - \omega^2)^2 + \gamma^2\omega^2]} \\ &= 2nk.\end{aligned} \tag{1.85}$$

1.5.1 Dielectrics

If γ is small

$$\varepsilon_1 \simeq n^2 = 1 + \frac{Ne^2}{\varepsilon_0 m(\omega_0^2 - \omega^2)} \tag{1.86a}$$

and $\varepsilon_2 = 0$, except when $\omega \simeq \omega_0 = \lambda/2\pi c$ when

$$\varepsilon_2 = \frac{Ne^2}{\varepsilon_0\gamma\omega_0} = \frac{Ne^2\lambda_0}{2\pi c\varepsilon_0\gamma}. \tag{1.86b}$$

Figure 1.4 shows the variation of n and k with ω for sodium chloride which illustrates the characteristic features of dielectric behaviour. The curve for n shows several resonances, i.e. different values of ω_0, those at about $2 \times 10^{16}\ \mathrm{s}^{-1}$ being due to different groups of electrons within the sodium and chlorine atoms whilst that at $4 \times 10^{13}\ \mathrm{s}^{-1}$ is caused by vibration of the polar molecule (phonon). For $10^{14} < \omega < 10^{16}\ \mathrm{s}^{-1}$, $n > 1$ and increases with frequency; this is called the normal dispersion part of the curve and is used in visible and infrared spectroscopy. The parts of the curve where n is decreasing with frequency are known as the anomalous dispersion regions. As predicted, k shows absorption peaks at frequencies corresponding to each of the resonances in n, but elsewhere its value is negligible.

Sellmeier evolved an arbitrary formula for the variation of refractive index with wavelength over the normal dispersion curve for a medium with several absorption bands; the above treatment accords with his formula which is, for b bands,

$$n^2 = 1 + \sum_{a=1}^{a=b} e^2/4\pi\varepsilon_0 m\,[N_a\lambda^2\lambda_a^2/(\lambda^2 - \lambda_a^2)] \tag{1.87}$$

where N_a is the number of electrons per unit volume resonating at wavelength λ_a.

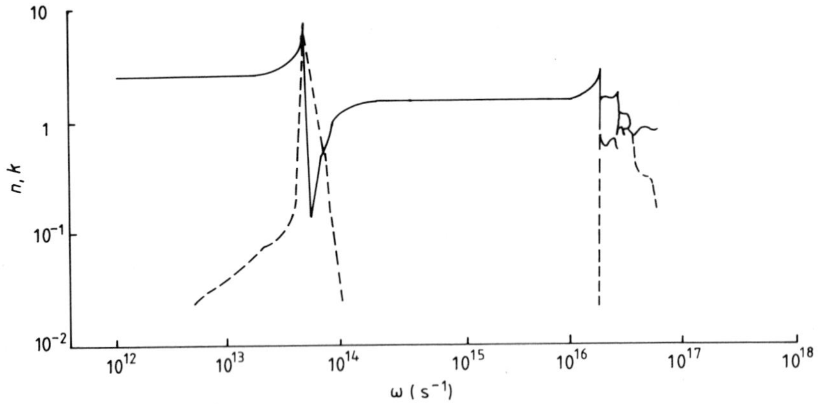

Figure 1.4 n (full curve) and k (broken curve) versus ω for sodium chloride. (From *Handbook of Optical Constants of Solids*, ed E D Palik (1985), courtesy Academic Press Inc.)

The treatment given above strictly only applies to a dielectric gas where the molecules are sufficiently far apart that, after being polarised by the applied electric field, they do not interact. In a solid dielectric account must be taken of the local electrical polarisation field due to the presence of the surrounding induced dipoles. The standard method of procedure is to imagine a small sphere drawn round the point at which the value of the local field is required and to calculate the force on a unit positive charge placed at the centre of the sphere (i) due to all the charges inside the sphere and (ii) due to all the remaining charges outside the sphere. In an isotropic medium the dipoles are symmetrically distributed and thus the contribution due to (i) is zero; that due to (ii) can be shown to be equal to $P/3\varepsilon_0$. The total local field is thus given by

$$E_{\text{loc}} = E_0 + P/3\varepsilon_0. \tag{1.88}$$

If E_{loc} is now used in equation (1.80) in place of E_0, the expression for P becomes

$$P = \frac{Ne^2(3\varepsilon_0 E_0 + P)}{3m\varepsilon_0[(\omega^2 - \omega_0^2) + \mathrm{i}\gamma\omega]} \tag{1.89}$$

$$P/(P+3E_0\varepsilon_0)=\frac{Ne^2}{3m\varepsilon_0[(\omega^2-\omega_0^2)+\mathrm{i}\gamma\omega]}. \tag{1.90}$$

On substituting $P=\varepsilon_0E_0(\varepsilon_r-1)$, equation (1.90) can be obtained in the form

$$\frac{\varepsilon_0E_0(\varepsilon_r-1)}{\varepsilon_0E_0(\varepsilon_r+2)}=\frac{(n^2-1)}{(n^2+2)} \tag{1.91}$$

$$=\frac{Ne^2}{3m\varepsilon_0[(\omega^2-\omega_0^2)+\mathrm{i}\gamma\omega]}.$$

This is the well known Clausius–Mosotti equation. Not only does it apply to dielectrics, but it can also be used when dealing with island or particulate films, which can often be approximated to the above situation (see Chapter 8).

1.5.2 Metals

In a metal some of the electrons are tightly bound to the atoms, as in the dielectrics, whilst some are very loosely bound and can easily become detached so that they are free to move through the lattice. These free electrons can be accommodated in the dispersion equations (1.83) to (1.85) by setting their natural frequency of vibration, ω_0, equal to zero; this is equivalent to equating the binding force for the free electrons, β, to zero. In addition, due to the presence of free electrons distributed uniformly throughout the material, there will be no electrical polarisation and the local field will be equal to the applied field, E_0. The equation for the complex dielectric constant of the material then reduces to

$$\varepsilon=1+\frac{Ne^2}{\varepsilon_0m\omega(\mathrm{i}\gamma-\omega)} \tag{1.92}$$

hence

$$\varepsilon_1=1-\frac{Ne^2}{\varepsilon_0m(\omega^2+\gamma^2)}. \tag{1.93}$$

In a free-electron metal, the major part of the damping is due to electron collisions. It is then the usual practice to equate γ with ω_c, the collision frequency. ω_c is equal to $1/\tau_c$, where τ_c is the average time between collisions, also known as the collision relaxation time. The free electrons may collide with lattice atoms, other electrons, grain boundaries or photons.
Thus

$$\varepsilon_1=1-\frac{\sigma_0\tau_c}{\varepsilon_0(1+\omega^2\tau_c^2)} \tag{1.94}$$

and

$$\varepsilon_2 = \frac{Ne^2\gamma}{\varepsilon_0 m\omega(\omega^2 + \gamma^2)} \tag{1.95}$$

$$= \sigma_0/\varepsilon_0\omega(1 + \omega^2\tau_c^2) \tag{1.96}$$

where σ_0 is the DC electrical conductivity, equal to $Ne^2/m\omega_c^2$.

These are the equations which were obtained by Drude (1887), and they have been extensively used in studies of the optical behaviour of metals.

1.5.3 Plasma oscillations

Before further discussion of the Drude equations, it is necessary to introduce the concept of plasma oscillations. A metal may be regarded as a plasma of the mobile free electrons inside the fixed positive field of the lattice ions. Normally the total net charge density in any small region of the metal is zero, but if a fluctuation in density does occur electrons will flow in or out of the region in an attempt to restore the electrical balance. Overshoot will occur, the process will go into reverse and an oscillation of the electrons will be set up. This would normally be damped out rapidly, but if an electric field of the same frequency is applied the oscillation will be maintained and strong absorption of the applied signal will take place. The resonant frequency as described above is called the plasma frequency, ω_p, and an expression for ω_p may be obtained as follows.

Let the electron become displaced along the x-direction through a distance, $s(x, t)$. The electrical polarisation produced is given by

$$P = -Nes. \tag{1.97}$$

From Maxwell's first equation, $\rho = 0$ and therefore

$$-\operatorname{div} \boldsymbol{P} = \varepsilon_0 \operatorname{div} \boldsymbol{E} \tag{1.98}$$

Thus,

$$\varepsilon_0 \frac{dE_x}{dt} = Ne \frac{ds}{dt} \tag{1.99}$$

or,

$$E_x = Nes/\varepsilon_0.$$

The force on the electron is given by

$$-eE_x = -Ne^2 s/\varepsilon_0 = m \frac{d^2 s}{dt^2}. \tag{1.100}$$

The force is thus proportional to the negative displacement, s, and the motion will be simple harmonic with angular frequency

$$\omega_p = (Ne^2/\varepsilon_0 m)^{1/2}. \tag{1.101}$$

A means of observing the absorption caused by the plasma in a metal is to measure the energy loss sustained by a beam of electrons passing through a thin sheet of the material; a sharp maximum in the electron energy loss is observed when the energy is equal to $h\omega_p/2\pi$. Electron-energy-loss measurements of this kind have been carried out for many metals, and are discussed in Chapter 4 when dealing with the experimental results for the optical constants of bulk materials.

It should be noted that plasma oscillations take place along the direction of the electric field produced by the fluctuations, and therefore appear as longitudinal waves. Normally they would not be expected to interact with the transverse waves of the electromagnetic spectrum, but the physical presence of the boundary at the surface of the metal allows coupling to take place and then electromagnetic energy may be absorbed. A resonance is produced at ω_p which is called a volume plasmon (Ferrell and Sterne 1962). If there is a regular pattern of roughness on the surface, another plasmon may be excited, known as a surface plasmon. It may be shown that the angular frequency of the surface plasmon, $\omega_s = \omega_p/\sqrt{2}$. It can also be demonstrated that a volume plasmon can be detected as a sharp peak in the function $\mathrm{Im}(1/\varepsilon)$ plotted against wavelength or photon energy, whilst a surface plasmon produces a peak in a similar plot of the function $\mathrm{Im}[1/(\varepsilon+1)]$.

Steinmann (1968) has discussed plasma resonances in solids. The free-electron model for plasmons, discussed above, works satisfactorily for the alkali metals and for aluminium when the plasma frequency is well clear of any other absorption effects. However, it is not sufficiently sophisticated for the majority of metals because of the interactions with bound electrons which occur in the same frequency range. The assumption of a homogeneous electrical background on which the simple theory is based can no longer be sustained and one effect of this is that the actual plasma frequency is smaller than that given in equation (1.101). Plasma frequencies for individual materials are discussed in Chapter 4.

Expressing Drude's equations, (1.93) and (1.95), in terms of the plasma frequency leads to

$$\varepsilon_1 = n^2 - k^2 = 1 - \omega_p^2/(\omega^2 + \omega_c^2) \tag{1.102}$$

$$\varepsilon_2 = 2nk = \omega_p^2\gamma/\omega(\omega^2 + \omega_c^2). \tag{1.103}$$

If $\omega \gg \omega_p$, i.e. in the x-ray region, ε_2 and k tend to zero so that ε_1 will approximate to n^2 and will be equal to $1 - [\omega_p/(\omega^2 + \omega_c^2)]$; consequently n will be positive with a value slightly less than 1 and the metal will behave similarly to a dielectric and will be transparent. As the frequency is reduced, ε_1 will also decrease and will become zero when $\omega^2 = \omega_p^2 - \omega_c^2$, this often taking place in the ultraviolet. When $\varepsilon_1 = 0$, $\omega_p^2 = \omega^2 + \omega_c^2$; in those materials where $\omega_p \gg \omega_c$, the plasma frequency occurs at the point when $\varepsilon_1 = 0$. It may be noted that for some metals there are several frequencies where $\varepsilon_1 = 0$ and so they support more than one plasmon. At still lower frequencies, in the visible and near-infrared regions, ε_1 will become negative so that k^2 is greater than n^2; this will lead to strong absorption and reflection of electromagnetic radiation.

When the frequency is even lower, in the far-infrared, microwave and radiowave regions, ω will eventually become less than ω_c. Under these circumstances, ε_1 will tend towards a value of $1 - (\omega_p/\omega_c)^2$, which is a constant quantity having a negative value. At the same time, ε_2 will become equal to $\omega_p^2/\omega\omega_c$, and so will continue to increase as ω decreases. Thus, both n and k will increase with decreasing ω, but the difference, $n^2 - k^2$, will remain constant; this implies that n and k become more nearly equal at very small ω. Hagen and Rubens (1900) showed that, under these conditions,

$$n \simeq k = (\omega_p^2/2\omega\omega_c)^{1/2} = (\sigma/2\varepsilon_0\omega)^{1/2}. \tag{1.104}$$

Confirmation of this theory has been partially obtained from measurements of the optical constants for gold, lead and aluminium at very long wavelengths. Values of n and k for lead have been published by Brandli and Sievers (1972) extending to 1600 μm and by Golovashkin and Motulevich (1968) at shorter wavelengths. Figure 1.5 shows these results plotted against ω on log scales, illustrating the approximation to the theory of Hagen and

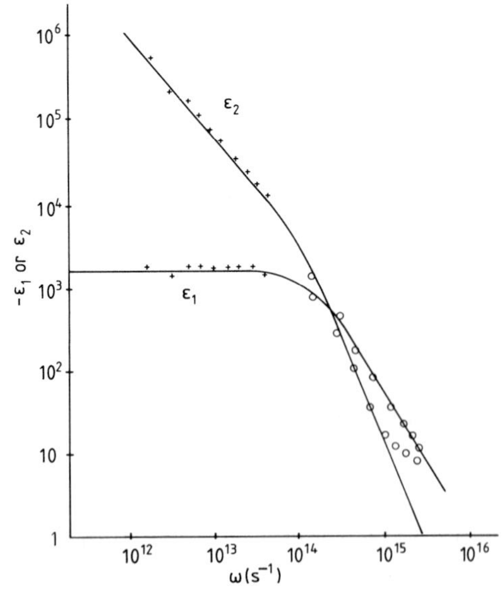

Figure 1.5 ε_1 and ε_2 versus ω for lead. The full curves are for the Drude model. + are the results of Brandli and Sievers, while ○ are those of Golovashkin and Motulevich. (From Ordal M A *et al* 1983 *Appl. Opt.* **22** 1100, courtesy the Optical Society of America.)

Rubens as the frequency decreases. Results for aluminium have been given by Bennett *et al* (1963) and by Shiles *et al* (1980) and these also show some convergence of the values of n and k at long wavelengths; extrapolated, they would become approximately equal at about 200 μm.

In addition to the free electrons, metals also contain electrons which are tightly bound to the atomic nucleus, and on the classical theory they would be expected to behave, under the influence of an electromagnetic field, in a similar manner to those in dielectrics; resonances would be expected leading to peaks in k and ε_2 at certain values of ω.

Figure 1.6 shows n and k for silver plotted as functions of λ and illustrates the various phenomena discussed in this section. At long wavelengths n and k are both large, with $k > n$ as predicted by the free-electron theory. As the wavelength decreases, both n and k also decrease and silver displays the characteristic metallic properties of high reflectance and low transmittance. A plasma resonance is indicated at approximately 3×10^{-1} μm when n and k are equal. For wavelengths below this point, $n > k$ and silver behaves similarly to a partially transparent dielectric. Absorption resonances due to the bound electrons produce subsidiary maxima in k in the ultraviolet and x-ray regions; at very short wavelengths the value of n tends to unity.

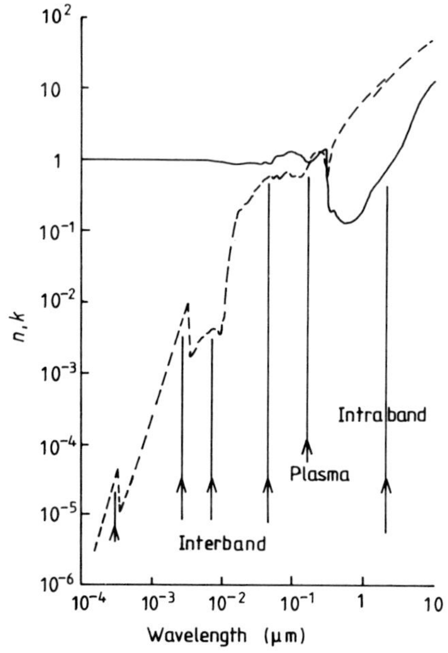

Figure 1.6 Log-log plot of n (full curve) and k (broken curve) versus wavelength for silver. (From *Handbook of Optical Constants of Solids*, ed E D Palik (1985), courtesy Academic Press Inc.)

Although the classical theories of bound and free electrons are partially successful in predicting the occurrence of the main features of the dispersion curves for both dielectrics and metals, correlation between the theoretical and experimental values of these features is not always satisfactory. In order to obtain better agreement it is necessary to use the quantum theory, which is the subject of the next section.

1.6 Quantum theory of electrons in solids

The application of the quantum theory to a material in the solid state is a highly specialised subject and for fuller details the reader is referred to some of the standard text-books on this subject such as Mott and Jones (1936), Sokalov (1967) and Kittel (1986). Only a brief summary of the aspects of solid state theory relevant to the optics of solids will be given here.

In the quantum theory, the wave nature of electrons is invoked and this leads to the concept that electron energies may only have certain quantised values. These allowed energy values are derived by considering the electron wavelengths which are capable of setting up stationary waves. In the case of the bound electrons, the stationary waves are set up in the electron's orbit, but the free electrons produce standing waves throughout the solid. The easiest case to deal with initially is the hypothetical one-dimensional solid when the lattice ions are ignored; the electrons are then confined inside a constant potential energy well whose length is that of the solid. It may readily be shown that the allowed energies, U, of the 'free' electrons are given by the equation

$$U = h^2/2mL^2 = j^2h^2/8mL^2 \tag{1.105}$$

where h is Planck's constant, L is the length of the 'solid' and j is an integer or quantum number. Equation (1.105) predicts a set of discrete values for the energy of the electron, and, when values for the universal constants are inserted, these energy levels are found to be closely packed and separated by about 10^{-33} J; i.e. they form a quasi-continuous 'band'. On a simple picture the electrons will occupy the lowest available energy levels, and, at the absolute zero of temperature, the band will be filled to a level known as the Fermi energy. An electron at the Fermi energy may easily move to unoccupied higher energy levels within this band by gaining the above very small amount of energy, for example in a collision, by acceleration in an electric field or by an increase in the temperature.

Expressing equation (1.105) in terms of the wavenumbers of the electrons, $K = 2\pi/\lambda$ leads to

$$U = h^2K^2/8m\pi^2. \tag{1.106}$$

Equation (1.106) indicates a parabolic relationship between U and K. If this simple model is extended to the three-dimensional case, the equation of allowed energy values becomes

$$U = h^2/8m\pi^2(K_x^2 + K_y^2 + K_z^2) \quad (1.107)$$

and thus all electrons having the same energy will lie on the surface of a sphere in K-space, i.e. a space with coordinate axes K_x, K_y and K_z. These correspond to three quantum numbers; a fourth is needed if the spin of the electron is taken into account. On the Pauli exclusion principle, no two electrons in the solid may have the same set of four quantum numbers. One of the implications of this principle and of equation (1.107) is that there can be several electrons having the same energy but with different sets of quantum numbers; the number of electrons for a given value of energy is called the density of states and this increases as the energy increases. The electrons lying in the highest occupied energy levels in the band (i.e. at the Fermi energy) lie on the Fermi surface, which in this case is a sphere.

The effects of a regular one-dimensional lattice of positive ions inside the metal may be included in the simple model by superimposing a periodic potential inside the potential energy well; it is found that the parabolic distribution of electron energies no longer applies. The indefinite range of allowable energies is split up into a number of bands separated by regions where electron energy levels are not permitted. These energy gaps occur at values of K given by $\pm j\pi/a$ where a is the lattice spacing. Another way of describing the situation is to regard the U–K curve as being divided into a number of zones known as Brillouin zones with discontinuities at each zone boundary. The first zone lies between $-\pi/a$ and $+\pi/a$, the second, in two parts, between $-2\pi/a$ and $-\pi/a$ and again between $+\pi/a$ and $+2\pi/a$, and so on. These concepts are illustrated in figure 1.7(a) which shows a typical U–K curve. In three dimensions, the gaps between the bands may occur at different values of K in different crystallographic directions, and consequently it is possible for some of the bands to overlap; this will also bring about modifications in the pattern of the density of states. If the electrons are weakly bound to the atoms, the theory predicts only narrow gaps at the zone boundaries and the material behaves in a very similar way to a classical free-electron metal. On the other hand, when tight binding occurs considerable distortion of the parabolic distribution takes place, and the classical theory can no longer be applied, even approximately.

Another useful concept is that of the reduced Brillouin zone. Because the electrons may be regarded as waves, any zone may be shifted along the K-axis through an integral number of units of $+2\pi/a$ so that all parts of the U–K curve can be accommodated within the first zone. This facility, which is also illustrated in figure 1.7(b), is extremely useful when all possible electron transitions in a solid are being considered.

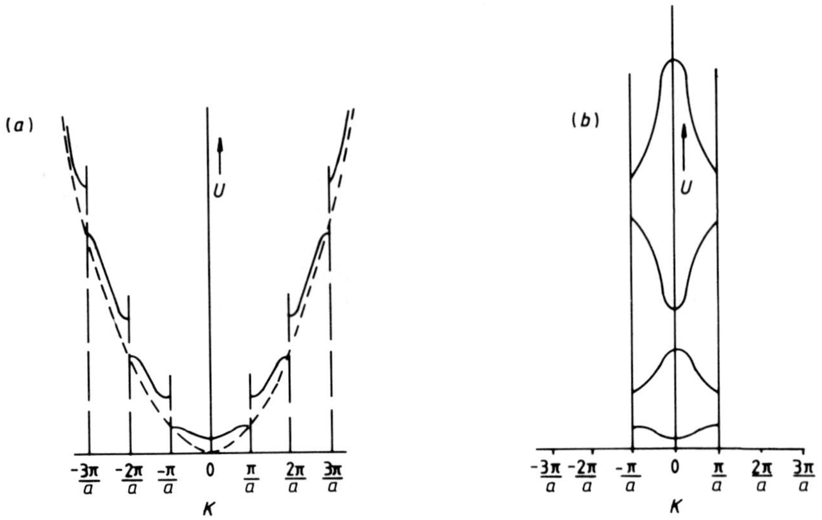

Figure 1.7 (*a*) Brillouin zones for a one-dimensional lattice. (*b*) Reduced Brillouin zones.

The model described above is the basis of the band theory of solids and it can be used to give a simple description of metals, insulators and semiconductors. Metals, which are characterised by the presence of free electrons, are represented by partially filled or overlapping bands (see figures 1.8(*a*) and (*b*)) which contain many unfilled energy levels into which the electrons may be easily excited by thermal energy and hence move readily throughout the solid, behaving similarly to the free electrons of the classical theory. Insulators are represented by energy bands which are completely filled with electrons and separated from the next higher, empty band by a large energy gap ($\sim$5 eV) so that free movement of the electrons throughout the solid is not usually possible (figure 1.8(*c*)); electron transitions can only take place when several electron volts of energy are absorbed by the electron. Semiconductors have a similar band structure to insulators but with a small energy gap ($\sim$1 eV) (figure 1.8(*d*)) so that at elevated temperatures some electrons can gain sufficient energy to make a transition to the next higher, empty band and thus initiate the conduction process. The departure of the electrons from the lower or valence band creates some unfilled levels which then behave as positively charged 'holes' and these also contribute to the conduction process.

It may readily be shown that each energy band can hold two electrons from each atom in the solid; thus each of the s-bands will be able to hold all the s-electrons from that shell in an atom, whilst three p-bands will be needed to contain all possible p-electrons in an atom. Similarly, five d-bands and seven f-bands are required for the respective electrons.

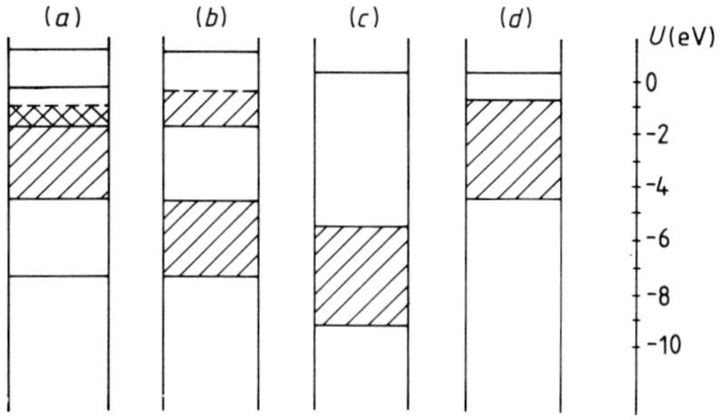

Figure 1.8 Band theory representation of metals. (a, b) Overlapping bands and partially filled band; (c) insulator; (d) semiconductor.

The same basic principles again apply when this model is translated into three dimensions. The Brillouin zones, whose boundaries mark the positions in K-space of the energy gaps, now become three-dimensional figures whose shapes depend on the crystal structure of the material; in fact the shapes of the Brillouin zones are the same as those of the reciprocal crystal lattices. Figure 1.9 shows the Brillouin zones for three of the most common crystal structures, (a) face-centred cubic (FCC), the Brillouin zone being a dodecahedron, (b) body-centred cubic (BCC), for which the Brillouin zone is a fourteen-sided figure known as a tetrakaihedron and (c) hexagonal, where the Brillouin zone is a hexagonal prism.

The relationship between U and K for the electrons becomes much more complex for a three-dimensional solid as the available energy levels are different for different crystallographic directions. Also shown in figure 1.9 are the main axes of the Brillouin zones which all start at the centre (Γ) corresponding to $K = 0$, and end at significant points on the zone boundaries. These points are designated H, P and N for the BCC lattice corresponding to the (0, 1, 0), (1, 1, 1) and (1, 1, 0) crystallographic axes. The corresponding axes for the FCC lattice are labelled L, (1, 1, 1), X, (1, 0, 0), and W, (2, 1, 0), whilst for the hexagonal lattice the axes are A, (0, 0, 1), K, (1, 0, 0) and M, (0, 1, 0). The usual method of presentation of this situation is to provide a number of U–K diagrams corresponding to these main crystallographic directions. Figure 1.10 shows the U–K diagrams for sodium, which is a comparatively simple metal having a BCC structure where the free electrons are contained within one s-band.

Within the Brillouin zone the electrons are again characterised by discrete energies. Near the centre of the zone the electron energies are distributed according to the parabolic relationship of equation (1.106) and the surfaces of equal energy are again spherical, but, as the energy increases, these surfaces depart markedly from the spherical form. The Fermi surface is often considerably distorted and may come into contact with the faces of the Brillouin zone or cross into higher zones. The optical and electrical

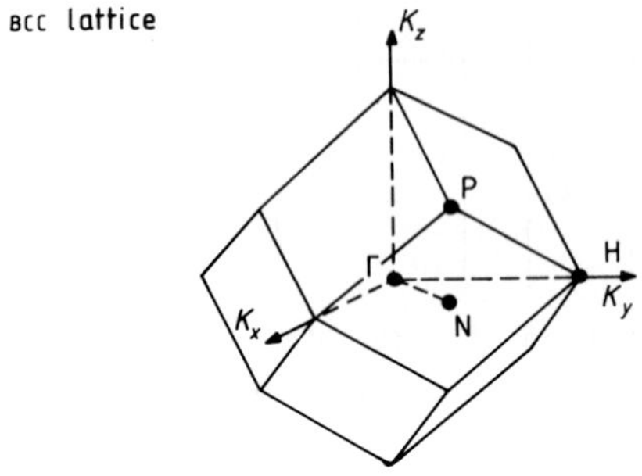

Γ: $\frac{\pi}{a}(0, 0, 0)$

H: $\frac{2\pi}{a}(0, 1, 0)$

P: $\frac{\pi}{a}(1, 1, 1)$

N: $\frac{\pi}{a}(1, 1, 0)$

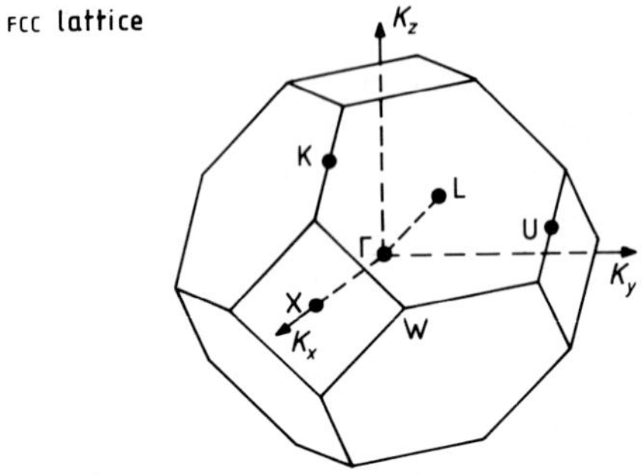

Γ: $\frac{\pi}{a}(0, 0, 0)$

L: $\frac{\pi}{a}(1, 1, 1)$

X: $\frac{2\pi}{a}(1, 0, 0)$

W: $\frac{\pi}{a}(2, 1, 0)$

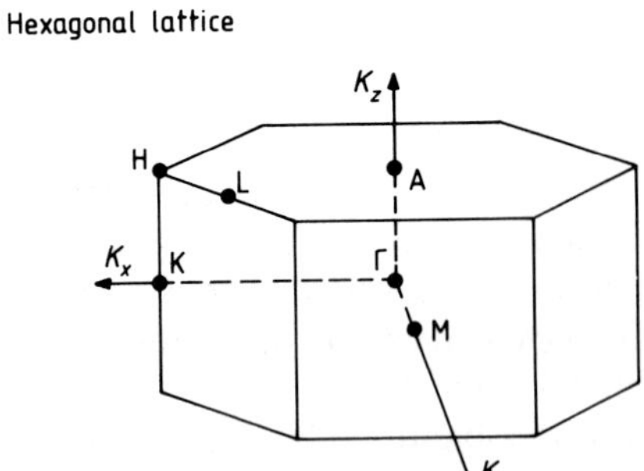

Γ: $(0, 0, 0)$

A: $\frac{\pi}{c}(0, 0, 1)$

K: $\frac{4\pi}{3a}(1, 0, 0)$

M: $\frac{2\pi}{a\sqrt{3}}(0, 1, 0)$

Figure 1.9 Brillouin zones for BCC, FCC and hexagonal lattices. (From *Physics Handbook* by Carl Nordling and Jonny Österman, Student litteratur ab, Sweden and Chartwell Bratt Ltd, UK, 1982.)

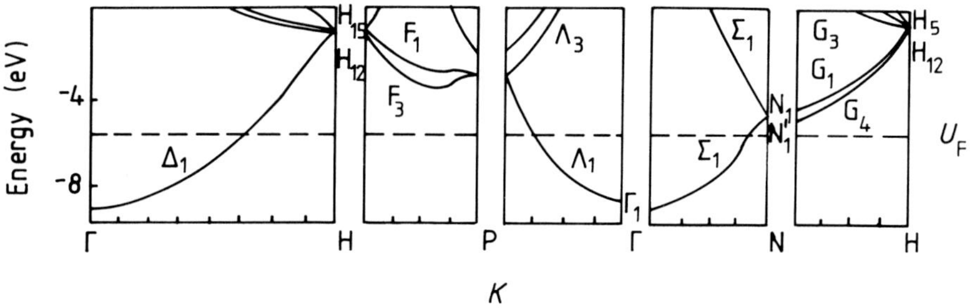

Figure 1.10 Three-dimensional band structure for sodium.

properties of materials are considerably influenced by the relationship between the Fermi surface and the boundaries of the Brillouin zone. Much experimental ingenuity has been used in determining the shapes of the Fermi surfaces of a great many materials; some of these for individual materials are discussed in Chapter 4.

Another way of looking at the departure of the U–K curve from the parabolic form for free electrons is to regard it as arising from a change in the effective mass of the electron. The effective mass, m^*, is defined by

$$m^* = \frac{h^2}{4\pi^2}\left(\frac{d^2U}{dK^2}\right)^{-1}. \tag{1.108}$$

m^* is a measure of the reaction of an electron in the solid to an externally applied electric field and takes account of the complex electric fields operating in the interior of solids. By appropriate differentiation of equation (1.106) it is easily shown that, for the parabolic free electron case, $m^* = m$.

All the equations which have been derived in this chapter for ε_1 and ε_2 contain a term involving the mass of the electron. Optical experiments can be performed from which it is possible to calculate the value of this mass, which is then referred to as the optical mass, m_0. m_0 may be greater or smaller than m and might not be equal to m^* as defined in equation (1.108). Some of the values of m_0 for various materials are considered in Chapter 4.

In addition, particularly in the cases of metals whose Fermi surfaces depart markedly in shape from the free-electron sphere, there is the possibility that electrons at different points on the Fermi surface may be characterised by different relaxation times. This effect will cause broadening of the absorption band produced by the outermost electrons and has been cited by some authors in explaining the optical properties of several of the transition metals.

1.6.1 Effects of band theory on optical properties

The concepts of the band theory of solids can be applied to the interactions between electromagnetic waves and the electrons in materials. Two basic types of electron transitions are possible when a photon is absorbed. The first is an intraband transition in which the electron moves to a higher energy level within the same band; a prerequisite for this type of transition is that free electron energy levels are available in the band. This type of transition is similar to the effect of electromagnetic radiation on free electrons as previously considered, and will produce strong absorption at long wavelengths (low frequencies) in the same position as that predicted by Drude.

The second possibility is an interband transition where the electron gains sufficient energy from the photon to jump into the next higher band; when this happens electrical conduction becomes possible for insulators and semiconductors as well as for metals, and an absorption band will be produced whose lower frequency edge (absorption edge) is the critical frequency to initiate such electron excitation. These absorption bands which can occur anywhere between the visible and the x-ray regions are similar to those produced by bound electrons in the classical theory. Two forms of interband transitions are possible. In the direct mode, the electron makes an energy transition at constant K, i.e. parallel to the U-axis. In the indirect mode, the electron simultaneously experiences a change in K; this is usually regarded as a two-stage process, a change in U at constant K followed by a change in K at constant U (phonon). Direct and indirect transitions are illustrated in figure 1.11.

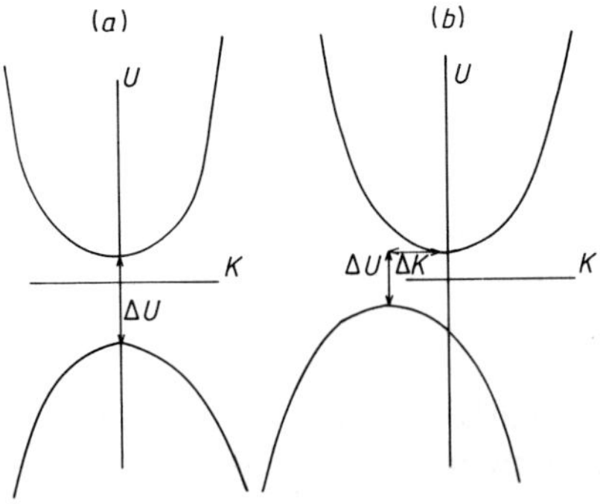

Figure 1.11 Direct (a) and indirect (b) interband transitions.

The quantum treatment of dispersion (Nozieres and Pines 1958) shows that the interaction between an applied electric field and an absorbing atom in a solid can be represented by a series of linear oscillators, the resonant frequency of each corresponding to an interband transition. The contribution of each oscillator to the optical properties of the material is then defined by a factor f_j called the oscillator strength. The equations for ε_1 and

ε_2 for an insulator are as follows:

$$\varepsilon_1 = \frac{Ne^2 f_j(\omega_j^2 - \omega^2)/\varepsilon_0 m}{(\omega_j^2 - \omega^2)^2 + \omega^2\gamma_j^2} \tag{1.109}$$

$$\varepsilon_2 = \frac{Ne^2 f_j \omega\gamma_j/\varepsilon_0 m}{(\omega_j^2 - \omega^2) + \omega^2\gamma_j^2}. \tag{1.110}$$

In the case of a metal the same equations apply for the interband transitions; in addition there is absorption due to the intraband transitions when ω_j is put equal to zero and the plasma resonances have also to be taken into account. It will be seen that there is considerable similarity between equations (1.109) and (1.110) and those obtained when applying the classical theory.

1.6.2 Other absorption mechanisms

The previous discussion, on the possible energy levels in a solid, applies for perfect crystal lattices; there are a number of other mechanisms connected with imperfections in the lattice, which can give rise to the absorption of radiation in a solid. One of these is the presence of impurities which will cause distortion of the lattice and so affect the band structure. Impurity atoms have their greatest effect in semiconductor materials where they generate additional energy levels which can be located in the previously forbidden energy gap, so enhancing the conduction process by electrons and/or holes. Electrons and holes can also form a second kind of absorption mechanism which is particularly important in insulators. An electron-hole pair, known as an exciton, can be maintained as a dipole element by the polarisation field inside the insulator; this dipole will resonate at a particular frequency thus causing strong absorption. The energy of the exciton is always less than that needed to excite an electron across the band gap.

Colour centres are formed by the addition of impurities such as transition metals or rare earths to an insulator. A single 'spare' electron from an impurity atom becomes trapped inside a vacancy due to a missing lattice ion and gives rise to broad absorption and emission bands in the visible spectrum, hence the terminology. In a similar manner, holes may be trapped in alkali halides and produce absorption bands in the ultraviolet region.

1.6.3 Anomalous skin effect

As the frequency of electromagnetic radiation incident on a metal increases, its depth of penetration into the metal decreases and the skin effect becomes the most important feature in determining the optical behaviour. The classical theory of the so-called normal skin effect leads to the following relationship between the depth of penetration, d, and the frequency

$$d = 1/(\mu_0\mu_r\sigma\omega/2)^{1/2}. \tag{1.111}$$

This relationship is based on the assumption that the wavelength of the radiation is long compared with the mean free path (MFP) of the electron, so that, between successive collisions, the electron can be regarded as travelling in a constant electric field. Only those electrons within the skin depth will be accelerated by the electric field associated with the radiation and this will result in a reduction in electrical conductivity. Because the extinction coefficient, k, is directly related to the electrical conductivity (see equation (1.44)), the optical properties of the metal will also be influenced at high frequencies.

Experimental measurements of the variation of the conductivity with frequency, however, revealed that the simple theory outlined above was not followed, except at temperatures greater than 300 K when the mean free path was very small; the discrepancy became greater as the temperature was reduced and the mean free path became larger. This effect is known as the anomalous skin effect. The correct explanation of this effect was given by London (1940) and takes account of the fact that, at high frequencies, the time taken by the electron to traverse one mean free path length will be much greater than the period of the radiation and so the electron will be subject to a varying electric field during its traverse.

Three frequency regions may be distinguished in which the anomalous skin effect may be discussed.

(1) Very small frequencies: here $d >$ MFP and the normal skin effect theory will still be adequate.

(2) Very high frequencies: here $d <$ MFP and the electron will spend only a very small part of its MFP within the electromagnetic field so that, once again, the electron may be regarded as being in a constant field during this time. The normal skin effect theory will again hold.

(3) Intermediate frequencies which cover the visible and near infrared regions, when neither of the above conditions hold and the effects of the varying electric field must be considered. The mathematics of the anomalous skin effect are complicated and will not be given here; further details may be obtained from the article by Reuter and Sondheimer (1949) who have dealt with the effect of this phenomenon the optical properties in this region.

Dingle (1953) has also given a good account of the application of the anomalous skin effect in optical work. He found that it was necessary to take into account the change in the electron collision frequency, ω_c, and produced the following equation for the new frequency, ω_c':

$$\omega_c' = \omega_c + \tfrac{3}{8}(v_F/c)\omega_p(1-p) \qquad (1.112)$$

where v_F is the velocity of the electron at the Fermi energy and p is a factor between 0 and 1. For specular reflections of the electrons $p = 1$, whilst for diffuse reflections $p = 0$. Beattie (1957) has derived relatively simple expressions for ε_1 and ε_2 in the visible and near-infrared spectral regions taking into account the anomalous skin effect and making use of the factor p.

1.7 Excitons

These are elementary localised excited states which can occur in solids and are able to propagate through a crystal lattice; two limiting cases are of interest in this work. Frenkel excitons can occur in some alkali halides when a particular atom, ion or molecule may be in an electronically ionised state; coupling to the adjacent lattice is strong so the excitation can be rapidly transferred from site to site without any motion of the atoms. This occurs by the propagation of a wave of electrical polarisation, the excitons becoming apparent as optical absorption bands or as characteristic luminescence. The Mott–Wannier excitons tend to occur in semiconductor crystals when optical excitation promotes electrons from the lower energy valence band to an upper conduction band, leaving a hole in the valence band. The electrons and holes produced are independently mobile and give rise to electrical conduction. However, they also have mutual electrical attraction and may combine into a hydrogen-like particle the radius of which is much greater than the crystal lattice spacing. In the simplest cases, the exciton has a set of energy levels similar to those of the hydrogen atom i.e. E proportional to n^{-2}, but corresponding to small binding energies. It is also possible for excitons to become trapped on impurity atoms or lattice defects. These excitons become observable as additional narrow absorption bands close to the main band in the ultra-violet spectrum.

References

Beattie J R 1957 *Physica* **23** 898
Bennett H E, Silver M and Ashley E J 1963 *J. Opt. Soc. Am.* **53** 1089
Born M and Wolf E 1975 *Principles of Optics* 5th edn (Oxford: Pergamon)
Brandli G and Sievers A J 1972 *Phys. Rev.* **B 5** 3500
Dingle R B 1953 *Physica* **19** 311, 348, 729, 1187
Ditchburn R W 1976 *Light* (London: Academic)
Drude P 1887 *Ann. Phys., Lpz.* **32** 584
Ferrell R A and Sterne E A 1962 *Am. J. Phys.* **30** 810
Golovashkin A I and Motulovich G P 1968 *Sov. Phys.-JETP* **26** 88
Hagen E and Rubens H 1900 *Ann. Phys., NY* **1** 352
Heavens O S 1965 *Optical Properties of Thin Solid films* (New York: Dover)
Kittel C 1986 *Introduction to Solid State Physics* 6th edn (New York: Wiley)
Lipson S G and Lipson H 1969 *Optical Physics* (Cambrigde: Cambrigde University Press)

London H 1940 *Proc. R. Soc.* A **176** 522
Lorentz H A 1906 *Theory of Electrons and its Applications to the Phenomena of Light and Radiant Heat* (New York: Columbia University Press)
Mott N F and Jones H 1936 *Theory of the Properties of Metals and Alloys* (Oxford: Oxford University Press)
Nozieres P and Pines D 1958 *Phys. Rev.* **109** 741
Reuter G E H and Sondheimer E H 1949 *Proc. R. Soc.* A **195** 336
Shiles E, Sasaki T, Inokuti H and Smith D Y 1980 *Phys. Rev.* B. **22** 1612
Sokolov A V 1967 *Optical Properties of Metals* (London: Blackie)
Steinmann W 1968 *Phys. Status Solidi* **28** 437
Ward L 1993 *Opt. Laser Techol.* **25** 393–5

2 Experimental Methods for Measuring the Optical Constants of Bulk Materials

2.1 Introduction

This chapter is devoted to the consideration of experimental methods for the determination of the optical constants of materials in bulk form. These techniques can be divided into four categories, namely:

(i) photometry, in which the ratio of the intensities of reflected to incident radiation in the s- and/or p- states of polarisation is measured, the angle of incidence being an additional parameter;
(ii) polarimetry where the phase changes on reflection in the s- and/or p-states are measured also possibly at different angles of incidence;
(iii) a mixture of polarimetric and photometric functions;
(iv) ellipsometry, where plane polarised light, on reflection from the surface, becomes elliptically polarised and the constants of the ellipse are determined.

These classifications are by no means exclusive, even between photometry and ellipsometry, but any method which measures both the light intensities and the phase changes, however indirectly, will be classed as ellipsometric.

2.2 Surface of specimen

In all optical measurements a factor of paramount importance is the condition of the surface under investigation. Imperfections, such as non-flatness, scratches and roughness, will render the surface nonspecular whilst contamination and surface films introduce layers having different optical properties from those of the specimen itself. In either case, the equations derived in Chapter 1 will be invalidated and suspicion will be thrown on the quality of the results obtained. The preparation of the surface of the specimen must therefore be carried out with great care to ensure that it is optically smooth and free from contamination. This may be achieved by one of the following techniques

(i) mechanical polishing by grits and powders,
(ii) electropolishing in a suitable chemical bath at a prescribed current density,
(iii) cutting or cleaving in a vacuum,
(iv) deposition of very thick films under vacuum conditions.

2.2.1(a) Mechanical polishing

The exact mechanism of polishing by abrasive materials has still not been fully explained. One of the most important contributions in this field was made by Beilby (1921) who argued that two processes take place during polishing. Firstly, the peaks on the surface are ground away, the valleys being filled with the eroded material and, secondly, this heavy working of the surface produces a layer, known as the 'Beilby layer', whose structure is amorphous. Support for Beilby's theory was adduced from the effect of chemical etching on a smooth mechanically polished surface; scratch marks reappeared, apparently those due to earlier polishing which had subsequently been filled in by material from the peaks and were thus more readily etched away. French (1932, 1933) used x-ray diffraction to study the surfaces of metallic specimens polished by emery papers. He obtained diffuse diffraction patterns indicating that the crystal size had been reduced by the polishing process; further polishing with magnesium oxide in water or rouge in alcohol increased the diffuseness of the patterns. French therefore concurred with Beilby's ideas that the polishing mechanism involved more than a mere levelling of the surface. Rooksby and Randall (1932) confirmed that the most probable surface structure was of very small crystals, but they were critical of French's claim that an amorphous layer was produced. Bowden and Hughes (1937) suggested that, in the polishing process, the real areas of contact between the surface and the abrasive were very small ($\sim 10^{-8}$ m^2) and that the frictional forces involved could generate high transient temperatures at these points so that local melting could take place. There has, however, been little evidence to support this idea.

The process of abrasion by grits (e.g. emery) involves the gouging of the surface by the sharp corners of the particles. The rate of removal of abraded material is much less than would be theoretically predicted as only about 10% of the grits are orientated at the correct 'attack angle' to be effective in gouging. In addition, deformation of the crystal structure takes place because of the high local forces; the depth of this is much greater than had been previously estimated.

The traditional method of polishing is to commence with coarse grits (grade 00) and then to move on to a finer grit, polishing at right angles to the previous scratches until they can no longer be seen; the process continues in this way up to the finest available grits and then a final polish may be given using rouge or magnesia powders. There will, however, be considerable deformation underneath the surface and this will be revealed by etching. It is currently suggested that the production of deep deformation in the form of dislocations and multiple crystal boundaries, rather than the filling of the valleys, is the correct explanation of Beilby's observation of the reappearance of scratches.

Thus, if each stage of the polishing process is carried out for a much longer period than that necessary simply to remove the previous scratches, the underlying deformed material will also be abraded and replaced by a thinner layer of deformation due to the finer abrasive being used. In this way it is now possible to produce mechanically polished specimens which are largely free from the surface layers which affected the results of many of the earlier workers. Many optical specimens receive a final polish with diamond abrasives; the polishing mechanism is the same as with emery but produces effects on a smaller scale, including the depth of the underlying deformation. All specimens should be well washed in distilled water and carefully dried as soon as the polishing process has been completed.

Archard (1985) has reviewed this field; in addition to the conclusions above, he also speculated that chemical reactions with the abrasive might be involved in the polishing process, these being particularly noticeable in the final stages of the polishing. Samuels (1971) also gives a detailed description of the polishing process.

It is most important to make sure that any deformed surface layer has been removed before making optical measurements and it is now common practice to examine all polished surfaces of solid specimens under a phase contrast microscope to ensure that the crystal structure is well exposed.

2.2.1(b) Roughness

The roughness of the surface clearly has an important influence on its optical properties and, before making measurements, every effort should be made to produce as smooth a surface as possible. Roughness is measured as the RMS deviation from the mean surface and is usually expressed in Å; typical values for an optically smooth surface are of the order of 10 Å. Bennett and King (1970) have described the effect of polishing techniques on the roughness and residual surface film of fused quartz, reaching the conclusion that the bowl feed technique produced better results than the fresh feed method. Using float polishing, Bennett *et al* (1987) reduced the roughness of glass surfaces from 10 to 2 Å.

Techniques for measuring surface roughness may be divided into two types—contact and non-contact. The contact method (Bennett and Darcy 1981) involves scanning the surface at a fixed rate with a very light stylus which is attached via a lever system to a small coil located in a magnetic field. Vertical movement of the stylus causes the coil to move in the field generating a small voltage in the coil; the magnitude of this voltage can be interpreted as a height. The instrument can be made accurate to 1 nm of vertical movement and 1 μm of lateral movement. The advantage of this method is that, after calibration, no reference surface is required; its disadvantage is the possibility of scratching the test surface.

Many of the non-contact methods involve interferometry; the field has been covered in books by Bennett and Mattson (1989) and Dyson (1970). Bennett (1976) has described the use of fringes of equal chromatic order (FECO) which are formed when a collimated beam of light undergoes multiple reflections between two partially silvered surfaces one of which is that whose profile is being measured and the other is as near a perfect surface as possible. When the emergent light is dispersed, interference fringes are produced in the form of irregular lines superimposed on a continuous spectrum. In the Mirau interference method (Bhushan *et al* 1985, a beam of white light passes through a red filter and a microscope before being split into two portions; one half is then reflected from the test surface and the other from a reference sample which is oriented at a small angle to it. The two beams are then recombined when straight line interference fringes are produced if the two surface are equally smooth; deviations are shown up as waviness in the fringes. This device can detect roughness of 1 Å.

Another instrument, the common path heterodyne interferometer (Sommargren 1981), uses a He–Ne laser as source; the red line is divided into two components differing in frequency by about 2 MHz by Zeeman splitting in an axial magnetic field, these being polarised in the s and p directions. The two beams pass through a Wollaston prism and then fall on the specimen a short distance apart ($\sim$ 160 μm); the specimen is rotated so that one beam moves in a circle of circumference 1 mm about the other. The two beams are then recombined and the beat frequency between them gives a measure of the height differences along the circle. In an alternative method the second beam can be directed on to the surface by a rotating polygonal mirror.

The surface may also be surveyed by a scanning electron microscope, a probe beam of electrons forming a raster pattern on the surface. The secondary electrons produced, whose intensity is a function of the emission angle, are collected by a detector which then gives a height profile over the raster. This technique has been described by Guenther and Pulker (1976).

2.2.2 Thick films

In order to avoid the problems encountered in mechanical polishing, some workers have used very thick films as close approximations to solids. The techniques for the deposition of films are discussed in Chapter 5, but it must be pointed out that there is some doubt that the structures of these films are identical with those of the solid materials. In particular, it has been found that the state of aggregation of the films depends critically on the evaporation conditions.

2.2.3 Cutting and cleaving

It is possible to cut soft materials, such as the alkali metals, inside a vacuum system and so produce a smooth, contamination-free surface. Wood (1933) used this technique with the metal sodium. Harder materials may be cut by means of an acid string saw. Alternatively, some crystalline materials, such as rock-salt, may be cleaved to produce a clean, flat surface.

2.2.4 Electropolishing

A metal can be electropolished by using it as the anode in a suitable electrolytic cell. The action of the electric current causes metallic ions to be removed from the surface and this effect is greatest at the high spots; hence the hills tend to be removed in preference to the valleys and a smoother surface is created. It is necessary for the specimen to be made reasonably flat by mechanical polishing prior to electropolishing. In addition the current density must be carefully controlled to lie on the plateau of the voltage–current characteristic. The electrolytes must also be kept cool as some of these mixtures cause violent reactions if the temperature is allowed to rise. West (1965) has given a concise account of this technique, including the appropriate polishing solutions and current densities, for a number of metals. Electropolishing is a good method of removing deformed layers produced in mechanical polishing. However, the electrolytes used may cause contamination of the surface unless careful washing takes place immediately the process is finished.

2.2.5 Ion polishing

This method of polishing avoids the use of both grits and chemicals which may damage the surface; a suitable apparatus has been described by Hoffmann *et al* (1975). Polishing is produced by the sputtering action due to bombarding the surface by gas ions at a glancing angle (see Chapter 5). The ions, usually of xenon, are accelerated towards the specimen by a large electric field and will preferably strike and remove the high spots on the surface. The sputtering beam can be scanned across the surface which is simultaneously rotated. It is possible to cause damage by the actual bombardment or by the entrapment of xenon molecules, but both effects can be eliminated by annealing at 450 °C. This technique is usually applied after initial polishing by more conventional methods.

2.3 Radiation detectors and noise

The next important factor in making optical measurements is the adequacy of the detecting and measuring system. The history of the development of optical measurements has been marked by the increasing sensitivity and spectral range of the detectors available and also by the reduction of the background noise which limits the use of all the devices.

2.3.1 Noise

Since noise is often the greatest source of error in optical measuring systems, it is useful for it to be considered first of all. Fuller descriptions of the various forms of noise and their reduction have been given in a number of books, for example Dereniak and Crowe (1984), to which the reader is referred. Two main categories of noise are recognised; photon noise and detector (system) generated noise. Photon noise arises from the bombardment of the detector by photons. Any fluctuations in the rate of arrival of the photons will give rise to a random production of current carriers in the detector. The problem is most acute with low-powered sources, where the rate of production of photons is small, and at high frequencies, where the individual photon energies are large and thus the rate of production of the photons may again be small so that any fluctuations will be more serious. Another source of photon noise is radiation from the surroundings falling on the detector; this can only be minimised by keeping the ambient temperature as low as possible.

There are several types of detector-generated noise. Johnson noise is caused by the thermal motion of the charged particles within a resistive element in the detector, leading to spurious electric currents. A closely related effect is called temperature noise, and is caused by changes in temperature of the detector other than those caused by the signal. Shot noise, which occurs in photodiodes, is a result of the photon nature of radiation; if the arrival of the photons is in the form of a Poisson distribution, so is the generation of the photoelectrons. An effect which occurs in photoconductors and photodiodes is generation-recombination noise which is caused by fluctuations in the rates of generation and recombination of the electrons and holes and hence of the carrier densities. The type of noise known as '$1/f$' is very commonly encountered. No exact explanation has been given for this source of noise, but it is possibly linked with ohmic contacts inside the detector. It is always present in current-biased detectors, such as the bolometer and the photoconductive cell, but not in photovoltaic cells. It is clearly most serious at low frequencies. Finally, mention must be made of microphonic noise which is generated by mechanical vibration of the detector, causing small changes in resistance or capacitance which will cause corresponding fluctuations in the output.

All the types of detector-generated noises discussed above can also be produced in the amplifier and recorder stages so it is important to pay just as much attention to noise suppression in these parts of the detector system.

2.3.2 Detectors

The development of the linear thermopile will serve as an illustration of the way in which radiation detectors have been steadily improved with time. In its original form this device consisted of a large number of junctions between bismuth and antimony wires connected in series; alternate junctions were attached to small blackened discs whilst those in between were kept cool. When the discs were exposed to radiation, a small thermo-EMF was generated. Although the thermopile was capable of absorbing and detecting radiation of all wavelengths, its relatively massive construction led not only to a small temperature rise, and hence to low sensitivity, but also to a long response time of several seconds. In consequence, optical measurements made with this device were tedious and were also limited by the background (Johnson) noise in the thermopile and by fluctuations in both the intensity of the source and the ambient temperature. Improvements were effected by constructing thermopiles of smaller thermal capacity and hence more rapid response (~ 1 s) and also by using two devices connected back to back thereby eliminating the drift caused by changes in the ambient temperature. A Wheatstone bridge circuit using four thermopiles, two exposed to the radiation and two covered, brought about a further increase in sensitivity. In spite of these improvements, however, the thermopile has been overtaken in sensitivity and speed of response by many other radiation detectors. Nevertheless, modern thermopiles are available, formed by the evaporation of very thin layers of antimony and bismuth and having response times of about 50 ms. With such a relatively short response time, the incident radiation can be interrupted at a frequency of about 10 Hz and the resulting signal processed by an amplifier tuned to the same frequency. In this way a much larger signal can be obtained from the detector unit and at the same time most of the background 'white' noise generated by the cell will not be passed by the tuned circuit thereby improving the signal/noise ratio of the detector. The thermopile is most useful in the near infrared when devices such as the photocell become very insensitive.

Another detector which is sensitive at all wavelengths is the bolometer. In its original form this device consisted of a thin foil of metal, coated black so that it would be an efficient absorber of radiation, which caused an increase in temperature and hence in the electrical resistance of the foil. This parameter could then be used as a quantitative measure of the radiation. Bolometers also suffered from long response times and were subject to both

Johnson and $1/f$ types of noise, but, again, modern devices can be made in thin-film form having small response times and hence usable with chopped radiation and tuned amplifiers. Bolometers may also be made from oxide or semiconductor materials which have large negative temperature coefficients of resistance (Low 1961, Clark *et al* 1977). By using an extrinsic semiconductor material with a very narrow energy gap measurements can be made up to wavelengths of 100 μm; such devices are called thermistors.

The Golay cell (Golay 1947) is another device which is sensitive to all wavelengths of radiation. It consists of a cylindrical vessel having on one end a blackened radiation-absorbing disc and on the other a membrane. Absorption of radiation causes an increase in the gaseous pressure inside the cylinder and the membrane is distorted, tilting a mirror and so moving a spot of light on a scale. There is a small hole in the cylinder to prevent effects due to changes in barometric pressure. Although the response time is quite long ($\sim$20 ms) the cell is operated with an alternating signal. The cell is subject to thermal noise but a sensitivity of 5×10^{-8} W can be obtained. However the cell is delicate and needs very careful setting up and operation as it is susceptible to microphonic vibrations.

The photoemissive cell has been used in a very large number of studies of optical constants. When photons of sufficient energy fall on the sensitive area of the cell electrons are emitted and may be collected by a positively charged electrode; at a constant wavelength, the resulting current is proportional to the radiation intensity. In order to cause emission of electrons, the photon energy must be greater than the workfunction of the photosurface ($\sim$1.5 eV) and so these cells are not effective in the infrared region. In addition their quantum efficiency, i.e. the electron emission per quantum of radiation, is reduced at short wavelengths. The useful wavelength range is therefore approximately 0.3 to 0.8 μm. Over this spectral range photoemissive cells are more sensitive and rapid than thermopiles; for example the caesium cell, which is useful between 0.3 and 0.7 μm, has a sensitivity of 50 mA W^{-1} and a response time of about 5 ns. Modern photocells have cathodes of the bialkali (antimony–potassium–caesium or antimony–rubidium–caesium) or trialkali types (antimony–sodium–potassium–caesium) with sensitivities up to 100 mA W^{-1}. Photocells are subject to both shot and Johnson noise.

A still greater sensitivity over the same spectral range can be obtained with the photomultiplier tube which consists of ten or more photosensitive surfaces (dynodes). The electrons ejected by the radiation from the first dynode are attracted to the second by a potential difference of about 100 V gaining sufficient kinetic energy in the field to cause multiplication of the ejected electrons. This process continues at each dynode and an overall gain of 100 000 000 can be achieved. A very stable power supply is required as any fluctuations will be similarly magnified.

A further significant development in radiation detection techniques occurred with the introduction of photoconductive devices. These cells are made from compounds such as PbS, InSb and GaAs and have a high electrical resistance which decreases on irradiation due to the production of additional electron-hole pairs across the narrow energy gap. They have very short response times ($\sim 50\ \mu s$) and this enables the use of chopped radiation at a frequency of about 1000 Hz, the output being connected to an amplifier tuned to the same frequency. The spectral range of these cells extends into the near-infrared region; thus lead sulphide is sensitive from 0.3 to 2.5 μm, lead selenide up to 4.0 μm and lead telluride up to 7.0 μm. The photoconductive cell is subject to Johnson, generation–recombination and $1/f$ noise, Johnson noise being most serious at high frequencies, generation–recombination noise in the middle frequencies and $1/f$ at the low frequencies. The first two types of noise can be reduced by cooling the cell to liquid nitrogen temperature and this also increases the spectral range on the long-wavelength side.

A radiation-sensitive device which has been introduced comparatively recently is the pyroelectric cell first described by Cooper (1962). This cell consists of a collection of minute crystallites (a suitable material being triglycine sulphite) each of which behaves as a small electric dipole; the dipoles are initially aligned by cooling the cell below its Curie temperature in an electric field. If the temperature is raised, for example by the absorption of radiation, the electrical polarisation of the cell undergoes a reversible decrease, the magnitude of which is proportional to the radiation intensity. The cell is usable between 1.0 and 15.0 μm and has a time constant of about 75 ms so that it can be used with a radiation chopper at 10 Hz. The main source of noise is Johnson noise.

The photodiode is a device such as a p–n junction which has an internal potential barrier; irradiation causes the generation of electron–hole pairs and these produce a current in an external circuit. The energy of the incident photon must be greater than the potential barrier, which can be made small by appropriate doping of the basic semiconductor material. This device is also susceptible to generation–recombination noise, particularly if the energy gap is very small when it is necessary to operate the cell at low temperatures. The photodiode is also affected by Johnson, shot and $1/f$ noise. The p–n junction may be used over the visible spectrum and up to about 1.0 μm. The phototransistor operates on the same basic principle as the photodiode but incorporates internal amplification and is therefore more sensitive.

A suitable detector for microwaves is a crystal rectifier whilst x-rays can be measured by an ionisation chamber.

The characteristics of the various radiation detectors are summarised in table 2.1.

As indicated earlier, one of the main lines of development in optical measurements has been the improvement of the signal/noise ratio for all types of measuring systems, and a further step in this direction was the introduction of phase-sensitive detection. This operates in conjunction with a radiation chopper by comparing the signal to be measured with another signal generated at exactly the same frequency by a second photodetector receiving light from an independent source through the same chopper disc. The phase difference between the two signals can be adjusted so that the amplifier is only operating for a small fraction of the cycle and this again reduces the total noise transmitted, whilst not affecting the signal.

Table 2.1 Characteristics of radiation detectors.

Type	Range (μm)	Response time(s)	Maximum sensitivity	Noise
Thermopile	1–25	5×10^{-2}	$50\ V\ W^{-1}$	Johnson
Bolometer	1–25	5×10^{-2}	$50\ V\ W^{-1}$	Johnson
Thermistor	1–100	5×10^{-2}	$500\ V\ W^{-1}$	$1/f$
Golay	1–2000	2×10^{-2}	—	Microphonics
Pyroelectric	6–1500	10^{-1}	$500\ V\ W^{-1}$	Johnson
Photocell	0.3–0.8	5×10^{-9}	$50\ mA\ W^{-1}$	Shot, Johnson
Photomultiplier	0.3–0.8	5×10^{-9}	$5 \times 10^4\ A\ W^{-1}$	Shot, Johnson
Photoconductive	0.3–10	2×10^{-4}	$10^5\ V\ W^{-1}$	Johnson, $1/f$, G-R
Photovoltaic	0.5–1.0	10^{-7}	$0.5\ A\ W^{-1}$	Johnson, $1/f$, shot, G-R

2.4 Radiation sources and optical windows

The radiation sources available for optical measurements include Nernst filaments and incandescent filament lamps and strip lamps which produce copious outputs in the visible and near-infrared regions. Commercially manufactured black bodies are useful as sources for the far infrared. The laser is a powerful source, particularly at discrete wavelengths in the visible and ultraviolet regions. The ultraviolet region can also be covered by fluorescent lamps, whilst for the soft x-ray region, as well as x-ray tubes, the powerful synchrotron radiation from particle accelerators is now more readily accessible. In the microwave region a suitable source is a cavity oscillator.

Many optical measurements are carried out inside closed vessels under vacuum conditions and it is necessary to use optical windows to admit the radiation. The choice of appropriate window material is important to ensure that the window causes the least possible modification to the passing radiation. The window material should therefore be optically flat and have a very low absorption coefficient over the spectral range being investigated. In addition it should not be birefringent or introduce rotation of the plane of polarisation of the radiation. Suitable materials are quartz (cut so that the radiation travels along the optic axis) for ultraviolet radiation, glasses for visible light, and quartz and rock-salt for infrared radiation.

2.5 Techniques for non-absorbing materials

The characteristic optical properties of insulators consist typically of values of n lying between 1 and 4 and values of k which are very small or even negligible. The main exceptions to this group are the ferroelectric materials, where n can have values above 100.

The refractive indices of dielectrics and semiconductors may be determined from a prism of the material of angle A, by measuring the angle of minimum deviation, D, of radiation passing through the prism. The refractive index, n, is given by the equation

$$n = \sin \tfrac{1}{2}(A + D)/\sin \tfrac{1}{2}A. \qquad (2.1)$$

This method has been used by Briggs (1950) for measurements on silicon and germanium, and by Malitson (1964) on barium fluoride.

Measurement of the critical angle, θ_c, may also be used to determine the refractive index using the relationship $n \sin \theta_c = n_2$, where n_2 is the known refractive index of the second medium. For materials where k is negligible, this method gives a sharp cut-off at the critical angle and n may be determined accurately. Meeten *et al* (1984) have discussed the situation where k is finite when the sharp critical angle is replaced by a region where R and T are continuously changing. It then becomes much more difficult to decide the cut-off position and considerable errors may occur in the values of n.

Another technique for obtaining the optical constants of insulators, particularly in the microwave region, is to measure their complex dielectric constant, $\varepsilon = \varepsilon_1 - i\varepsilon_2$, which is equal to $(n - ik)^2$. Maier (1970) has described a rectangular waveguide technique by which ε can be determined with great accuracy.

One of the most accurate techniques for measuring the value of n is the Rayleigh refractometer. In this instrument, two sets of interference fringes are produced between light passing through the specimen and a reference beam. A small change in the refractive index of the specimen, dn, causes one of the fringe patterns to be shifted relative to the other; the change in optical path difference being given by $dn\ d$, where d is the thickness of the specimen. This path change is also equal to $ds\ \lambda$, ds being the fringe shift and λ the wavelength of the light. Ditchburn (1976) has stated that the minimum detectable change in n is about $\lambda/40d$ and the maximum $200\lambda/d$.

Johnson *et al* (1969) used another interference method to study the optical properties of CdTe and GaAs. The beams reflected from the top and bottom of a parallel sided block of the material of thickness d will interfere constructively when $2nd$ is equal to an odd number of half wavelengths. By scanning a range of wavelengths, a series of such null points could be found and thus the refractive index could be determined as a function of wavelength.

The refractive index of a dielectric may also be found by reflection photometry as described in §2.6; this method has been used by McCarthy (1963).

2.6 Photometric methods for absorbing materials

The optical constants of absorbing materials may be determined from measurements of the reflectance R_s and R_p of solid specimens at known angles of incidence. Equations for these two optical functions were developed in Chapter 1 (equations (1.53) and (1.61)) and are as follows, using the same notation:

$$R_s = \frac{(Y_1 - u_2)^2 + v_2^2}{(Y_1 + u_2)^2 + v_2^2} \tag{2.2}$$

$$R_p = \frac{(e_1 - Z_1 u_2)^2 + (e_2 - Z_1 v_2)^2}{(e_1 + Z_1 u_2)^2 + (e_2 + Z_1 v_1)^2} \tag{2.3}$$

where

$$Y_1 = n_1 \cos\theta_1 \qquad Z_1 = n_1/\cos\theta_1 \qquad \varepsilon_1 = n_2^2 - k_2^2$$

$$\varepsilon_2 = 2n_2k_2 \qquad \varepsilon_3 = (n_1 \cos\theta_1)^2$$

$$2u_2^2 = \varepsilon_1 - \varepsilon_3 + [(\varepsilon_1 - \varepsilon_3)^2 + \varepsilon_2^2]^{1/2} \qquad \text{and}$$

$$2v_2^2 = \varepsilon_3 - \varepsilon_1 + [(\varepsilon_1 - \varepsilon_3)^2 + \varepsilon_2^2]^{1/2}.$$

These equations are not only non-linear but also transcendental and it is difficult to rearrange them in the form of analytical solutions for n and k without making simplifying assumptions some of which will be discussed later. Solutions can be obtained by graphical techniques and several such methods have been evolved, amongst which are those due to Collins and Bock (1943) and Avery (1952); these methods are extremely laborious and tedious but the authors have produced sets of working graphs which may be useful to other workers. Since the advent of the digital computer, however, it has become possible to obtain numerical solutions to these equations, especially if reasonable starting values based on previous results can be used. The values of n and k are determined by an iterative process which seeks to find the best fit between the experimental values of the optical functions and those calculated from the entered values of n and k.

Approximate solutions for n_2, and k_2 may be obtained if it can be assumed that ε_3 is negligible compared with ε_1. Then, $u_2 = n_2$ and $v_2 = k_2$, and this leads to the following expressions for R_s and R_p:

$$R_s = \frac{Y_1^2 - 2n_2Y_1 + n_2^2 + k_2^2}{Y_1^2 + 2n_2Y_1 + n_2^2 + k_2^2} \qquad (2.4)$$

and

$$R_p = \frac{Z_1^2 - 2n_2Z_1 + n_2^2 + k_2^2}{Z_1^2 + 2n_2Z_1 + n_2^2 + k_2^2}. \qquad (2.5)$$

These equations may be firstly solved for n_2, giving

$$2n_2 = \frac{(Z_1^2 - Y_1^2)(R_s + R_p - R_sR_p - 1)}{(R_s - R_p)(Y_1 + Z_1) + (1 - R_sR_p)(Y_1 - Z_1)}. \qquad (2.6)$$

k_2 can then be calculated from

$$k_2^2 = [R_s(Y_1 + n_2)^2 - (Y_1 - n_2)^2]/(1 - R_s). \qquad (2.7)$$

The experimental techniques for the measurement of R_s and R_p are straightforward, bearing in mind the precautions necessary to eliminate the various sources of error discussed below. There are two main techniques for the determination of reflectance: direct and comparison. Figure. 2.1 shows a schematic layout of one experimental arrangement for the direct method. The radiation from the monochromator may be unpolarised or can be made plane polarised by an appropriate device (see §2.8) before falling on the reflecting surface of the specimen. The detector must be mounted on a rotating arm so that it can be swung accurately to measure the intensities of both the incident and reflected beams. This experimental necessity, although it allows different angles of incidence to be used, is one of the weakest features of this type of photometric measurement as the repositioning of the detector is difficult to perform with consistent accuracy. This limits the overall accuracy attainable and many of the earlier results were

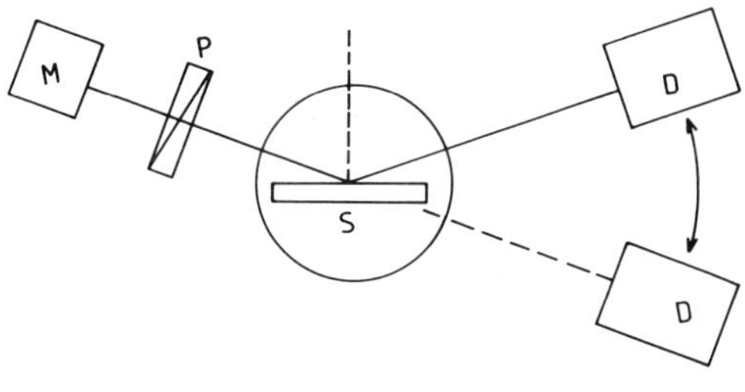

Figure 2.1 Apparatus for direct measurement of reflectance. M, monochromator; P, polarising device; S, specimen on table; D, detector.

only accurate to about 1%. In some of the more recent papers, e.g. Okoomian (1969), authors claim accuracies ten or more times better than this mainly due to efforts to eliminate the necessity to change detector positions.

Hunter (1960) describes the design of an accurate reflectometer of this type having the following features: an adjustment to obtain coaxial rotation of the detector and the specimen; the ability to rotate the detector separately; the ability to measure the reflectance on either side of the axis of the instrument; and the facility to measure the reflectance at different azimuth angles.

An experimental arrangement for the comparison method due to Strong (1938) is illustrated in figure 2.2. The beam of radiation is allowed to fall either on a comparison mirror, A, of reflectance R_A, from which there is one reflection, or on the surface under investigation, of reflectance R, from where it passes to a second comparison mirror, B, of reflectance R_B, and then back for a further reflection at the surface. Comparing the intensities of the two reflected beams yields the quantity R^2R_B/R_A. The ratio R_B/R_A may be found by comparing the radiation intensities when (i) the specimen is replaced by mirror A and (ii) A and B are interchanged. The apparatus still suffers from the difficulty of aligning the mirrors between readings and it is also not easy to alter the angle of incidence during the course of the experiment.

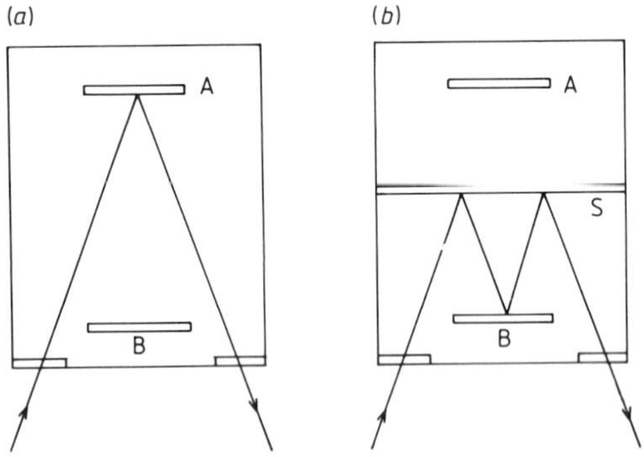

Figure 2.2 Strong's comparison method for reflectances. S, specimen; A, B, comparison mirrors.

The measurement of absolute reflectance presents a number of problems other than those outlined above, and these have been discussed by Kuhn and Wilson (1950, 1958). They may be summarised as follows:

(1) The apertures of both the light source and the detector must be sufficiently small that any ray passing directly from one to the other does so outside the specimen.

(2) The light source must emit uniformly into the solid angle subtended by the aperture of the measuring system (i.e. uniformly across its cross section). This condition is very important because, in reflectance measurements, the reversed virtual image of the source formed by the specimen becomes the object for the detector and so the source must be identical when viewed from both sides.

(3) The same argument applies to the detector as its sensitivity may not be uniform across its surface. Ground glass screens placed in front of both the source and the detector can be used to ensure the uniform situations described in (2) and (3) although they will lead to a loss of light intensity.

(4) The light scattered by any system component including the specimen must be negligibly small.

(5) The surface of the specimen must be planar.

(6) All adjustments in the system must be capable of high accuracy.

In addition, Hunter (1960, 1965) has considered the errors due to the following.

(1) Non-parallelism of the incident radiation which, in general, is not serious.

(2) An unknown amount of polarisation in the incident beam; if neglected, this can cause considerable errors but, if known, its effect can be taken into account.

(3) Slow fluctuations in the intensity of the source; this may be minimised by taking the intensity measurements in a systematic pattern and then averaging.

Even after all these precautions have been taken, there are still further difficulties with photometric measurements, particularly at large angles of incidence. The main problem is associated with the decrease in the effective aperture of the specimen as the angle of incidence increases; it is important that the specimen itself does not become the limiting aperture of the system as this would then change with angle of incidence. Consequently, as many of the specimens used are of small area, the aperture of the light beam must be reduced when large angles of incidence are used with a consequent loss of intensity and a worsened signal/noise ratio.

A modification of the photometric technique which is very useful for materials of high reflectance is that employing multiple reflections of the beam. Gates *et al* (1958) have described equipment which uses 2, 4, 6, 8 or 16 reflections and Arnon and Baumeister (1978) have used this technique to measure accurately reflectance larger than 0.99.

Collins and Bock (1943) measured the reflectances of linearly polarised light at a number of different angles of incidence, claiming an accuracy of ±1%. They calculated the optical constants of the material by a graphical process, the most reliable results being obtained with materials of low reflectance.

Avery (1950, 1952) attempted to overcome the difficulties of absolute reflectance measurements by measuring the ratio R_p/R_s at a number of different angles of incidence. A polarising device was introduced into the incident beam and, by rotating this through 90°, values of the intensities of the incident and reflected beams in the s- and p- states could be measured. Provided the rotation of the polariser did not cause a lateral shift of the light beam, the ratio R_p/R_s was independent of the uniformity of the beam or the position of the photodetector. It was not even necessary to assume that the s- and p- components of the incident beam were equal, as their ratio could be determined from a preliminary measurement which was also independent of the beam uniformity or the position of the detector. Having determined R_p/R_s for at least three different angles of incidence, Avery was able to calculate n and k by a graphical process.

Another photometric method is the measurement of the reflectances of unpolarised radiation at two or more different angles of incidence, e.g. Simon (1951). Russo (1985) has described a novel technique to measure the ratio R_p/R_s by reflecting the radiation from two faces of a pyramid-shaped specimen, polarising one beam in the s-direction and the other in the p-direction and then measuring their intensities simultaneously.

In recent years the quality of reflectance photometry as a method for the determination of n and k has been greatly improved by the use of the technique of multiple angle photometry in which the reflectance is measured at several angles of incidence; these readings can be performed rapidly and accurately with modern equipment. n and k are then computed by one of the iterative processes already mentioned (section 2.6) using Fresnel's equations as standards. Thc accumulation and use of so much data enables the location of n and k to be made with much greater precision than with the single value reflectance technique and also tends to pick up any 'missing' solutions neglected by the simpler methods. Miller *et al* (1988) have given a description of the use of this technique which they employed to measure the optical parameters of $Au/SiO_2/Si$ films.

2.7 Polarimetric methods

The polarimetric technique involves the measurement of the phase difference on reflection, Δ, between the s- and p-states of polarisation. An interferometer, such as Michelson's, may be used for this purpose; figure 2.3 shows such an arrangement. The beam is polarised in either the s- or p-plane and interference fringes are produced after two reflections at an angle θ at the surface of the specimen, S. On rotating the polariser P through 90°, a fringe shift is produced from which Δ can be calculated. This technique has been used by Bennett (1966) and also by Shklyarevskii (1957). It is necessary to measure Δ at two different angles of incidence in order to solve for both n and k. The phase changes can be determined to an accuracy of about one degree which is comparable with that for the photometric methods. The main limitation to accuracy in polarimetry is caused by the lack of sharpness of the interference fringes.

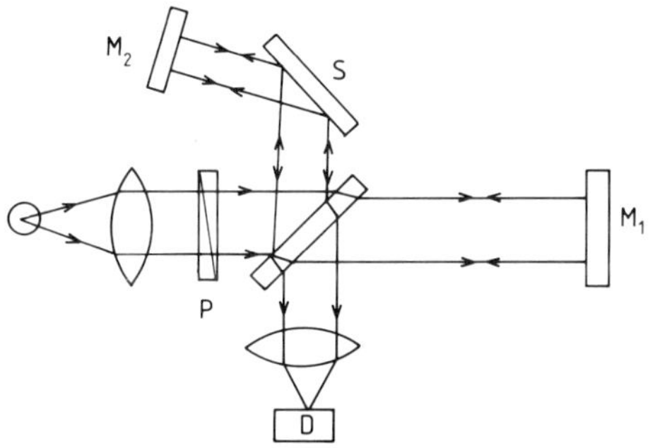

Figure 2.3 Interferometric method for measuring phase differences on reflection. M_1, M_2, mirrors; P, polarising device; S, specimen; D, detector.

2.8 Ellipsometric methods

Reflection ellipsometry has been extensively used to measure optical constants for almost a century since it was first introduced by Drude (1894, 1959).

As was shown in §1.3, when plane polarised light is reflected from a surface, changes are produced in both its amplitude and phase, these changes being different for the s- and p-states of polarisation. Consider a beam of radiation of amplitude A, plane polarised at an angle Q to the plane of incidence on the specimen. The incident beam may be resolved into components $A_s = A \cos \mathrm{A}$ and $A_p = A \sin Q$ and these will be in phase. After reflection the amplitudes and phases of each component will be altered and the reflected beam will become elliptically polarised. If the constants of the ellipse can be determined it is possible to calculate the optical constants of the specimen.

In practice the optical constants are generally determined from the measurement of two angles. Firstly, the phase difference, Δ, between the p- and s-components of the reflected beam must be found, and this may be accomplished by introducing a compensating device into the reflected beam which produces an exactly equal and opposite phase difference to that produced by the reflection (suitable compensating devices are discussed later in this section). After compensation the beam again becomes plane polarised but at a new azimuth angle, Ψ, to the plane of incidence. Ψ depends on the ratio of the square roots of the reflectance in the s- and p-states of polarisation as follows: $\Psi = \arctan \sqrt{R_p/R_s}$. Ψ may be determined by orienting a second polarising device at right angles to the new azimuth so that the transmitted radiation is totally cut off.

2.8.1 Instruments of ellipsometry

The most commonly used instrument in ellipsometry is the null ellipsometer which operates as described above. A typical experimental arrangement for this instrument is shown in figure 2.4. The incident beam is plane polarised at a suitable angle to the plane of incidence by means of the polariser, P. After reflection from the specimen, S, at an angle of incidence of θ_1, a compensator, C, is placed in the beam so that the phase of the s-wave is retarded relative to the p-wave. The beam then passes through a second polarising element called the analyser, A. Finally, the emergent beam is detected by a suitable device, D. An aid to judging the exact extinction, or null, point such as a half shade may also be included in the optical system in front of the detector. The angle between the polarising and analysing arms (twice the angle of incidence) can be read by means of two scales fitted with verniers; similarly the azimuths of the polarising, analysing and compensating devices can also be determined by scales equipped with verniers.

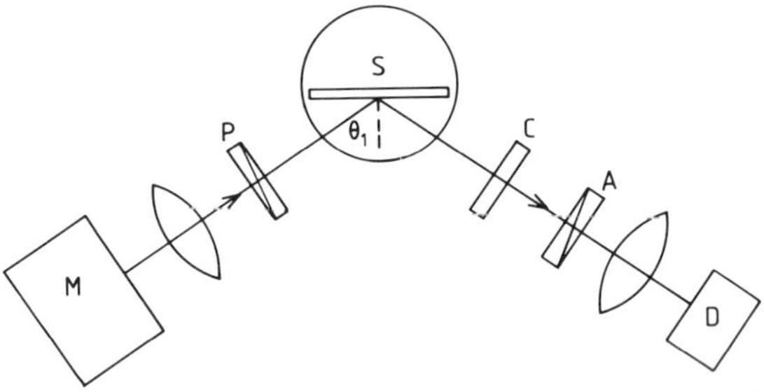

Figure 2.4 Null ellipsometer. M, monochromator; P, polariser: S, specimen; C, compensator; A, analyser; D, detector.

The polarising and analysing devices fall into three classes: those which depend on double refraction; those which use direction-dependent absorption (dichroism); and those using reflection. In the first category the best-known devices are the Nicol and the Glan–Thompson prisms. The Nicol prism is shown in figure 2.5(*a*) and consists of a crystal of calcite (calcium carbonate) cut along the shorter diagonal plane and usually with its end face ground to an angle of 71°; the prism is recemented with Canada balsam which has a refractive index between n_e and n_o. The angle of incidence of radiation on the ground face of the crystal is such that the ordinary ray is totally internally reflected at the cement but the extraordinary ray is transmitted, this having its vibrations polarised in the plane vertical to the paper. The geometry of the crystal limits the angle of divergence of the transmitted beam to about 24°. The Glan–Thompson prism, illustrated in figure 2.5(*b*), is also made from calcite. Two crystals are cemented together with their optic axes parallel and again the extraordinary ray is transmitted. As the end faces of the Glan–Thompson prism are flat, a larger angle of divergence is possible and this prism is now more commonly used.

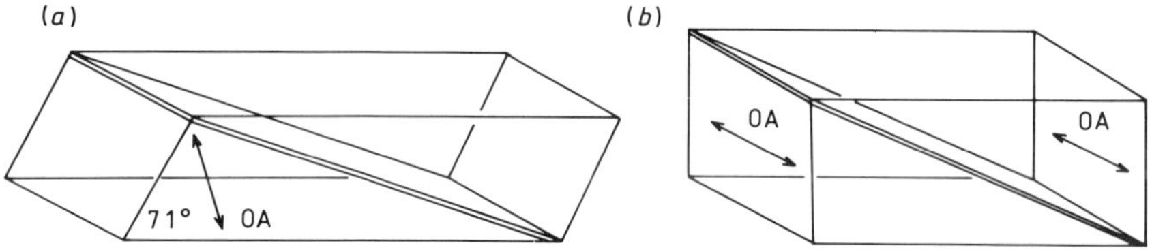

Figure 2.5 (*a*) Nicol prism. (*b*) Glan–Thompson prism.

Dichroic devices are represented by tourmaline (a complex silicate of boron and aluminium) and Polaroid which are anisotropic in their absorbing properties. Polaroid consists of a sheet of iodine polyvinyl alcohol supported on a gelatine film and oriented by stretching the film. Radiation is totally absorbed when its *E*-vibrations are in the stretch axis but only partially absorbed when the vibrations are at right angles to this; there is thus always a partial loss of intensity when using these devices. A range of Polaroids is available for use in the visible and near-infrared regions. Polaroids, although cheap, are not considered to be as effective as the prisms described above. A fine wire grid can be used to produce polarisation in the far-infrared and microwave regions (see Auton 1967).

Reflection devices make use of the Brewster's angle phenomenon already discussed in §1.3.2. For a non-absorbing material at a certain angle of incidence the p-component of the reflected radiation becomes zero so that the reflected beam contains only s-polarised radiation. In order to preserve the collimation of the radiation it is necessary for three reflections to take place in the device. On the other hand, the transmitted component is only partially p-polarised and a succession of parallel plates is needed to ensure that this polarisation is almost complete. Glass plates may be used in the ultraviolet and visible regions whilst selenium plates are effective in the infrared.

Devices called compensators or retarders produce relative phaseshifts between the two orthogonally polarised components preferably without affecting the relative amplitudes. There are two main classes of retarders, birefringent retarders and total internal reflection retarders. Birefringent retarders use a uniaxial anisotropic crystal such as quartz or mica with its optic axis cut parallel to the end faces of the device. Radiation incident normally on the retarder is split into two linearly polarised components, one with vibrations parallel to the optic axis and the other at right angles to this. These two waves propagate through the crystal with speeds of c/n_o and c/n_e respectively, n_o and n_e being the refractive indices of the ordinary and extraordinary rays respectively, and, if the crystal thickness is d, they will emerge with a phase difference of $2\pi d/\lambda(n_o - n_e)$. Depending on the thickness of the retarders, they are known as quarter- or half-wave plates for a particular wavelength. The device can be made tunable to different wavelengths or to alternative phase differences by varying the effective thickness. This is accomplished in the Babinet–Soleil compensator shown in figure 2.6. There are two main parts; a single block of quartz with its optic axis vertical and a pair of wedge-shaped quartz pieces with their optic axes at right angles to that of the block. When the block and the wedges are of equal thickness, the total retardation is zero. The retardation may be altered by sliding one of the wedges to change the effective thickness of the retarder. The relative retardation between the s- and p-components of the radiation may also be varied by rotating the retarder about an axis parallel to the radiation path.

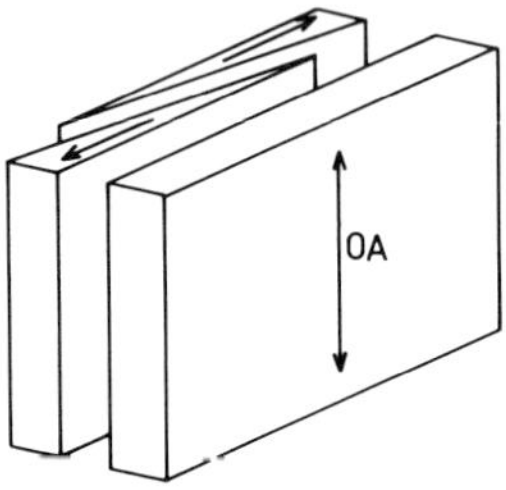

Figure 2.6 Babinet–Soleil compensator.

The second class of retarders makes use of the different phase changes introduced into the s- and p-components of radiation on internal reflection. Figure 2.7 shows a form of the Fresnel rhomb in which there are again three reflections so that the collimation of the beam is not affected. If the reflecting faces of the rhomb are made highly reflecting by coating with aluminium the device is almost independent of wavelength.

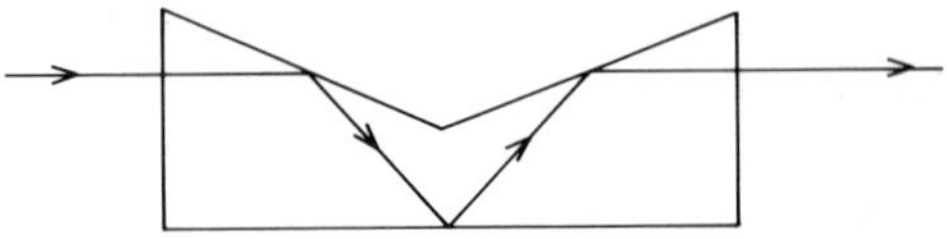

Figure 2.7 Fresnel rhomb.

2.8.2 Half shade

Half shades are often placed directly in front of the detector in an ellipsometer to enable the optical null point to be determined with more precision. They consist of two thin semicircular pieces of quartz mounted side by side and having their optic axes inclined at a small angle, a. When radiation polarised in a plane bisecting a is incident, the two sides are equally illuminated, but a small change of angle causes one side to become bright and the other dark. The angle a should be less than one degree when the polariser can be set to about 4″ of arc.

2.8.3 Depolarisers

Depolarisers are sometimes introduced between the radiation source and the ellipsometer in order to ensure that the incident beam is completely unpolarised; this is important if comparisons are to be made at different azimuths. A similar device may be used for the same purpose between the ellipsometer and the detector, and removes any errors caused by the sensitivity of the detector being dependent on the polarisation angle. Depolarisers consist of thick plates of highly birefringent materials so that random phases are introduced into the various components of the beam.

2.8.4 Sources and detectors for ellipsometry

One of the most important factors limiting the use of ellipsometers is the availability of polarising components which will operate in the required region of the spectrum. For this reason most of the published work has been in the near-ultraviolet, the visible and the near-infrared regions of the spectrum, where the Glan–Thompson prism and Polaroids are most effective. The best source of radiation is the laser, a wide range of lasers being available to cover wavelengths in the visible and ultraviolet regions. An alternative source in this region is the high-pressure gas discharge tube.

The eye can be used to detect the ellipsometer balance point but increasingly the tendency has been to make use of photoelectric devices for this purpose. Photomultipliers can be employed in the ultraviolet and visible regions, especially if they are cooled. In the near infrared silicon photodiodes or triodes are good detectors. In using any of these detectors it is important to pay attention to the noise characteristics of the devices as discussed in §2.3.2.

2.8.5 Operation of a null ellipsometer

A full description of the setting up and operation of an ellipsometer has been given by Azzam and Bashara (1977, pp 153–268) to which the reader is referred for details, only a brief account being given here. The first step is the alignment of the two arms of the instrument which must be carried out and checked carefully as otherwise all the subsequent measurements will contain systematic errors. Next, the scales of all the optical components must be calibrated: one common method of calibrating the polariser–analyser combination is to make use of Malus' law which states that the intensity of radiation passing through two polarising devices whose axes are at an angle θ is proportion to $\cos^2\theta$. Bennett (1966) describes a calibration procedure using three polarising devices; the two outer devices are oriented with their axes in the same direction whilst the third is aligned at an angle θ. The transmitted radiation is then proportional to $\cos^4\theta$. By using three polarising elements any polarisation of the source or the detector will be eliminated. McCrackin *et al* (1963) have also described a calibration technique for a polariser–analyser system. Firstly the polariser is rotated in small steps and, for each step, the angle of the analyser to give minimum output is determined; then the procedure is repeated, now rotating the analyser in small steps and finding the corresponding positions of the polariser. When plotted the two sets of readings produce two lines whose point of intersection gives the correct calibration relationship. This technique may also be used with an imperfect polariser which produces slightly elliptically polarised light. A further check on calibration can be made by employing the Brewster's angle reflection from a non-absorbing dielectric material of known refractive index, when the p-component should be zero.

An ellipsometer may be operated by varying the orientations of any two of the polariser, P, compensator, C, or analyser, A. For example. P may be set at 45° to give equal s- and p-components, C and A then being varied until a null is obtained. Alternatively the compensator may he fixed and P and A adjusted. The null situation may be determined either by eye, aided by a half shade, or by a photoelectric detector. There are a number of ways of carrying out the nulling procedure in order to ensure that maximum accuracy is achieved. For example. if C is set at $+45°$, there will be two conjugate positions of P and A which give extinction, P_1, A_1 and P_2, A_2: then, $\Delta = -(P_1 + P_2)$ and $\Psi = -(A_2 - A_1)/2$. A similar conjugate pair, P_3, A_3 and P_4, A_4 are obtained when C is set at $-45°$: these give $\Delta = (\mathrm{P}_3 + P_4)$ and $\Psi = -(A_4 - A_3)/2$. Using either of these two settings of C the method of nulling is known as two-zone averaging. If all four sets of data are used the method is called four-zone averaging and gives $\Delta = (P_3 + P_4 - P_1 - P_2)/2$ and $\Psi = -(A_4 - A_3 + A_2 - A_1)/4$. It is generally considered possible to measure azimuths to an accuracy of $\pm 0.01°$ whilst the compensator can be set to 0.1–1.0°. Consequently, if possible, it is better practice to keep C fixed and to vary P and A. The effects on the values of n and k caused by errors in the setting of these components is considered in more detail in §2.9.

According to Azzam and Bashara (1977, pp 213–33) the main sources of error in the determination of the ratio of the reflected amplitudes, Ψ, are:

(1) errors in the azimuth angles of the various components;
(2) the deviation of the compensator from orthogonality;

(3) p-s cross scatter due to surface roughness or anisotropy;
(4) imperfections in the analyser;
(5) cell window birefringence effects;
(6) polariser ellipticity;
(7) imperfections in the compensator; and
(8) depolarisation of the polariser output.

The main errors in the determination of the phase difference, Δ, are:

(9) errors in the analyser azimuth;
(10) ellipticities in the polariser and analyser;
(11) non-orthogonality in the compensator;
(12) depolarisation of the polariser output;
(13) window birefringence;
(14) azimuth errors in the polariser and the compensator; and
(15) imperfections in the compensator.

Two-zone averaging removes errors (1)–(4) in Ψ and (9)–(13) in Δ; four-zone averaging removes all errors except (8) for Ψ and (14) for Δ.

Riedling (1981) has also dealt with the systematic errors in ellipsometry due to flaws in optical components and alignment.

There are two main methods of operating an ellipsometer. In the first the polariser is set at 45° and the compensator and the analyser are rotated until a null is obtained. The direction of the compensator then gives the angle x between the major axis of the ellipse and the plane of incidence, whilst the difference between the settings of the analyser and the compensator gives the ellipticity angle, y. Then it may be shown that

$$\cos 2\Psi = \cos 2y \cos 2x \tag{2.8}$$

and

$$\tan \Delta = \tan 2y/\tan 2x. \tag{2.9}$$

The second method of operation involves setting the compensator so that its direction of fast vibrations is at 45° to the angle of incidence and then rotating the polariser and analyser until a minimum is obtained.

2.8.6 Mathematics of ellipsometry

A simplified treatment of the mathematics of ellipsometry is as follows. Starting with Fresnel's equations, (1.27) to (1.39), the ratio of the amplitudes of the reflected p- and s-components can be written

$$-a_{1s} \cos(\theta_1 + \theta_2)/a_{1p} \cos(\theta_1 - \theta_2) = a \exp(i\Delta) \tag{2.10}$$

where a_{1s}, a_{1p} are the amplitudes of the incident s- and p- components, θ_1 and θ_2 are the angles of incidence and refraction at the specimen, Δ is the phase angle between the reflected s- and p-components and a is the ratio of their amplitudes; $a = \tan\Psi$. Setting $a_{1s}/a_{1p} = b$ gives

$$a \exp i\Delta = \frac{-b(\tan\theta_1 \tan\theta_2) - 1}{(\tan\theta_1 \tan\theta_2) + 1} \tag{2.11}$$

and a rearrangement of this equation leads to

$$\frac{b + a \exp i\Delta}{b - a \exp i\Delta} = \tan\theta_1 \tan\theta_2 \tag{2.12}$$

$$= \frac{\tan\theta_1 \sin\theta_1}{[(n_2 - ik_2)^2 - \varepsilon_3^2]^{1/2}}. \tag{2.13}$$

A common practice is to polarise the incident radiation at an angle of 45° to the plane of incidence so that the p- and s- components are equal, i.e. $b = 1$.

In addition ε_3^2 can often be neglected compared with $(n - ik)^2$, and then equation (2.13) becomes

$$\frac{1 + a \exp i\Delta}{1 - a \exp i\Delta} = \frac{\tan\theta_1 \sin\theta_1}{n_2 - ik_2}. \tag{2.14}$$

Separating the real and imaginary parts of (2.14) gives, for the real part

$$n_2(1 + a\cos\Delta) + ak_2 \sin\Delta = \tan\theta_1 \sin\theta_1(1 - a\cos\Delta) \tag{2.15}$$

and for the imaginary part

$$an_2 \sin\Delta - k_2(1 + a\cos\Delta) = -a\tan\theta_1 \sin\theta_1 \sin\Delta. \tag{2.16}$$

Equations (2.15) and (2.16) may now be solved for n_2 and k_2, giving

$$n_2 = \frac{\tan\theta_1 \sin\theta_1(1 - a^2)}{1 + a^2 + 2a\cos\Delta} \tag{2.17}$$

and

$$k_2 = \frac{2a\tan\theta_1 \sin\theta_1 \sin\Delta}{1 + a^2 + 2a\cos\Delta} \tag{2.18}$$

2.8.7 Principal incidence and azimuth

Further simplification of equations (2.17) and (2.18) can be achieved if the phase difference between the reflected p- and s-components is arranged to be 90°. This can be brought about by varying the angle of incidence, the condition being known as principal incidence, θ_{pr}.

If a compensator is placed in the reflected beam such that the phase difference between the p- and s-components is cancelled, the reflected beam becomes plane polarised with its electrical vibrations in a plane making an angle Ψ_{pr} with the plane of incidence; Ψ_{pr} is called the principal azimuth.

Then,

$$n_2 = \tan\theta_{pr}\sin\theta_{pr}(1-a^2)/(1+a^2) \quad (2.19)$$

$$= \tan\theta_{pr}\sin\theta_{pr}\cos 2\Psi_{pr} \quad (2.20)$$

and

$$k_2 = 2a\tan\theta_{pr}\sin\theta_{pr}/(1+a^2) \quad (2.21)$$

$$= \tan\theta_{pr}\sin\theta_{pr}\sin 2\Psi_{pr}. \quad (2.22)$$

The method of ellipsometry developed by Drude (1894) and used subsequently for many years was that associated with the principal angle of incidence. The polariser was set at an angle of 45° to the plane of incidence and a quarter-wave plate was incorporated in the reflected beam. A null output was obtained by successive adjustments of the analyser and the angle of incidence; principal azimuth was obtained from the final setting of the analyser and principal incidence from the final setting of the angle of incidence. The technique was reputed to be exceedingly tedious. Nevertheless, this method was used extensively in the early years of the present century by some of the better-known practitioners in this field, for example Weniger and Pfund (1919), Fragstein (1933) and Forsterling and Friederickz (1913). Another group which used this technique in the 1930s was that led by Lowery (Lowery *et al* 1935, 1937).

Another technique which uses angle of incidence nulling is that described by O'Bryan (1936) and shown in figure 2.8. Polarised radiation, after its first reflection at the specimen, S, is retroreflected by a metallic mirror, M, so that it suffers a second reflection at the specimen before again passing through the polariser, P. By altering the angle of incidence, the total phase difference produced by the two reflections at the specimen can be adjusted so that the beam, on reaching the polariser for the second time, is polarised orthogonally to its original azimuth and is thus in the nulling condition, this having been achieved without using a compensator. Assuming that the normal incidence reflection at M causes equal phase and amplitude changes in the s- and p-components, it may be shown that nulling occurs when

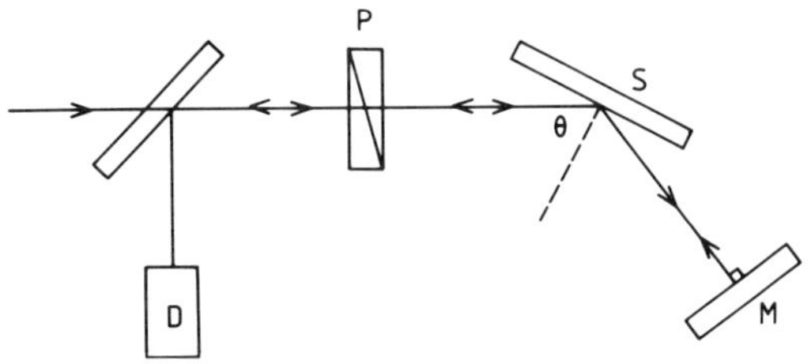

Figure 2.8 O'Bryan's ellipsometer. M, mirror; P, polariser; S, specimen; D, detector.

$\Delta = \pm 90^\circ$ and the azimuth of the polariser is equal to $\pm\Psi_{pr}$. Noskov and Charikov (1956) also used a similar method, but employing four reflections of the beam. These are examples of the use of multiple reflections in ellipsometry, others being those of Fragstein and Kampermann (1963) and Motulevich and Shubin (1957) in which four reflections were used, and Shkylarevskii and Padalka (1959) when eleven reflections were used.

The angle of incidence tuning technique described above is suitable for the determination of optical constants in the ultraviolet and visible regions of the spectrum and is still often used at the present time. However, it has one very considerable drawback when applied to metals in the infrared, where the values of both R_s and R_p tend towards unity. Under these conditions it is readily shown that (i) the angle of principal incidence tends towards 90° where the tangent is changing rapidly and (ii) the angle of principal azimuth tends towards 0, i.e. the azimuth is almost identical with the initial plane of polarisation of the incident radiation so that the rotation of the analyser to obtain the azimuth will be extremely small. The accuracy of determination of n and k is thus considerably reduced and consequently few of the results for wavelengths greater than 5 μm published in the first part of this century are now regarded as reliable.

Beattie and Conn (Beattie 1955, 1957) overcame this particular difficulty by measuring the ratio $(R_p/R_s)^{1/2}$ i.e tan Ψ at a number of predetermined angles of incidence. Dispensing with the compensator, and with the polariser set at 45°, the reflectances were determined with the analyser at 0°, 45°, 90° and 315°. Both Δ and Ψ could be calculated from these intensity ratio measurements, and hence the optical constants. They were able to show that in order to obtain well-defined values of n and k the quantities $\tan^2\Psi \sin^2\Psi/k$ and $2A\cos\theta$ should be of comparable magnitude, A being given by the ratio $(1 + R_sR_s^*)/(1 - R_sR_2^*)$. $A\cos\theta$ is roughly proportional to $(n_2^2 + k_2^2)/n_2$ whilst $\tan\Psi \sin\Psi$ is proportional to $(n_2^2 + k_2^2)/k_2$ and hence discrimination should be possible. The experimental arrangement is shown in figure 2.9.

2.8.8 Developments in ellipsometry

The earliest ellipsometers were all dependent on the final null position being determined by the human eye. One obvious limitation was that such measurements had to be confined to the visible region of the spectrum. In addition the eye is subject to fatigue and consistent results may not be obtained. The eye was therefore replaced by a photocell, one of the earliest instruments to incorporate this device being that of Bor *et al* (1939). A further step towards greater accuracy and consistency was taken by the introduction of automatic adjustment; the output from the photocell was used in a feedback system to make adjustments in the nulling process by the mechanical operation of the optical elements. An example of this type of instrument was described by Roberts and Meadows (1974). Another means of automation is to employ the (amplified) photocell current to operate a Faraday cell. This is a rectangular prism of heavy glass which causes rotation of the plane of polarisation of the radiation proportionately to a magnetic field acting along its axis. The photocell current can thus be used to rotate the plane of polarisation of the incident or reflected radiation so as to bring about a null situation. Mattieu *et al* (1974) have described an ellipsometer operating on this principle. Automatic ellipsometry was also used by Takasaki (1966) and by Ord and Wills (1967).

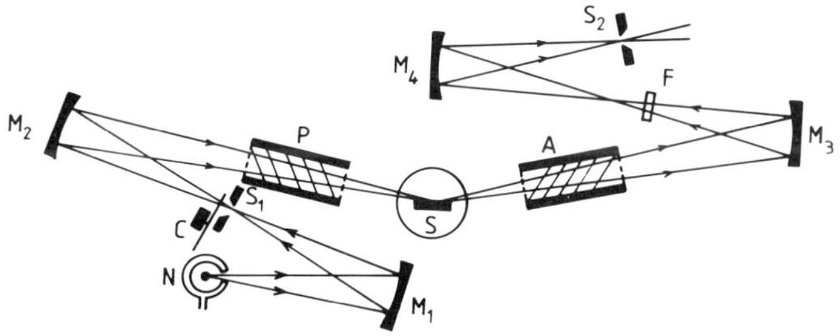

Figure 2.9 Beattie and Conn's photometric ellipsometer. M, etc, mirrors; P, polariser; S, specimen; A, analyser; S_1, S_2, slits; C, compensator; N, Nernst filament, F, filter. (From Beattie J R 1955 *Phil. Mag.* **46** 236.).

A further advantage of the automatic adjustment of ellipsometers is their rapid speed of operation which enables surface reactions to be followed or readings to be taken quickly at several wavelengths or angles of incidence.

Another variation on this instrument is the interference ellipsometer described by Hazebroek and Holscher (1973), which is a modification of Michelson's interferometer. A beam splitter directs one beam to the specimen and back again whilst the second (reference) beam is sent along another retroreflecting arm containing a standard reflector. Both returning beams pass through a Woolaston prism which directs the s- and p-components separately to two matched detectors. By adjusting the length of the retroreflecting arm the two beams may be combined either in or out of phase. Measurement of these two components enables the ratio of the s- and p-reflectances to be calculated. Azzam and Bashara (1977, pp 263–4) give details of the appropriate mathematics of this calculation.

A further group of ellipsometers is known as photometric ellipsometers. The principle of their mode of action is the measurement of the change in output intensity as a function of the angle of incidence for a given setting of the azimuth angle and the compensator. In static ellipsometers the output intensity is measured for a number of settings of the chosen optical component and the quantities Ψ and Δ deduced from them as in Beattie and

Conn's method. In dynamic photometric ellipsometers, continuous modulation of one of the polariser, analyser or compensator angles is carried out, and the output signal subjected to Fourier analysis in order to extract n and k. The rotating analyser instrument has been the most popular of these possibilities and examples may be found in the work of Archard *et al* (1952), Aspnes (1974) and Hauge and Dill (1973). The mathematical treatment of this technique has also been given by Azzam and Bashara (1977, p 412).

In the polarisation modulated ellipsometer the state of polarisation of the beam of radiation is modulated by a Faraday cell fed by an AC current of the required frequency (Jasperson *et al* 1973). The output from the detector then has to be processed by a harmonic analyser in order to recover n and k. One advantage of this technique is that all the optical components remain stationary, thus leading to high-speed operation.

A recent addition to the range of techniques for measuring optical constants is AIDER (angle of incidence derivatives in ellipsometry and reflectance). In this system measurements are made not only of the ellipsometric functions, Ψ and Δ, and the reflectance, R, but also the derivatives of these functions with respect to the angle of incidence, θ. This is achieved by modulating the angle of incidence by a mechanical system. The extraction of optical data from the output of the spectrometer is complicated and Bermudez (1980) may be consulted for further details: Azzam (1983) provides mathematical expressions for the derivatives of the appropriate optical functions.

As in photometry, it is now a common practice to use multiangle versions of ellipsometry. Modern techniques allow such measurements to be made rapidly and the corresponding values of n and k can be determined by a least-squares fit computer program. An example of the use of this method has been given by Paneva and Ohlidal (1986) and the techniques have reviewed by Tronon *et al* (1990).

2.9 Accuracy studies

It is useful to make estimates of the inherent accuracy which is achievable in each of the techniques of photometry, polarimetry and ellipsometry so that the most suitable technique may be chosen for a particular material.

In order to determine both n and k it is necessary to measure two optical functions, for example R_p and R_s or Ψ and Δ. Let them be called X and Y. The accuracy with which n and k can be calculated is related to both the accuracy to which X and Y can be measured and also to the form of the intersections of the contour lines of constant X and constant Y; these intersections will produce a network of parallelograms in n, k space. Intersections of X and Y at small angles lead to long, narrow error parallelograms generating large uncertainties in n and k, particularly along the major diagonal. When the X and Y contours are widely spaced considerable uncertainties will be possible in any direction in the n–k plane. The most accurate results will be obtained with closely spaced contours intersecting at or near right angles.

The form and shape of the error parallelograms may be analysed in more detail in the following way. The experimental errors in the optical functions X and Y may be written

$$\Delta X = \frac{\partial X}{\partial n}\Delta n + \frac{\partial X}{\partial k}\Delta k + \frac{\partial X}{\partial \theta}\Delta\theta \qquad (2.23)$$

and

$$\Delta Y = \frac{\partial Y}{\partial n}\Delta n + \frac{\partial Y}{\mathrm{d}k}\Delta k + \frac{\partial Y}{\mathrm{d}\theta}\Delta\theta \qquad (2.24)$$

where $\Delta\theta$ is the error in the angle of incidence.

It will be assumed initially that $\Delta\theta$ is zero and that only the variations of X and Y with n and k need be considered. A method of dealing with errors in θ will be described later in this section.

Equations (2.23) and (2.24) may be solved for Δn and Δk to give

$$\Delta n = \left(\Delta X \frac{\partial Y}{\partial k} - \Delta Y \frac{\partial X}{\partial k}\right)\left(\frac{\partial X}{\partial n}\frac{\partial Y}{\partial k} - \frac{\partial X}{\partial k}\frac{\partial Y}{\partial n}\right)^{-1} \qquad (2.25)$$

and

$$\Delta k = \left(\Delta Y \frac{\partial X}{\partial n} - \Delta X \frac{\partial Y}{\partial n}\right)\left(\frac{\partial X}{\partial n}\frac{\partial Y}{\partial k} - \frac{\partial X}{\partial k}\frac{\partial Y}{\partial n}\right)^{-1} \qquad (2.26)$$

When both positive and negative values of ΔX and ΔY are considered there are four possible solutions to equations (2.25) and (2.26) representing the four corners of a parallelogram whose centre point is the original value of n and k. Using expressions for the partial derivatives of the optical functions, a computer program can be devised to calculate and plot these error parallelograms for a range of values of n, k and θ. These partial derivatives may be obtained from the exact equations for reflectance and phase changes on reflection given in §1.4.1.

In solving equations (2.25) and (2.26) it is necessary to specify values of ΔX and ΔY. Convenient values are ± 0.01 for the photometric functions and ± 0.01 rad for the polarimetric functions. Considerably greater accuracy than this has been claimed, particularly for the polarimetric functions, but the values chosen are not unreasonable and give a range of conveniently sized error parallelograms. In any case, the sides of the parallelograms may be scaled proportionately up or down if levels of accuracy different from those specified above are obtained.

A second arbitrary quantity is the definition of an 'accurate' parallelogram. For convenience this has been chosen as one whose longer diagonal was less than or equal to 0.6 units of n or k, corresponding to $\Delta n = \Delta k = \pm 0.2$ for a square error figure.

The procedure described above has been applied to each of the groups of techniques for n and k described earlier in this chapter. In all cases, very large error parallelograms have been omitted to avoid confusion.

2.9.1 *Photometry*

(a) *Combinations of* R_s *and* R_p

The contours of constant R_s and R_p are both roughly elliptical in shape with the major axes lying approximately on the line $n = k$ (see figure 1.2). The intersections of these two sets of contours will therefore generally take place at very small angles, but when n and k are small the contours are sufficiently close to generate a roughly square region of high accuracy near the origin of coordinates. Figure 2.10(a) shows the error parallelogram array for the combination (R_s, R_p), both functions being at an angle of incidence of 30°. This basic pattern expands or contracts as the angle of incidence is changed. R_s shows a slow increase with θ and this causes its contour pattern to contract. On the other hand R_p decreases with θ up to Brewster's angle and then increases rapidly towards a value of unity at $\theta = 90°$. Since the change in R_p with angle is the greater, the accurate region has its maximum extent at the Brewster's angle. Figure 2.10(b) shows the accuracy pattern when the angle of incidence is 75°. Although this analysis implies that the most accurate results with this combination should be obtained near Brewster's angle, consideration should also be given to the problems of reflectance measurements at high incidence angles discussed by Kuhn and Wilson (p 48).

The combination ($R_{s\theta 1}$, $R_{s\theta 2}$) produces no regions of high accuracy. R_s changes only slowly with θ and the two sets of contours always intersect at very acute angles producing long, narrow error parallelograms. ($R_{p\theta 1}$, $R_{p\theta 2}$), however, is a viable combination under certain circumstances, as the variation of R_p with θ can be large. The largest accurate regions are obtained when one of the angles of incidence is near normal and the other is close to the Brewster's angle.

(b) $(R_p/R_s)_{\theta_1}$, $(R_p/R_s)_{\theta_2}$

Measurements of the ratios of s- and p-reflectances at two different angles of incidence have been much used in order to calculate the optical constants, it being claimed that high experimental accuracy can be obtained by the elimination of errors due to the positioning of the detector. The inherent accuracies of these functions have also been tested and the square accuracy pattern was again observed. The accurate areas, however, were generally small and these functions would only be viable when one of the angles of incidence was less than 30° and the other between 60° and 75°.

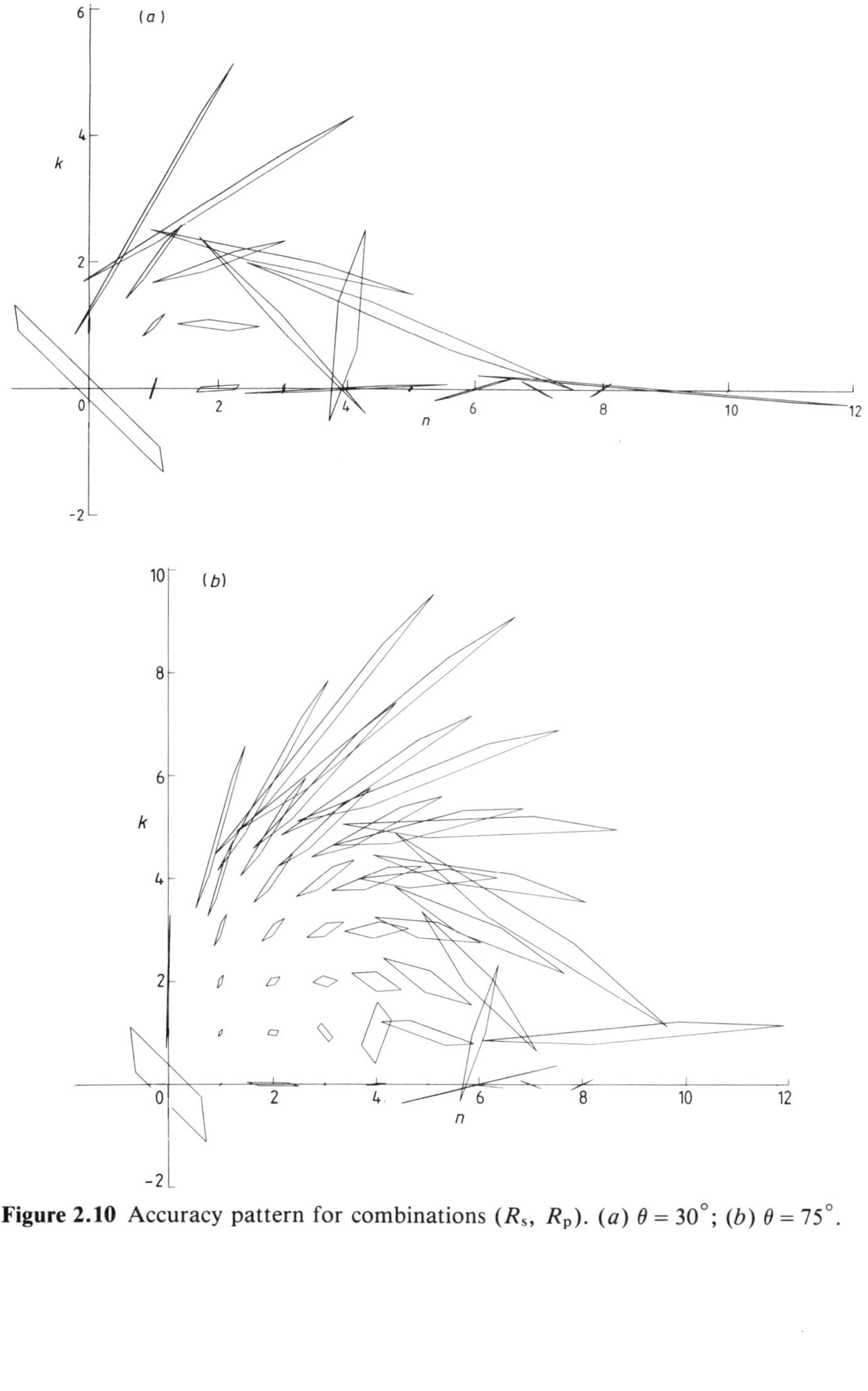

Figure 2.10 Accuracy pattern for combinations (R_s, R_p). (*a*) $\theta = 30°$; (*b*) $\theta = 75°$.

Table 2.2 Ranges of areas of accuracy in n and k for the following combinations of optical functions: (a) ($R_{s\theta_1}$, $R_{p\theta_2}$); (b) ($R_{p\theta_1}$, $R_{p\theta_2}$) and (($R_p/R_s)_{\theta_1}$, $(R_p/R_s)_{\theta_2}$).

	$\theta_p =$ 15°	45°	60°	75°	85°
$\theta_s =$ 15° (a)	—	—	0–3	0–4	0–3
(b)	—	0–2	0–3	0–4	0–3
45° (a)	—	—	0–2	0–3	0–4
(b)	0–2	—	0–1	0–2	0–3
60° (a)	0–3	0–2	0–2	0–3	0–3
(b)	0–3	0–1	—	0–1	0–2
75° (a)	0–4	0–3	0–2	0–3	0–3
(b)	0–4	0–2	0–1	—	0–1
85° (a)	0–5	0–4	0–3	0–2	0–3
(b)	0–4	0–3	0–2	0–1	—

Ward (1982, 1983) has given detailed results of the accuracy studies for the photometric functions, a summary of which is given in table 2.2.

Humphreys-Owen (1961) has also considered the accuracies obtainable in several reflectance techniques for determining n and k; examples are (i) unpolarised light at two different angles of incidence, (ii) R_s and R_p at the same angle, (iii) R_s/R_p at two different angles, (iv) R_s at two angles, and (v) R_p at two angles. In addition, he investigated methods which used a determination of Brewster's angle together with either (R_s, R_p) or R_s/R_p at the same or a different angle. He presented plots such as figure 2.11

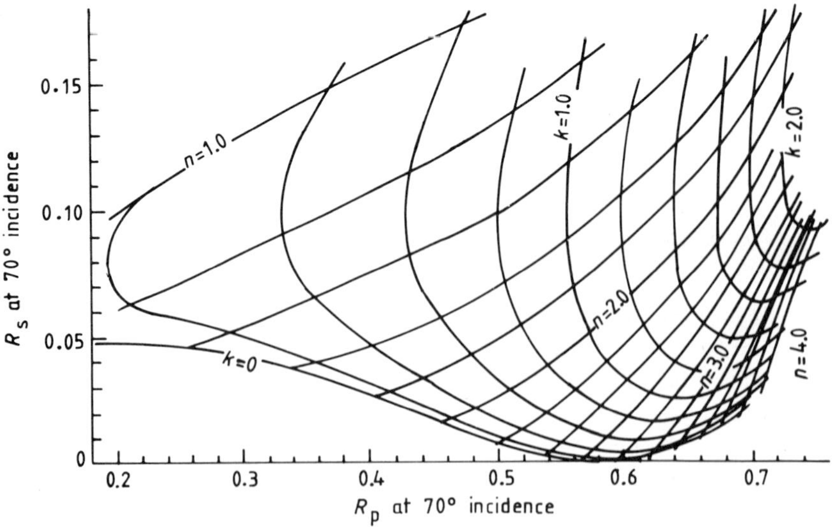

Figure 2.11 R_s (70°) versus R_p (70°) showing contours of constant n and constant k. (From Humphreys-Owen S P F 1961 *Proc. Phys. Soc.* **77** 953.).

which shows contours of constant n and k for the functions R_{p70}, R_{s70}. In this method of plotting, a wide spacing between the contours is indicative of good sensitivity which is seen to occur at small values of n and k.

Hunter (1965) also investigated the accuracy of reflectance versus angle of incidence techniques. Isoreflectance curves were plotted in the n–k plane, the angles of intersection of the curves providing a measure of the accuracy available. The most accurate results were obtained when using R_p; unpolarised radiation gave the next best alternative whilst the worst conditions were found with R_s. A typical set of isoreflectance curves for R_p is shown in figure 2.12.

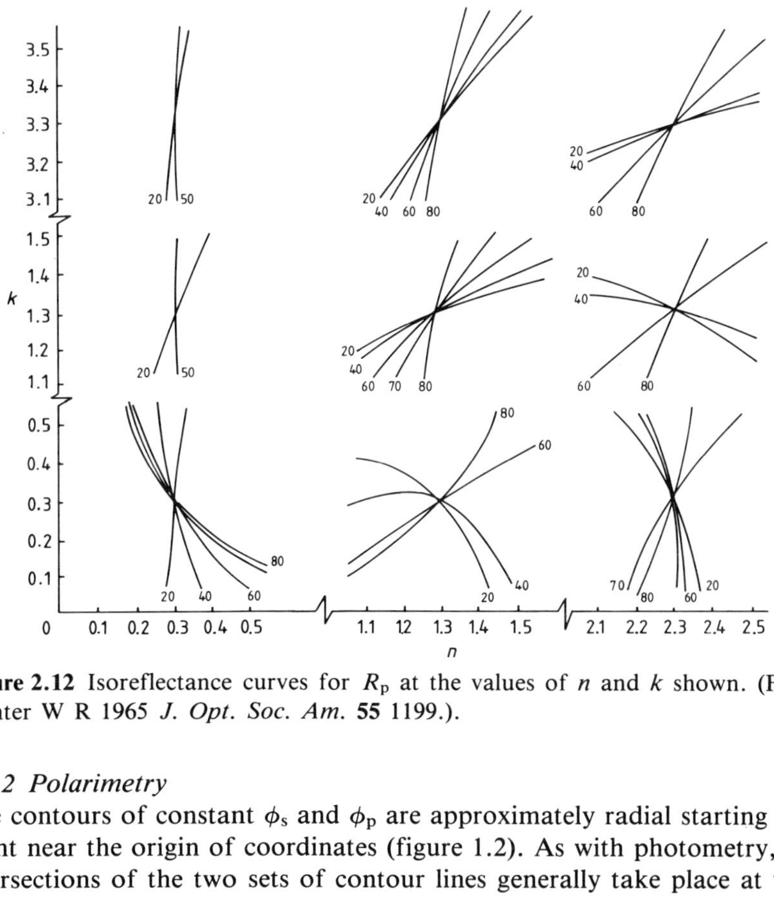

Figure 2.12 Isoreflectance curves for R_p at the values of n and k shown. (From Hunter W R 1965 *J. Opt. Soc. Am.* **55** 1199.).

2.9.2 Polarimetry

The contours of constant ϕ_s and ϕ_p are approximately radial starting at a point near the origin of coordinates (figure 1.2). As with photometry, the intersections of the two sets of contour lines generally take place at very small angles leading to the production of long, thin error parallelograms, particularly along the line $n = k$. Near to the origin, however, the intersecting angles of the contours are sufficiently large to generate small regions of high accuracy. A typical error parallelogram array for (ϕ_s, ϕ_p) is illustrated in figure 2.13(a); this is for an angle of incidence of 30°.

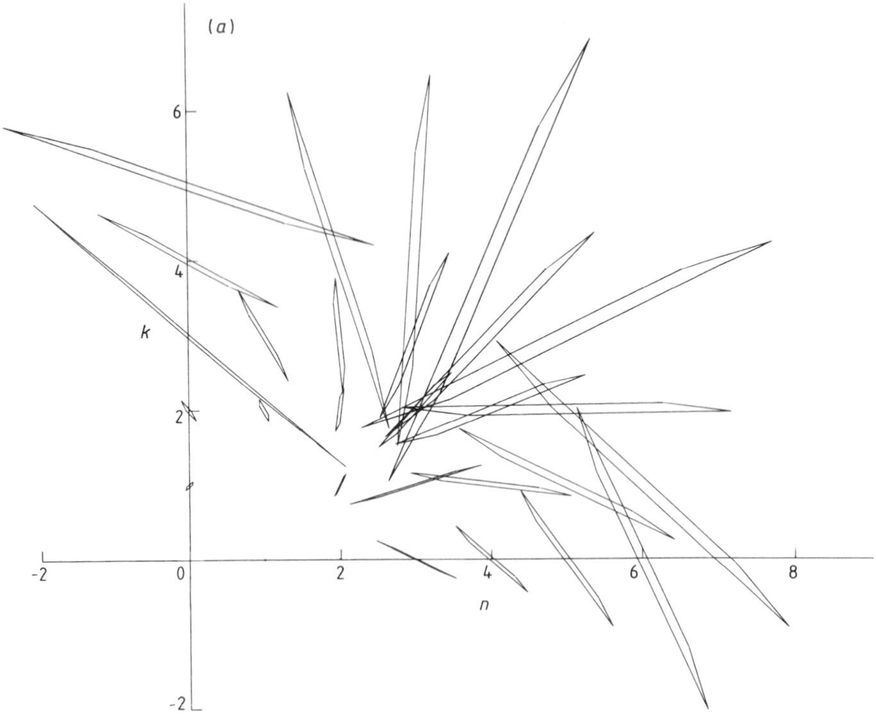

Figure 2.13(*a*) Accuracy pattern for (ϕ_s, ϕ_p) at 30°.

As the angle of incidence increases, the accurate region moves outwards from the origin and its area becomes larger, because of the expansion of the contours due to ϕ_p. Of more significance, however, is the appearance, at angles of incidence greater than 60°, of a singularity in the pattern of the error parallelograms characterised by a set of narrow parallelograms radiating from a particular point in the *n*, *k* array (figure. 2.13(*b*)). This singularity is brought about by the shape of the contours in ϕ_p which converge close to the *n*-axis (p 12) so that, when combined with those of ϕ_s, a set of error parallograms oriented towards this region is produced. The occurrence of this singularity clearly represents another restriction on the accuracy available when using this pair of optical functions which must therefore be used with care at angles of incidence of 60° and more. For solid and thick-film specimens, the singularity occurs at $k = 1$; its *n* coordinate as a function of angle of incidence is given in table 2.3.

(a) $((\phi_s)_{\theta_1}, (\phi_s)_{\theta_2})$ *and* $((\phi_p)_{\theta_1}, (\phi_p)_{\theta_2})$
The combination $((\phi_s)_{\theta_1}, (\phi_s)_{\theta_2})$ shows no regions of high accuracy, but $((\phi_p)_{\theta_1}, (\phi_p)_{\theta_2})$ produces the same pattern of accurate parallelograms as (ϕ_s, ϕ_p). The accurate region is largest when one of the angles is small and

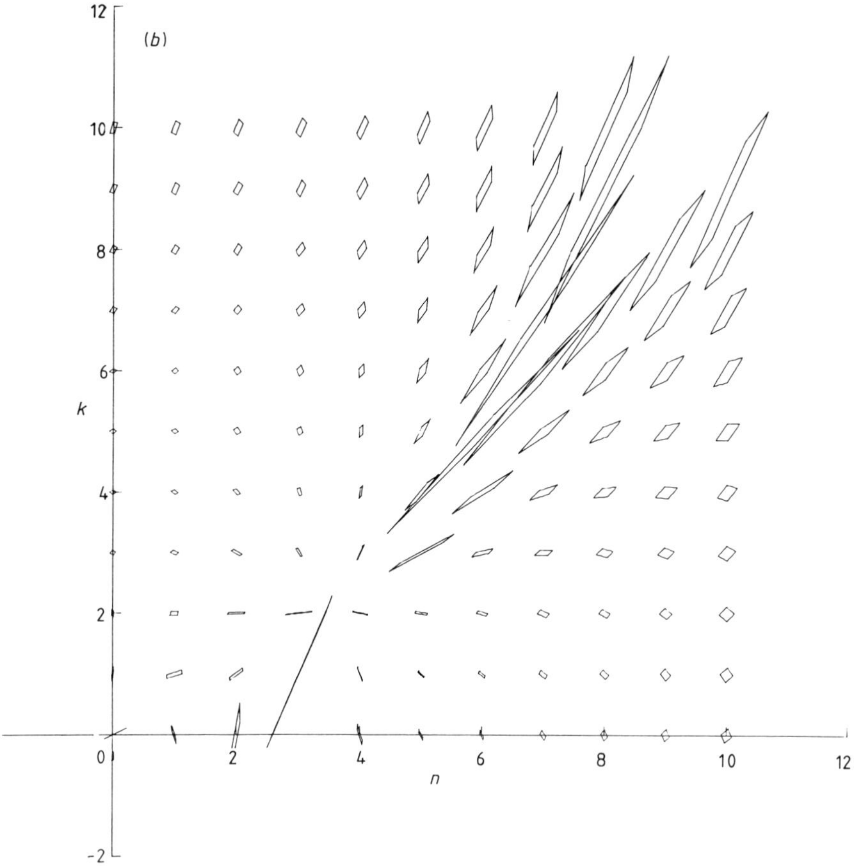

Figure 2.13(*b*) Accuracy pattern for (ϕ_s, ϕ_p) at 75°, showing singularity.

Table 2.3 *n*-coordinate of singularity as a function of angle of incidence θ.

θ(deg.)	n
60	2.0
75	3.8
80	5.5
85	11.5

the other is close to 90°; the presence of the singularity may also affect the shape of the accurate region.

Techniques such as Shkylarevskii's (1957) measure the phase difference, $\Delta = \phi_p - \phi_s$, at two different angles of incidence. Δ is dominated by ϕ_p and the pattern of accurate parallelograms is thus similar to that for $((\phi_p)_{\theta_1}, (\phi_p)_{\theta_2})$.

2.9.3 *Mixed photometric and polarimetric functions*

The contour lines of constant ϕ are roughly radial whilst those of constant R are elliptical and so it would be expected that combinations of functions of the type (R, ϕ) should produce contour patterns intersecting at large angles, leading to the possibility of broad regions of high accuracy. Figure 2.14(*a*) illustrates the case for (R_s, ϕ_s) at an angle of incidence of 15° where a region of high accuracy is present in the form of an annulus spreading

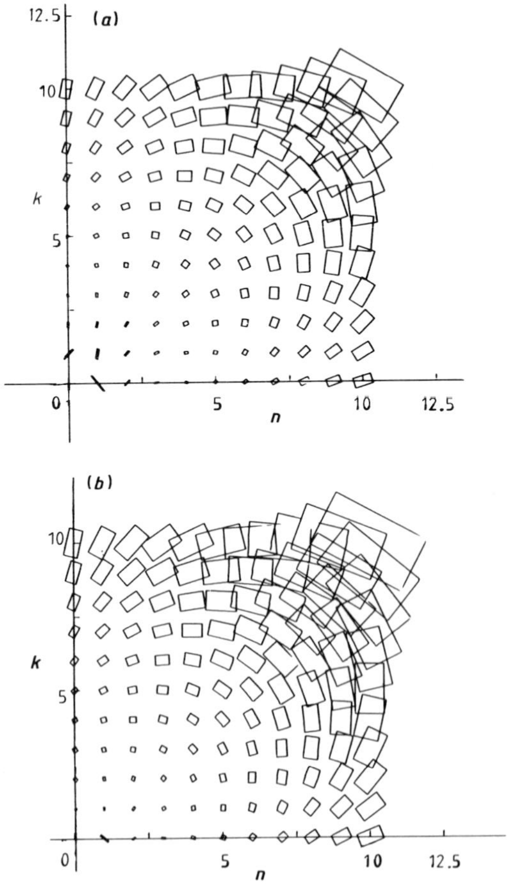

Figure 2.14 Annular accuracy patterns due to (R_s, ϕ_s) at (*a*) 15° and (*b*) 60°.

right across the n, k quadrant. As the angle of incidence is increased, the annulus contracts towards the origin following the contraction of the contour pattern due to R_s (figure 2.14(*b*)).

A similar pattern of error parallelograms is produced by the combination (R_p, ϕ_p) at small angles of incidence; as the angle of incidence increases the annulus of high accuracy now expands outwards in accordance with the expansion of the contour pattern due to R_p. However, when the angle of incidence reaches 60°, the singularity in ϕ_p makes its appearance and renders this combination less suitable for accurate work. Ward (1984) has carried out detailed studies on the accuracy of these combinations of mixed optical functions and has given tables of the accurate regions. Table 2.4 gives a sample of these results.

There appear to be few experimental techniques which use such a mixture of optical functions, but the accuracy studies indicate that they would be useful particularly for materials having roughly equal values of n and k.

Table 2.4 Annular regions of high accuracy: (*a*) ($R_{s\theta_1}$, $\phi_{s\theta_2}$); (*b*) ($R_{p\theta_1}$, $\phi_{p\theta_2}$). The numbers give the radii of accurate regions.

$\theta_1 =$	30°	60°	75°	85°
$\theta_2 = 30°$(*a*)	1–5	0–4	0–3	0–2
(*b*)	1–5	0–7	0–8	0–10
60°(*a*)	0–4	0–4	0–3	0–2
(*b*)	0–7	0–7	0–8	0–10
75°(*a*)	0–3	0–3	0–3	0–2
(*b*)	0–8	0–8	0–8	0–10
85°(*a*)	0–2	0–2	0–2	0–2
(*b*)	0–10	0–10	0–10	0–10

2.9.4 Reflection ellipsometry

The same procedure may be used to study the inherent accuracy available when using reflection ellipsometry by investigating the optical functions Ψ and Δ. A typical array of error parallelograms is shown in figure 2.15(*a*) where the accurate region is again in the form of an annular quadrant. At small angles of incidence the accurate region is almost non-existant; this is because R_s and R_p are almost equal under these conditions leading to widely spaced contours in Ψ. As the angle of incidence increases, the accurate region grows rapidly in size, but, once again, the singularity due to ϕ_p, now in the function Δ, becomes obtrusive for angles greater than 60° (figure 2.15(*b*)). Many of the reported studies of optical constants using ellipsometry have been carried out at angles of incidence greater than 75° (the principal angle of incidence tends towards 90° when n and k are large, for example for the transition metals in the infrared spectrum) and, consequently, there is a possibility that this singularity may have intruded inadvertently in the measurements. On the other hand, in the infrared, many materials such as the noble metals have large values of k but only small values of n (see figure 4.2) and these values of the optical constants would be well clear of the singularity even for angles of incidence above 75°.

Dignam and Moskvits (1970) nave investigated the errors in ellipsometry due to errors in Ψ and Δ; they give a graph showing the variation of these errors with the ratio n/k. When using the compensator technique, $\Delta k/k$ increased with n/k whilst $\Delta n/n$ decreased. The situation was reversed with the intensity ratio method. The analysis was refined in a later paper (Dignam and Moskvits 1971) to give, for the compensator method, $\Delta n = 2gk$ and $\Delta k = -2gn$, where g is the error in the azimuth. These values of Δn and Δk generate rectangular error figures parallel to the axes and becoming square when $n = k$; they are not dissimilar to the results of Ward which predict squares at all values of n and k but oriented at 45° to the axes. For the intensity ratio method the corresponding results were $\Delta n = 2g(1 - k^4/n^4)^{1/2}$ and $\Delta k = 4g(1 + k^2/n^2)^{1/2}$. These values also generate rectangles as before.

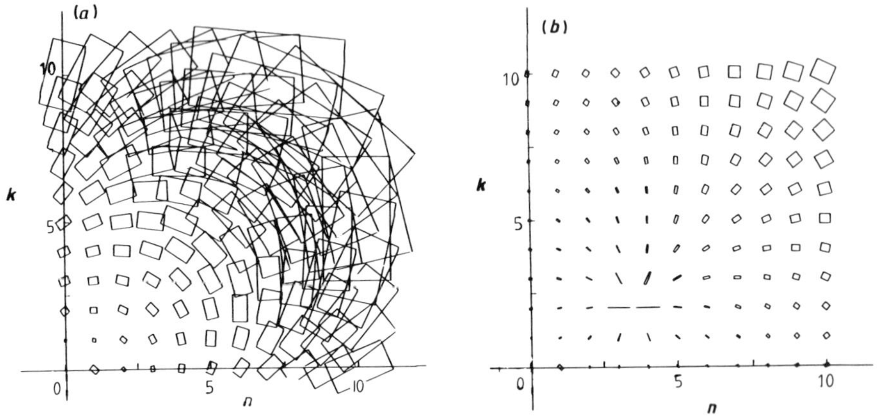

Figure 2.15 Accuracy patterns for reflection ellipsometry at (*a*) $\theta = 30°$ and (*b*) $\theta = 75°$.

Zendler *et al* investigated the sensitivity of the ellipsometric functions to the angle of incidence (Zendler *et al* 1974a) and also to beam deviation (Zendler *et al* 1974b). They published a series of graphs which showed the angular regions where large errors might occur. Errors due to beam deviation were often significant in zone-averaged measurements when using standard optical components.

Zaghloul and Azzam (1982) have published contour maps of constant Ψ versus constant Δ, which, over a limited range, can be applied to ellipsometry and reflectivity measurements.

2.9.5 Errors due to uncertainties in the angle of incidence

Errors in the angle of incidence can be taken into account by rearranging equations (2.23) and (2.24) in the form

$$\Delta X\left(1 \pm \frac{\partial X}{\partial \theta}\frac{\partial \theta}{\Delta X}\right) = \frac{\partial X}{\partial n}\Delta n + \frac{\partial X}{\partial k}\Delta k = \Delta X' \tag{2.27}$$

and

$$\Delta Y\left(1 \pm \frac{\partial Y}{\partial \theta}\frac{\partial \theta}{\Delta Y}\right) = \frac{\partial Y}{\partial n}\Delta n + \frac{\partial Y}{\partial k}\Delta k = \Delta Y'. \tag{2.28}$$

Considering the worst possible situations. i.e. that the signs of ΔX and $(\partial X/\partial\theta)(\Delta\theta/\Delta X)$ are the same and also those of ΔY and $(\partial Y/\partial\theta)(\Delta\theta/\Delta Y)$, the solutions for Δn and Δk become

$$\Delta n = \left(\Delta X' \frac{\partial Y}{\partial k} - \Delta Y' \frac{\partial X}{\partial k}\right)\left(\frac{\partial X}{\partial n}\frac{\partial Y}{\partial k} - \frac{\partial X}{\partial k}\frac{\partial Y}{\partial n}\right)^{-1} \tag{2.29}$$

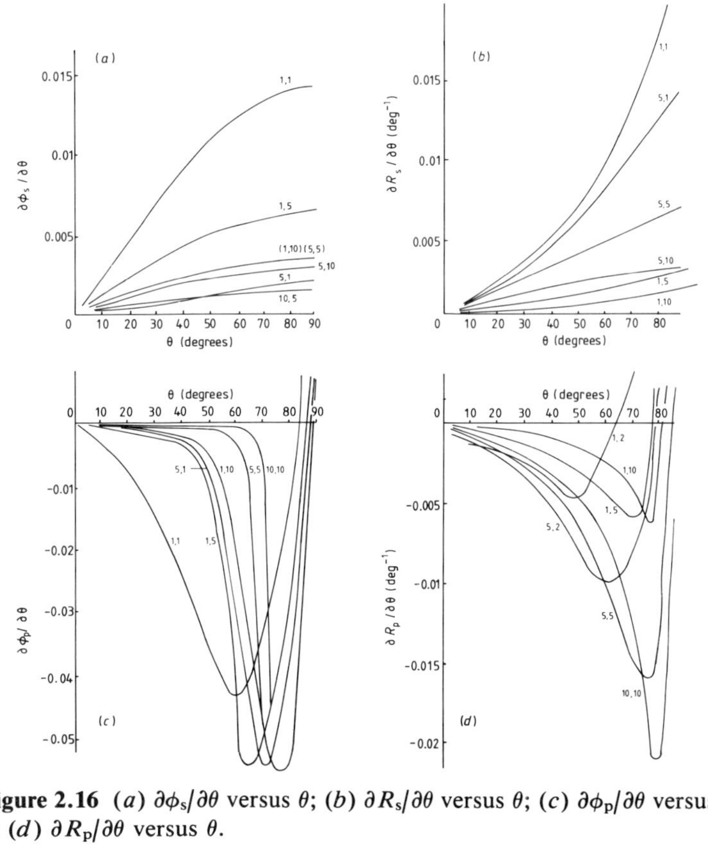

Figure 2.16 (a) $\partial\phi_s/\partial\theta$ versus θ; (b) $\partial R_s/\partial\theta$ versus θ; (c) $\partial\phi_p/\partial\theta$ versus θ; (d) $\partial R_p/\partial\theta$ versus θ.

and

$$\Delta k = \left(\Delta Y' \frac{\partial X}{\partial n} - \Delta X' \frac{\partial Y}{\partial n}\right)\left(\frac{\partial X}{\partial n}\frac{\partial Y}{\partial k} - \frac{\partial X}{\partial k}\frac{\partial Y}{\partial n}\right)^{-1}. \tag{2.30}$$

Consequently, the uncertainties in θ may be included in the analysis by scaling the uncertainties in X and Y by the factors $1 + (\partial X/\partial\theta)(\Delta\theta/\Delta X)$ and $1 + (\partial Y/\partial\theta)(\Delta\theta/\Delta Y)$ respectively. This may be done by scaling the sides of the parallelograms by these same factors. Figure 2.16 shows value of (*a*) $\partial R_s/\partial\theta$, (*b*) $\partial R_p/\partial\theta$, (*c*) $\partial\phi_s/\partial\theta$ and (d) $\partial\phi_p/\partial\theta$ plotted against θ for various combinations of n and k; in all four cases the largest values of the partial differentials occur at large angles of incidence. $\partial R_s/\partial\theta$ and $\Delta\phi_s/\Delta\theta$ are largest at small value of n and k whilst $\partial R_p/\partial\theta$ and $\partial\phi_p/\partial\theta$ have their biggest values at large n and k. As an example of the errors likely to be produced, consider a set of measurements at an angle of incidence of 60° on a material where $n = 1$ and $k = 5$. An error of 1° in the angle of incidence would generate the following values of $(\partial X/\partial\theta)(\Delta\theta/\Delta X)$: $R_s = 0.15$; $R_p = 0.49$; $\phi_s = 0.58$ and $\phi_p = 1.23$.

References

Archard J F 1985 *Phys. Bull.* **36** 212–14
Archard J F, Clegg P L and Taylor A M 1952 *Proc. Phys. Soc.* B **65** 758
Arnon O and Baumeister P 1978 *Appl. Opt.* **17** 2913
Aspnes D E 1974 *J. Opt. Soc. Am.* **64** 639
Auton J P 1967 *Appl. Opt.* **6** 1023
Avery D G 1950 *Phil. Mag.* **41** 1028
—— 1952 *Proc. Phys. Soc.* B **65** 425
Azzam R M A 1983 *Opt. Acta* **30** 1113
Azzam R M A and Bashara N M 1977 *Ellipsometry and Polarised Light* (Amsterdam: North-Holland)
Beattie J R 1955 *Phil. Mag.* **46** 222, 989
—— 1957 *Physica* **23** 898
Beilby G 1921 *Aggregation and Flow of Solids* (London: Macmillan)
Bennett H E 1966 *Appl. Opt.* **5** 1265
Bennett J M 1976 *Appl. Opt.* **15** 2705
Bennett J M and Darcy J H 1981 *Appl. Opt.* **20** 1785
Bennett J M and King R J 1970 *Appl. Opt.* **9** 236
Bennett J M and Mattson L 1989 *Introduction to Surface Roughness and Scattering* (Washington, DC: Optical Society of America.)
Bennett J M, Sheffer J J, Shibanio Y and Namba Y 1987 *Appl. Opt.* **26** 696
Bermudez M 1980 *Surf. Sci.* **94** 29
Bhushan B, Wyant J C and Koliopoulos C L 1985 *Appl. Opt.* **24** 1489
Bor J, Hobson A and Wood B 1939 *Proc. Phys. Soc.* **51** 932
Bowden F P and Hughes T P 1937 *Proc. R. Soc.* A **160** 575
Briggs H B 1950 *Phys. Rev.* **77** 287

Clark J, Richard P L and Yeh N-H 1977 *Appl. Phys. Lett.* **30** 664
Collins J R and Bock R O 1943 *Rev. Sci. Instrum.* **14** 135
Cooper J 1962 *J. Sci. Instrum.* **39** 467
Dereniak E L and Crowe D G 1984 *Optical Radiation Detectors* (New York: Wiley)
Dignam M J and Moskvits M 1970 *Appl. Opt.* **9** 1868
—— 1971 *Appl. Opt.* **10** 679
Ditchburn R W 1976 *Light* (London: Academic)
Drude P 1894 *Ann. Phys., Lpz.* **51** 77
—— 1959 *Theory of Optics* (New York: Dover)
Dyson J 1970 *Interference as a Measuring Tool* (Brighton: The Machinery Publishing Co. Ltd.)
Forsterling K and Friederickz V 1913 *Ann. Phys. Lpz.* **40** 201
Fragstein C 1933 *Ann. Phys. Lpz.* **17** 1
Fragstein C and Kampermann H 1963 *Z. Phys.* **173** 39
French R C 1932 *Nature* **129** 169
—— 1933 *Proc. R. Soc.* A **140** 637
Gates D M, Shaw C C and Beaumont D 1958 *J. Opt. Soc. Am.* **48** 88
Golay M J E 1947 *Rev. Sci. Instrum.* **18** 346, 357
Guenther K H and Pulker H K 1976 *Appl. Opt.* **15** 2992
Hauge P S and Dill F H 1973 *IBM J. Res. Dev.* **17** 472
Hazebroek H F and Holscher A A 1973 *J. Phys. E. Sci. Instrum.* **6** 822
Hoffmann R A, Lange W J and Choyke W J 1975 *Appl. Opt.* **14** 1803
Humphreys-Owen S P F 1961 *Proc. Phys. Soc.* **77** 949
Hunter W R 1960 *Appl. Opt.* **6** 2140
—— 1965 *J. Opt. Soc. Am.* **55** 1197
Jasperson S N, Burge D E and O'Handley R C 1973 *Surf. Sci.* **37** 548
Johnson C J. Sherman G H and Weil R 1969 *Appl. Opt.* **8** 1667
Kuhn H and Wilson B A 1950 *Proc. Phys. Soc.* **48** 775
—— 1958 *Proc. Phys. Soc.* B **63** 745
Low F J 1961 *J. Opt. Soc. Am.* **51** 14
Lowery H, Bor J and Wilkinson H 1935 *Phil. Mag.* **20** 390
Lowery H, Wilkinson H and Smare D L 1937 *Proc. Phys. Soc.* **49** 345
McCarthy D E 1963 *Appl. Opt.* **2** 596
McCrackin F L, Passli E, Stromberg R R and Steinberg H L 1963 *J. Res. NBS* **67A** 363
Maier H G 1970 *Frequenz* **24** 303
Malitson I H 1964 *J. Opt. Soc. Am.* **54** 628
Mattieu H J, McClure D E and Muller R H 1974 *Rev. Sci. Instrum.* **45** 798
Meeten G H, North A and Willmouth F M 1984 *J. Phys. E: Sci. Instrum.* **17** 642
Miller L S, Walder A J, Linseil P and Blundell A 1988 *Thin Solid Films* **156** 11
Motulevich G B and Shubin A A 1957 *Opt. Spectrosc.* **2** 661
Noskov M M and Charikov B A 1956 *Opt. Spectrosc.* **1** 1007
O'Bryan O 1936 *J. Opt. Soc. Ann.* **26** 127
Okoomian X 1969 *Appl. Opt.* **8** 2363
Ord J L and Willis B L 1967 *Appl. Opt.* **6** 1673
Paneva A and Ohlidal I 1986 *Thin Solid Films* **145** 23
Riedling K 1981 *Thin Solid Films* **75** 355

Roberts E H and Meadows A 1974 *J. Phys. C: Solid State Phys.* **7** 1974
Rooksby H P and Randall J T 1932 *Nature* **129** 280
Russo O L 1985 *J. Phys. D: Appl. Phys.* **18** 1723
Samuels L E 1971 *Metallographic Polishing by Mechanical Methods* (London: Pitman)
Shklyarevskii I N 1957 *Opt. Spectrosc.* **3** 638
Shklyarevskii I N and Padalka V G 1959 *Opt. Spectrosc.* **6** 78. 776
Simon I 1951 *J. Opt. Soc. Am.* **41** 336
Sommargren G E 1981 *Appl. Opt.* **20** 610
Strong J 1938 *Modern Physical Laboratory Practice* (New York: Prentice Hall) p 376
Takasaki H 1966 *Appl. Opt.* **5** 759
Tronin A Y, Filippov V V, Konstantinova A F and Karpuk M M 1990 *Proc. SPIE Opt. Eng. (USA)* **1270** 151
Ward L 1982 *J. Phys. D: Appl. Phys.* **15** 1361
—— 1983 *J. Phys. D. Appl. Phys.* **16** 271
—— 1984 *J. Phys. D. Appl. Phys.* **17** 1781
Weniger W and Pfund A H 1919 *Phys. Rev.* **14** 427
West J M 1965 *Electrodeposition and Corrosion Processes* (London: Van Nostrand Reinhold) pp 115–16
Wood R W 1933 *Phys. Rev.* **44** 353
Zaghloul A-R M and Azzam R M A 1982 *Appl. Opt.* **21** 743
Zendler J R, Kohler R B and Bashara N M 1974a *Appl. Opt.* **3** 159
—— 1974b *Appl. Opt.* **3** 1938

3 Kramers–Kronig Methods for Calculating Optical Constants

3.1 Introduction

In the previous chapter consideration was given to the direct measurement of optical functions from which the optical constants can be calculated over limited ranges of wavelengths or frequencies. The Kramers–Kronig technique, which is the subject of the present chapter, makes use of a spectral distribution of one or more of these functions determined or estimated between zero and infinite frequencies to calculate n and k at any frequency within this range. As we have seen in Chapter 2, modern techniques enable optical-type measurements to be made over a wide range of frequencies ranging from x-rays to microwaves. Combining the results from a number of sources it is therefore possible to draw up a set of values of the required optical function over a very wide spectral range. Nevertheless, there are some extreme wavelength regions where optical measurements are not possible and it then becomes necessary to make extrapolations of the behaviour of the optical functions. These extrapolations are often the least accurate part of the Kramers–Kronig analysis process and may lead to errors being generated in the calculated values of n and k: some actual methods of extrapolation are discussed later in this chapter. The optical functions used in Kramers–Kronig-type calculations can be reflectance, transmittance, or phase change on reflection or transmission; any optical function may be used provided it is a function of frequency.

The technique was first introduced by Kramers (1929) and then improved by Kronig (1931). Subsequently the procedure has been considerably developed in sophistication and it is now one of the most powerful techniques for the accurate determination of optical constants. As a considerable amount of summation mathematics is involved, the technique is ideally suited for computer work.

Kramers–Kronig analysis is based on the premise that, for an optical function consisting of both real and imaginary parts, if the variation of the real part with frequency is known, the imaginary part, also as a function of frequency, may then be calculated, and vice versa.

3.2 Complex frequency plane

Dispersion occurs when a wavepacket containing a broad band of frequencies passes through a medium whose refractive index is a function of frequency; the different frequency components then travel through the medium at different speeds. A simple example of this effect is the dispersion of white light into the visible spectrum by a prism. The dispersive equation relates the angular frequency of the wave, ω, with the wave number, K, via the wave velocity, v

$$v = \omega / K. \tag{3.1}$$

The refractive index of an absorbing medium being complex in form, i.e. $\tilde{n} = n - ik$, leads to the concept of a complex wave velocity, $v = v_1 - iv_2$, from the relation $n = c/v$. The idea of complex frequency then follows from the relation in equation (3.1) above, i.e. $\omega = \omega_r - i\omega_i$. The wave equation (equation (1.10)) may then be rewritten in the form

$$E = E_0 \exp[i(\omega t - Kx)] = E_0 \exp(+\omega_i t)\exp[i(\omega_r t - Kx)]. \tag{3.2}$$

The imaginary exponential term represents the oscillating part of the wave as a function of both time and position whilst the real exponential is a time-dependent decay term.

A complex frequency plane may be constructed with the real part of ω, ω_r, as the ordinate and the imaginary part, ω_i, as the abscissa. A dispersive function may be represented by two sets of contours in this plane; for example figure 3.1 which shows contours of constant amplitude and constant phase for the function $(\varepsilon - 1)$ which is given by equation (1.83), i.e.

$$(\varepsilon - 1) = \frac{Ne^2}{\varepsilon_0 m} (\omega_0^2 - \omega^2 + i\gamma\omega)^{-1}. \tag{3.3}$$

The other significant features to be noted on this diagram are the two points where the amplitude term becomes infinite—these are known as poles. Points where the amplitude becomes zero are called zeros.

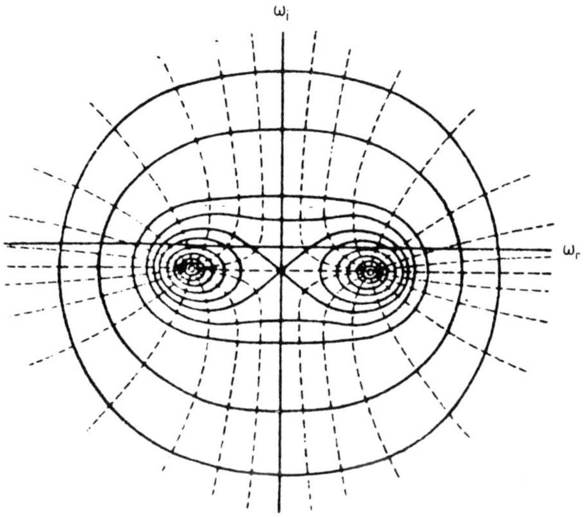

Figure 3.1 Complex frequency plane showing orthogonal contours of constant amplitude (full curves) and constant phase (broken curves) arising from the function in equation (3.3). (From Lipson S G and Lipson H *Optical Physics* (1969), courtesy Cambridge University Press.)

3.3 Causality

Causality plays a fundamental role in the Kramers–Kronig analysis as it is used to decide which part of the complex frequency plane is valid from the point of view of actual physical phenomena. Although a physical system may have a 'memory' of an event which has already happened, it cannot anticipate an event which has not yet taken place. This is a basic statement of the principle of causality which will be applied to the situation at the poles in the complex frequency plane. At a pole an infinite amplitude may be achieved in response to a finite driving force. Since the arrival of the driving force cannot be anticipated, the infinite amplitude of the system must be the result of the memory of a force which has decayed from an infinite amplitude occurring some time in the past, i.e. at some negative value of t. The only way that this statement can be reconciled with equation (3.2) is for ω_i to be negative so that, when $t = -\infty$, $\boldsymbol{E}$ is of infinite amplitude. Thus the poles must always lie in the lower half of the complex frequency plane. Alternatively, the principle of causality can be applied in a more practical form; all physical processes show natural decay but never exhibit natural growth, hence the real exponential in equation (3.2) must be negative, i.e. ω_i must be negative for positive values of t.

In some accounts of the complex frequency plane, ω is defined as $\omega_r + i\omega_i$;

the region of actual phenomena then falls in the upper half of the complex frequency plane.

3.4 Optical impedances

The Kramers–Kronig analysis deals with the response of an optical system to an imposed electromagnetic driving force, the relationship between these two quantities being called the response function. Examples of this function are the reflectance and the transmittance, and these may be expressed in terms of the optical impedances as was shown in Chapter 1. Bode (1945) developed the treatment of network impedances in terms of the concept of complex frequency space and his technique can also be applied to the optics of materials.

The optical functions may be expressed as either polynomials or factors in ω, for example

$$Z = P[(\omega - \omega_1)(\omega - \omega_2)(\omega - \omega_3)\ldots(\omega - \omega_m)]/ Q[(\omega - \omega_1')(\omega - \omega_2')(\omega - \omega_3')\ldots(\omega - \omega_m')]\,. \quad (3.4)$$

When $\omega = \omega_1, \omega_2, \ldots, \omega_m$ then $Z = 0$; these values of ω give the zeros. When $\omega = \omega_1', \omega_2', \ldots, \omega_m'$ then $Z = \infty$; these values of ω give the poles.

Normally the individual values of ω and ω' are all different and the poles and zeros are then said to be of the first order, or simple, but in special cases, several values of ω or ω' may coincide to give a multiple zero or pole. The poles and zeros are the analogues of resonances and anti-resonances in electrical circuits, but occurring at complex frequencies.

3.5 Contour integration

The above concepts may now be applied to the integration of a function round a closed loop in the complex frequency plane. In particular, it is necessary to investigate the restrictions which must be placed on the zeros and poles of the functions in order to obtain a real solution which can then be compared with actual measurements; a mathematical tool is therefore needed to transpose behaviour at complex frequencies into that at real frequencies. The real frequency axis may be regarded as the boundary of the lower half plane, so the broad mathematical problem becomes that of relating the behaviour of a function inside the lower half of the complex frequency space to its behaviour on the boundary of that region. This is achieved by means of contour integration which was originally formulated by Cauchy.

In ordinary calculus, an integral can be regarded as the area under a curve; integrals of complex variables may be defined in exactly the same way. In general, of course, these integrals will also be complex.

Now let us consider the integration of a complex function round a completely circular path, e.g. $z^x\,dz$ where z may be defined as $Y\exp(i\theta)$, Y being a real quantity. Thus:

$$\int z^x\,dz = \int_{\pi}^{-\pi} Y^x[\exp(ix\theta)]\ iY[\exp(i\theta)]\ d\theta \tag{3.5a}$$

$$= iY^{x+1}\int_{\pi}^{-\pi}[\cos(x+1)\theta + i\sin(x+1)\theta]\ d\theta. \tag{3.5b}$$

The integral of any sine or cosine function over one complete cycle is zero and so, provided x is not equal to -1, the integral of z^x is zero. If $x=-1$, the integral becomes simply $\int_{\pi}^{-\pi} d\theta$, i.e. $-2\pi i$ for the clockwise direction and $+2\pi i$ for the anti-clockwise direction.

These results can be expressed in a general form in Cauchy's theorem, which states that, if a function $f(z)$ is analytic (i.e. over the region concerned, it has a finite derivative) within a closed curve and also on the curve itself, $f(z)\,dz$ round the whole curve is equal to zero.

If $x<-1$, and for large values of z, the contour integral round any portion of the loop is also equal to zero, as the function itself tends to zero for large z. Also, if $x>-1$, and for small values of z, the contour integral again becomes zero round any portion of the loop as the path length tends to vanish at small z.

As an illustration of the above ideas, let us now consider $f(z)$ to be in the form of a polynomial and again integrate over a closed loop in the shape of a circle. Thus

$$f(z) = A_0 + A_1z + A_2z^2 + \cdots + A_nz^x. \tag{3.6}$$

This function is analytic, and from Cauchy's theorem its integral round any closed loop must be zero. Now, if a term A_k/z is added to the right-hand side of equation (3.6) a pole is produced at $z=0$ so the function is no longer analytic at all points within the loop. The integral then becomes $2\pi A_k i$ (anti-clockwise), the quantity A_k being known as the residue of the function at the pole.

Although these results have been obtained using a circular integration path, they are equally valid for a closed path of any shape.

Now, consider the integration of the function

$$f(z) = A_0/(z-z_a) + A_1 + A_2(z-z_a) + \cdots + \tag{3.7}$$

which has a pole at $z=z_a$. The integration loop may be drawn, as in figure 3.2, containing a subsidiary loop isolating the pole so that the function now becomes analytical inside the loop. The two integrations between the points P_1 and P_2 clearly cancel and the remaining integrals are $+2\pi iA_0$ in an anti-clockwise direction round the inner loop and $-2\pi i A_0$ in a clockwise direction round the outer loop; these again add up to zero. Where there are P poles and N zeros, the integral becomes $\pm 2\pi i(P-N)$ depending on whether it is clockwise or anti-clockwise.

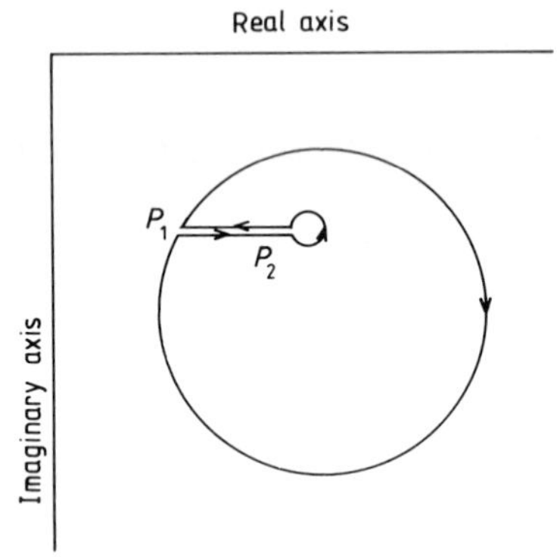

Figure 3.2 Integration loop round a pole.

3.5.1 Integration of the real components of a complex function

It is required to integrate the frequency-dependent optical function, X, around the path shown in figure 3.3; one example of X would be the complex reflection coefficient, $r\exp(i\phi)$. The semicircular integration path is made very large, and indentations are made along the ω_r axis to avoid any poles or zeros; it is also assumed that there are no singularities within the loop. If the integrand can be arranged to vanish at infinity more rapidly than ω^{-1}, the integral round the loop can be evaluated as shown above.

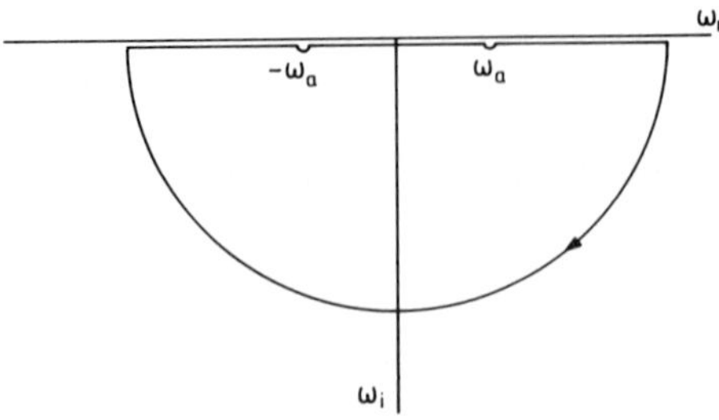

Figure 3.3 Integration of the frequency-dependent optical function.

X is assumed to be analytic at $\omega = 0$ and at $\omega = \infty$; it will be defined by the following equations

$$X = A + \mathrm{i}B \tag{3.8}$$

$$= A_\infty + \mathrm{i}B_\infty/\omega + A_1/\omega^2 + \mathrm{i}B_1/\omega^3 + \cdots + . \tag{3.9}$$

In these equations, the coefficients of the real terms (A) are even functions of ω, whilst those of the imaginary terms (B) are odd functions of ω. Starting with equation (3.9) let a new function be constructed: $X - A_\infty$. This will vanish more rapidly than ω^{-1} and thus, round the whole loop, $\oint (X - A_\infty)\, \mathrm{d}\omega = 0$. The integral may be split into two parts. that round the semicircle, $\S$, and that along the ω_r-axis, $\int_{\pm\infty}^{\pm\infty}$. Thus

$$\S\, (X - A_\infty)\, \mathrm{d}\omega + \int_{-\infty}^{+\infty} (X - A_\infty)\, \mathrm{d}\omega = 0. \tag{3.10}$$

Substituting for X from equation (3.8):

$$\int_{-\infty}^{+\infty} (A - A_\infty)\, \mathrm{d}\omega + \mathrm{i} \int_{-\infty}^{+\infty} B\, \mathrm{d}\omega + \S\, (\mathrm{i}B_\infty/\omega)\, \mathrm{d}\omega = 0 \tag{3.11}$$

as only the term $\mathrm{i}B_\infty/\omega$ makes any significant contribution to the semicircular integral when ω is large. Since B is an odd function of ω, the integral from $-\infty$ to $+\infty$ involving this term is zero. Also, since A is an even function of ω, the first term in (3.11) can be replaced by twice the integral from 0 to ∞. The integration of the third term is equal to $-\mathrm{i}(\mathrm{i}B_\infty\pi) = \pi B_\infty$. Thus

$$2 \int_0^\infty (A - A_\infty)\, \mathrm{d}\omega = -\pi B_\infty. \tag{3.12}$$

Consequently, the integration round the loop of the real part of X can be reduced to that along the real frequency axis as was the requirement at the beginning of this section. From now on, therefore, integrations involving ω will imply those along the ω_r axis. The result of the integration performed in equation (3.12) gives the value of the imaginary part of the function at infinity. In addition, the function has a residue of B_∞ at $\omega = \infty$ as shown in equation (3.9).

In practice, it is usually required to find the value of B at some finite frequency, say ω_a (see figure 3.3). This can be done by creating a pole at ω_a by dividing the function $X - A_a$ by $\omega - \omega_a$, where A_a is the value of A at ω_a. In order to maintain symmetry, a pole at $\omega = -\omega_a$ must also be created. Thus,

$$\oint \left(\frac{X - A_a}{\omega - \omega_a} - \frac{X - A_a}{\omega + \omega_a} \right) \mathrm{d}\omega = 0. \tag{3.13}$$

This integrand vanishes at least as rapidly as ω^{-2} at $\omega \gg \omega_a$ so the contribution of the semicircular part of the contour is zero. The rest of the path consists of the real frequency axis containing two small indentations to avoid the poles at $\omega = \pm\omega_a$. If the indentations are small, the values of X round them may be regarded as constants, $A_a + iB_a$ and $A_a - iB_a$ respectively. Thus the contour integral becomes

$$\int_{-\infty}^{+\infty} \left(\frac{2\omega_a(A - A_a + iB)}{\omega^2 - \omega_a^2}\right) d\omega + \int_{I_1} iB_a \left(\frac{1}{\omega - \omega_a} - \frac{1}{\omega + \omega_a}\right) d\omega$$

$$- \int_{I_2} iB_a \left(\frac{1}{\omega - \omega_a} - \frac{1}{\omega + \omega_a}\right) d\omega. \qquad (3.14)$$

The first term in equation (3.14) is the contribution along the ω_r axis whilst the second and third terms are due to the indentations I_1 and I_2. For the pole near $\omega = +\omega_a$, the second term in equation (3.14) can be evaluated as $i\pi(iB_a) = -\pi B_a$; similarly, the third term near $\omega = -\omega_a$ integrates to $-i\pi(-iB_a) = -\pi B_a$. In the first integral the imaginary term can be neglected because of its symmetry on the positive and negative axis. Consequently equation (3.14) may be written

$$B_a = \frac{\omega_a}{\pi} \int_{-\infty}^{+\infty} \frac{A - A_a}{\omega^2 - \omega_a^2} d\omega. \qquad (3.15)$$

$$= \frac{2\omega_a}{\pi} \int_0^{\infty} \frac{A - A_a}{\omega^2 - \omega_a^2} d\omega. \qquad (3.16)$$

When transposed in terms of the photon energy, this equation becomes:

$$B_a = \frac{2U_a}{\pi} \int_0^{\infty} \frac{A - A_a}{E^2 - E_a^2} dU. \qquad (3.17)$$

Equations (3.16) and (3.17) are two forms of one of the standard Kramers–Kronig equations and have been frequently employed in work of this kind. It may, however, be developed into another version as follows. If we set $Q = \ln(\omega/\omega_a)$, equation (3.16) may be reduced to

$$B_a = \frac{1}{\pi} \int_{-\infty}^{+\infty} \frac{A - A_a}{\sinh Q} dQ \qquad (3.18)$$

which can be integrated by parts. Integrating over positive values of Q gives

$$B_p = -\frac{1}{\pi} [(A - A_a) \ln \coth \tfrac{1}{2}Q]_0^{\infty} + \frac{1}{\pi} \int_0^{\infty} \frac{dA}{dQ} \ln \coth \tfrac{1}{2}Q \, dQ \qquad (3.19)$$

whilst over the negative values of Q we get

$$B_n = \frac{1}{\pi} [(A - A_a) \ln \coth \tfrac{1}{2}Q]_{-\infty}^{0} + \frac{1}{\pi} \int_{-\infty}^{0} \frac{dA}{dQ} \ln \coth \tfrac{1}{2}Q \, dQ. \qquad (3.20)$$

Thus,

$$B_a = B_p + B_n = \frac{1}{\pi} \int_{-\infty}^{+\infty} \frac{dA}{dQ} \ln \coth \tfrac{1}{2}Q \, dQ. \qquad (3.21)$$

Now, resubstituting for Q in terms of ω leads to the following equation for B_a:

$$B_a = \frac{1}{\pi} \int_{-\infty}^{+\infty} \frac{dA}{d\omega} \ln \frac{\omega + \omega_a}{\omega - \omega_a} \, d\omega. \tag{3.22}$$

This equation has also been frequently used in Kramers–Kronig analyses.

3.5.2 Integration of the imaginary component

A second contour integral theorem gives the analogous relation between the integration of the imaginary component over the real frequency axis and the behaviour of the real component at extreme frequencies.

A simple way to achieve this result is to consider the function $i\omega(X - A_\infty)$ which is equal to $i\omega(A - A_\infty - iB)$; this results in an interchange of the real and imaginary components. Using this function is equivalent to replacing the original A by $-B\omega$, which has even symmetry, and the original B by $i\omega(A - A_\infty)$, which has odd symmetry. Carrying through the arguments used in the previous sections leads to the following expressions:

$$A_a - A_\infty = -\frac{2}{\pi} \int_0^{\infty} \frac{\omega B - \omega_a B_a}{\omega^2 - \omega_a^2} \, d\omega \tag{3.23}$$

$$A_a - A_\infty = -\frac{1}{\pi} \omega_a \int_{-\infty}^{+\infty} \frac{d(\omega B)}{dQ} \ln \coth \tfrac{1}{2} Q \, dQ \tag{3.24}$$

$$A_a - A_\infty = -\frac{1}{\pi} \omega_a \int_{-\infty}^{+\infty} \frac{d(\omega B)}{d\omega} \ln \left(\frac{\omega + \omega_a}{\omega - \omega_a} \right) d\omega. \tag{3.25}$$

3.6 Practical applications of Kramers–Kronig relations

The expressions which have been most frequently employed when applying Kramers–Kronig analyses to sets of optical data are equations (3.16) and (3.17), a selection of authors being Andermann *et al* (1964), Jahoda (1957), Roessler (1965), Veleur (1968) and Bauer *et al* (1974). A technique which has often been used is to measure the reflectance, R, over as large a frequency range as possible and then extrapolate R to zero and infinite frequencies by means of plausible formulae. For example Rasigni and Rasigni (1977) extrapolated the reflectance to be asymptotic to 1.0 as long wavelengths and used a power law of the form $R_2 = R_1(\lambda_2/\lambda_1)^p$, where p is a real number, at short wavelengths. An alternative extrapolation at short wavelengths is as an exponential: $R_2 = R_1 \exp[C(\lambda_1 - \lambda_2)]$. Jezierski (1984) has developed an algorithm for extrapolations of the reflectance to both large and small values of the frequency. Now, $R^{1/2} = r \exp i\phi$, so that:

$$\tfrac{1}{2}\ln R = \ln r + \mathrm{i}\phi = A + \mathrm{i}B. \tag{3.26}$$

Thus B_a, i.e. the phase change on reflection ϕ_a at ω_a may be calculated by substituting ln r for A in equation (3.16). Once the corresponding values of R and ϕ are known, n and k may be computed from Fresnel's equations. Lichten (1985) and Berreman (1967) have described techniques of this kind whilst Quarry and Holland (1974) were able to calculate $\phi_p - \phi_s$ from measurements of the ratio R_p/R_s at oblique incidence. Kozima *et al* (1960) used measurements of the transmittance, T, in order to calculate χ.

An alternative process is to measure the optical absorption, ε_2, and then use equation (3.23) to calculate ε_1. In the case of conductors, there is a pole at $\omega = 0$ (see equation (1.95)), but ε_1 remains finite along the real ω-axis (equation (1.93)). Thus ε_1 is given by

$$\varepsilon_1 = 1 + \frac{2}{\pi}\int_0^\infty \frac{\omega e_2(\omega) - (\sigma/\varepsilon_0)}{\omega^2}\,\mathrm{d}\omega. \tag{3.27}$$

Hulthen (1982) has given details of a generalised system of Kramers–Kronig analysis in which it is not necessary to know the variation of one of the optical functions between $\omega = 0$ and ∞. However, partially overlapping knowledge of both the real and imaginary parts of the function is required.

Arkatova *et al* (1981) have produced a review of the literature on the usage of Kramers–Kronig relations.

3.7 Sum rules

Following on from the Kramers–Kronig analysis, a number of sum rules have been formulated which are useful in checking the calculated results for consistency. The first may be expressed as

$$\int_0^\infty (n(\omega) - 1)\,\mathrm{d}\omega = 0 \tag{3.28}$$

which implies that the average value of $n(\omega)$ over the whole spectral range is unity. It may be shown that this rule follows from the principle of causality (Smith and Mattheiss 1974).

A second rule is:

$$\int_0^\infty (\varepsilon_1(\omega) - 1)\,\mathrm{d}\omega = -2\pi\sigma_0. \tag{3.29}$$

For insulators $\sigma_0 = 0$, whilst in the case of metals this rule provides a means of assessing the intraband contribution.

A sum rule involving ε_2 is:

$$\int_0^\infty \omega\varepsilon_2(\omega)\,\mathrm{d}\omega = \tfrac{1}{2}\pi\omega_{\mathrm{p}}^2 \tag{3.30}$$

which can also be expressed in the forms:

$$\int_0^\infty \omega k(\omega)\,\mathrm{d}\omega = \tfrac{1}{4}\pi\omega_{\mathrm{p}}^2 \tag{3.31}$$

and

$$\int_0^\infty \omega\,\mathrm{Im}(\varepsilon(\omega)^{-1})\,\mathrm{d}\omega = -\tfrac{1}{2}\pi\omega_{\mathrm{p}}^2. \tag{3.32}$$

References

Andermann G, Caron A and Daws D A 1965 *J. Opt. Soc. Am.* **55** 1211

Arkatova T G, Gopshtein N M, Makarova E G and Mikhailov B A 1981 *Sov. J. Opt. Tech.* **48** 552

Berreman D W 1967 *Appl. Opt.* **6** 1519

Bauer R S, Spicer W E and White J J 1974 *J. Opt. Soc. Am.* **64** 830

Bode H W 1945 *Network Analysis and Feedback Amplifier Design* (Princeton, NJ: van Nostrand)

Hulthen R 1982 *J. Opt. Soc. Am.* **72** 794

Jahoda F C 1957 *Phys. Rev.* **107** 1261

Jezierski K 1984 *J. Phys. C: Solid State Phys.* **17** 475

Kozima K, Suetaki W and Schatz P N 1960 *J. Opt. Soc. Am.* **56** 181

Kramers H A 1929 *Z. Phys.* **30** 522

Kronig R de L 1931 *Proc. R. Soc.* **133** 255

Lichten W 1985 *J. Opt. Soc. Am.* **A2** 1969

Quarry M R and Holland W E 1974 *Appl. Opt.* **13** 595

Rasigni M and Rasigni G 1977 *J. Opt. Soc. Am.* **67** 54

Roessler D M 1965 *Br. J. Appl. Phys.* **16** 1359

Smith N U and Mattheiss L F 1974 *Phys. Rev.* B **9** 1341, 1365

Veleur H W 1968 *J. Opt. Soc. Am.* **58** 1356

4 Results for Bulk Specimens

4.1 Introduction

In this chapter short accounts are given of the results, which have been accumulated by many workers, for the optical constants and other related properties for metals, insulators and semiconductors in bulk form. The acquisition of this information has had two main purposes: to provide a source of data for use in the design of optical equipment of various kinds and to compare with theoretical models such as the free electron theory, the band theory of solids, the anomalous skin effect and resonances in electron plasma.

4.2 Metals

The numerous determinations of the optical constants of a wide range of solid metals which were made over the period between 1890 and approximately 1950 established the main features of the variation of n and k with wavelength or photon energy, especially in the ultraviolet, visible and infrared regions of the spectrum.

In this period measurements were generally carried out on bulk specimens which had been prepared either by mechanical or electropolishing as described in §2.2. It has subsequently been realised that the quality of the surface of the specimen plays a crucial role in determining the values of the optical properties of the material. The surfaces involved in these early investigations were generally neither sufficiently planar, contamination-free nor strain-free to enable some of the finer optical detail to be measured with any certainty. In addition, the imperfections themselves often produced spurious features in the optical properties.

Some of the typical optical properties of a metal are illustrated in figure

4.1(*a*) which shows the variation of n and k with wavelength for gold. Another way of displaying the same information is shown in figure 4.1(*b*) where ε_1 and ε_2 are plotted against the photon energy, U; this method of presentation is particularly suitable for energies up to the plasmon energy.

At very long wavelengths (far infrared and microwaves), the values of n and k are large (up to 1000, for gold for example) k being greater than n; thus, the real part of the dielectric constant, $\varepsilon_1 = n^2 - k^2$, is negative in this region. The imaginary part of ε, $\varepsilon_2 = 2nk$, is also large but positive, and corresponds to the low-wavelength side of the Drude absorption band produced by the free electrons (or, in terms of the band theory of solids, the intraband transitions of the electrons in the conduction band) which has its maximum at zero frequency, i.e. infinite wavelength.

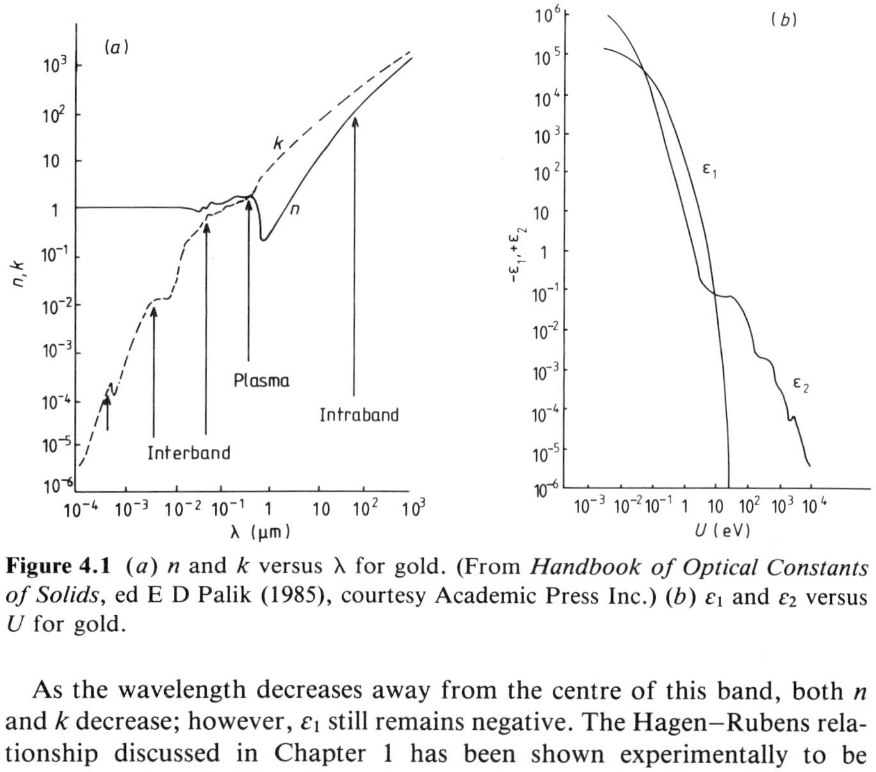

Figure 4.1 (*a*) n and k versus λ for gold. (From *Handbook of Optical Constants of Solids*, ed E D Palik (1985), courtesy Academic Press Inc.) (*b*) ε_1 and ε_2 versus U for gold.

As the wavelength decreases away from the centre of this band, both n and k decrease; however, ε_1 still remains negative. The Hagen–Rubens relationship discussed in Chapter 1 has been shown experimentally to be approximately valid for all metals at very long wavelengths and in certain cases, for example gold, silver and aluminium, down to wavelengths as small as 3 μm. As the quality of specimen surfaces has been improved, the predictions of the Hagen–Rubens theory have been investigated more thoroughly, and work on this topic is still being carried out, for example by Bennett and Bennett (1966).

In the near-infrared region a better fit between the experimental and the free-electron theoretical values of n and k can sometimes be obtained by taking account of the anomalous skin effect (§1.6.3).

In the visible and near-ultraviolet regions of the spectrum, the effects of the interband transitions of the electrons may become apparent in some metals, producing further absorption bands. Copper and gold are notable cases where this effect is observed as their absorption bands fall in the visible spectrum, giving rise to their characteristic colours. The interband transitions can also be observed as peaks in the reflectance and in ε_1 and as an abrupt increase in ε_2.

For other metals in the same wavelength region the values of k are very much smaller, and eventually become less than those of n; the value of ε_1 passes through zero and this is the condition for producing a plasma resonance which can also be observed as a narrow absorption band in the spectrum of Im $1/\varepsilon$. If ε_2 also falls to very low values the materials become transparent at wavelengths below the plasma resonance as in the alkali metals. The values of ε_1 become positive at these short wavelengths and the metal then behaves similarly to an absorbing dielectric. In some metals, however, the situation is more complicated as the frequency of the plasma resonance is modified by the proximity of interband transitions as described in §1.5.3.

Still further reduction of the wavelength allows additional interband transitions to be stimulated from the lower lying electron energy bands, and several absorption bands of this kind may be observed towards the x-ray region of the spectrum.

Henke (1981) has devised a method of calculating n and k at very short wavelengths based on X-ray scattering coefficients; this works well for both elements and compounds in this spectral region where the value of n is close to unity and k is small. He uses the following equations:

$$n = 1 - r_0\lambda^2 nf_1/2\pi : nf_1 = \Sigma\, n_q f_{1q}$$

$$k = r_0\lambda^2 nf_2/2\pi : nf_2 = \Sigma\, n_q f_{2q}$$

where λ = wavelength in Å, r_0 = classical electron radius = 2.818×10^{-15} m, f_{1q} and f_{2q} are the scattering coefficients for a given element q. For a compound, each element must be summed separately; n_q is the number of atoms of type q in the compound.

Tables of the coefficients f_1 and f_2 have been published (Henke *et al* 1982) for many of the elements in the periodic table and the values of n and k calculated by this method are good up to wavelengths of about 100 nm; peaks in k caused by interband transitions of the M and L electrons are readily shown up.

Since about 1955 it has become possible to effect an enormous improvement in the quality of metal surfaces (see §2.2). The broad optical features described above have been confirmed and more accurately catalogued and, in addition, other absorption phenomena have been observed which may be directly related to (i) the condition of the surface, i.e. roughness and contamination, (ii) internal vacancies and flaws within the material, and (iii) grain boundaries. In order to avoid the problems presented by mechanical polishing, many of these more recent measurements have been performed on very thick films ($d > 100$ nm) of extremely pure materials deposited rapidly in a vacuum of about 10^{-10} Torr and maintained at this pressure throughout the optical measurements which are completed as quickly as possible to minimise the contamination of the surface. It is still likely, however, that such films will not have the same surface finish as solid specimens and, in addition, their internal structure may contain many more dislocations and stacking faults. Considerable progress has been made, however, in producing films whose optical properties are very close to those of bulk specimens. Very fast deposition rates (several tens of nanometres per second), together with heated substrates, are usually necessary in order to produce large microcrystals and reduce grain boundaries which are the requirements towards this end. It is still necessary to use bulk specimens for anisotropic materials, although carefully prepared thick films may be preferable for polycrystalline specimens.

It is now almost standard practice to examine all optical specimens by x-rays and/or Auger spectrometry to ensure that the crystal structure is fully revealed, and also to detect any surface contamination.

Figure 4.2 shows plots of k versus n for a number of materials in their bulk form, the wavelengths in micrometres being given alongside each plot. In the infrared, the noble metals are characterised by large values of k and small n, whilst for the ferromagnetics the values of n and k are roughly equal; for all metals both n and k tend to increase with wavelength over the spectral region covered. Dielectrics have small n and negligible k whilst ferroelectrics have large n and also small k.

Ordal *et al* (1983, 1985, 1987, 1988) have published collated sets of optical constants for aluminium, cobalt, copper, gold, iron, lead, nickel, palladium, platinum, silver, titanium and tungsten and these tables, which cover a wide range of wavelengths, are a useful reference source. In addition, two handbooks edited by Palik (1985, 1991) contain detailed values of the optical constants for many materials, also over a wide spectral range. Another useful source of data on the optical constants of a wide range of materials is Landolt-Börnstein (1985). It is, however, more instructive to consider the optical behaviour in terms of groups of metallic elements of similar properties such as the alkalis, the noble metals, the transition elements and also alloys, and this practice is followed in the present chapter. Weaver *et al* (1981) have given detailed results for the transition metals, promising further publications dealing with the noble metals, aluminium and the rare earths and the alkalis and the alkaline earths. An article by

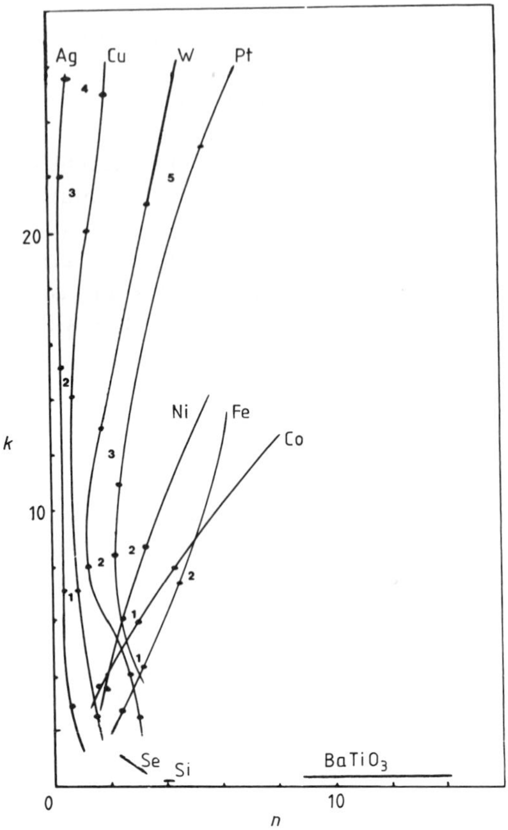

Figure 4.2 Values of the optical constants n and k for various materials. (Numbers alongside each curve represent the wavelength in micrometres) (From Ward L 1985 *Optica Acta* **32** 156.)

Nilsson (1974) gives a good background to the optical properties of metals and alloys.

In discussing the optical properties of metals over a range of wavelengths, it is customary to attempt to identify the features caused by free electrons, interband transitions and plasma resonances. For each metal this involves detailed analysis of the features in the spectra of n and k, combined with reference to the appropriate electron band structure diagram. One example of this process is described in some detail for copper, but, for the remaining metals, only the more important optical features are pointed out.

4.2.1 Alkali metals

The alkali metals, lithium, potassium, sodium, caesium and rubidium, are monovalent and have body-centred cubic structures; the first Brillouin zone is thus in the form of a dodecahedron (see figure 1.9). The outermost conduction electrons occupy the lower part of the broad s-bands and are loosely bound to the parent atoms; the Fermi surfaces are nearly spherical and do not come into contact with the boundaries of the Brillouin zone. The solid state situation is thus fairly straightforward and approximates closely to the classical free-electron model. Nevertheless, the experimental results obtained for these metals have proved difficult to interpret and there are two main reasons for this. Firstly, all the metals are chemically extremely reactive and rapidly become contaminated when exposed to gases particularly oxygen. Secondly, it is extremely difficult to prepare surfaces of these materials which are free from microscopic roughness.

Wood (1933), in a classical series of measurements on these metals when specimens were cut and measured inside the vacuum system, observed that they became transparent in the ultraviolet spectrum. This was one of the first experimental observations of plasma frequencies which, in this group of metals, occur at energies well clear of the interband transitions so that the free-electron values of the plasma frequencies are obtained. Wood also noted that the upper wavelength edge of the transparent region moved towards longer wavelengths with increasing atomic weight (see table 4.1). This is to be expected as the plasma wavelength, being proportional to $1/\sqrt{N}$, is thus also proportional to the square root of the atomic weight divided by the density, and this quantity increases with the atomic weight.

Table 4.1 Plasma wavelengths (experimental and theoretical) for the alkali metals.

Metal	λ_p, expt. (nm)	Atomic weight, A	Density, ρ (kg m^{-3})	$\sqrt{A/P}$	λ_p, theory (nm)
Li	205	7	534	0.114	155
Na	210	23	970	0.154	210
K	315	39	860	0.213	290
Rb	360	85.5	1530	0.236	322
Cs	440	133	1870	0.267	364

Some of the better known of the earlier investigations of the optical properties of the alkali elements was carried out by Ives and Briggs (1936, 1937, 1938) using ellipsometry over the range 250 to 580 nm. For potassium, a maximum was expected in the absorption at 340 nm using p-polarised light, but this was actually observed at 360 nm. Agreement with theory was better at the shorter rather than the longer wavelengths and this may have been a consequence of the difficulty in obtaining clean surfaces.

From more recent times, the work of Mayer and his collaborators (Mayer and El Naby 1963a, b, c) on these materials is well known. Very pure samples of the metals were obtained by repeated distillation in a vacuum of about 10^{-10} Torr and the final optical specimens were made by evaporating on to a cup-shaped substrate. The optical technique used for the determination of n and k was that of Beattie and Conn (see Chapter 2), the readings being taken as quickly as possible after deposition to minimise contamination of the specimen surface. The temperature of the substrate was varied between 90 and 370 K so that some of the results obtained refer to the metals in their molten state.

Mayer and Heitel (1966) and also Hodgson (1963) found a very strong absorption band for photons having energies less than the interband threshold (410 nm) in a region where Drude free-electron absorption is weak. For example, in sodium the peak was at 690 nm (Mayer) and 890 nm (Hodgson); it was later suggested that this peak might have been caused by a surface film of the metallic hydroxide, although Overhauser and Butler (1976) found no evidence that the hydroxide itself produced such absorption. They believed the phenomenon could be explained in terms of optical anisotropy, perhaps induced by the presence of the hydroxide. However, many investigators have failed to find Mayer's absorption band. For example Smith (1969, 1970) using multiple reflection ellipsometry, although agreeing with Hodgson's results in the infrared and noting the presence of the interband transitions, found no evidence for the anomalous band. His explanation was that he was working with thick films rather than the solid-like materials used by Mayer. Monin (Monin 1973, Monin and Boutry 1974) deposited the alkali metals at 71 K and then reheated them before carrying out ellipsometric measurements on the metal/vacuum interface. His results were close to those obtained by Smith who had used a metal/ quartz interface. Measurements at 77 K, when surface microcrystallites would have been present, showed a strong absorption peak which was not coincident with Mayer's, and it would therefore appear that both Monin's and Mayer's peaks were caused by the optical properties of the micro-crystals (Meesen 1972). This phenomenon is considered in more detail in Chapter 8 when dealing with discontinuous films. Another investigation on the alkali metals was carried out by Palmer and Schnatterley (1971) using ellipsometry on the metal/vacuum surface for films deposited on sapphire at 77 K. Although their results for sodium were in agreement with those of Ives and Briggs, they did not tally with the results of Smith, Mayer or Hodgson. However, when using the metal/substrate interface, which is less subject to contamination, the agreement was closer. In the case of potassium films, the agreement was better for both interfaces.

Payan and Roux (1969) discovered two anomalous bands; the first at 1.9 eV (0.65 μm) only occurred with discontinuous films and was believed to be caused by the state of the surface. The second band was only observed when the $\boldsymbol{E}$ vector of the radiation was parallel to the plane of incidence, i.e. the p-state of polarisation; it was probably due to surface plasmons at right angles to the surface.

In the near-infrared region, many authors have attempted to fit their experimental results into Drude-like formulae. Thus, Cohen (1958) used the relationship

$$\varepsilon_1 = A - B/\omega^2$$

where $A = 1 + 4\pi Na$ and $B = \omega_p^2$. This relationship may be derived by putting $\omega \gg \omega_c$ in equation (1.103). If the free-electron theory is valid, a plot of $-\varepsilon_1$ versus $1/\omega^2$ or λ^2 should be linear with a slope of ω_p^2; the intercept on the ε_1 axis then gives the value of A from which the polarisability, a, may then be determined.

Mayer and Heitel (1966) used Cohen's formula with the alkali metals, and in general the results, shown in figure 4.3(a), were in agreement with these ideas; the slope, and hence ω_p, increased on proceeding towards elements of lower atomic numbers as with Wood's results. The values of a were about 0.15 for all the alkali metals.

Considering the imaginary part of the dielectric constant, ε_2, there will be absorption of energy due to the damping of electron oscillations by a number of causes such as interactions with phonons, lattice defects, interband transitions and other mechanisms. For free electrons, the electrical conductivity is given by

$$\sigma = \frac{\omega_p^2 \omega_c}{\omega^2 + \omega_c^2}.$$

When $\omega \gg \omega_c$, σ is proportional to $1/\omega^2$. Figure 4.3(b) shows σ versus λ or U for sodium; the Drude formula is a good fit up to energies of 1 eV but above that there is a wide divergence. Between 1 and 2.5 eV there is a peak in ε_2 which is probably that found in Mayer's earlier work, whilst a further peak at 3 eV may be attributed to direct transitions between the conduction band and higher unfilled bands. For potassium and caesium, the plots of σ versus U showed unexpectedly strong absorption in the spectral region between the Drude and interband effects.

Cohen himself (1966) attempted to interpret Mayer's results for the optical behaviour of the alkali metals. For sodium at 90 K there was a smooth changeover from the Drude to the interband absorption mechanisms instead of the sharp absorption edge which would be expected. At higher temperatures, the strength of the extra absorption band at 1.7 eV grew with temperature at the expense of the interband transitions. The Drude band and the residual interband absorption were still present in the liquid state.

Caesium behaved similarly to sodium except that the extra absorption band was found overlapping with both the Drude and interband regions; this band was highly asymmetrical with a pronounced dip below the estimated Drude absorption on the long wavelength side followed by a peak at about 1.0 eV reaching above the Drude value. The peak was present and

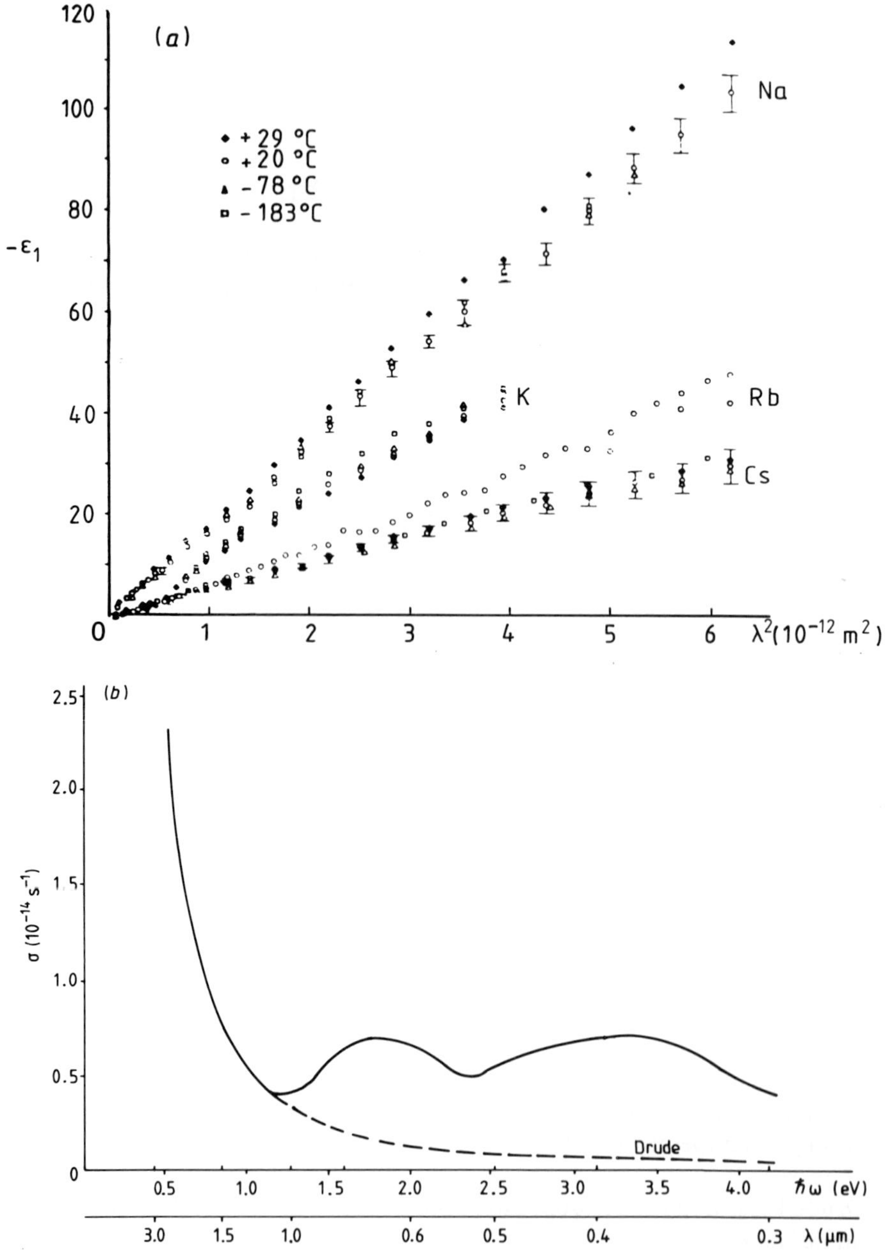

Figure 4.3 (*a*) Real part of the dielectric constant, $-\varepsilon_1$ versus λ^2 for the alkali metals. (*b*) Absorption of sodium versus wavelength. (From *Optical Properties and Electronic Structure of Metals and Alloys*, ed F Abelès (1966), courtesy North-Holland Publishing.)

strong at all temperatures. An increase in absorption was also observed at the highest photon energies used (4.2 eV). Potassium also behaved in a similar manner with a large peak at 1.4 eV followed by a second absorption peak at 3.0 eV.

It was not possible to account for these extra absorption effects in terms of Drude, interband or intraband transitions. In addition explanations were needed for the following effects:

(i) the strong asymmetry in the curves for potassium and caesium;
(ii) the strength of the band was comparable to that of the interband absorption line at all temperatures in potassium and caesium and, in sodium, at all temperatures where the extra band was fully developed;
(iii) some extra incoherent background which apparently interfered with the absorption process responsible for the peaks in potassium and caesium;
(iv) in sodium as the peak grew in intensity there was a corresponding decrease in intensity from the interband transitions.

Cohen attempted to explain these phenomena in terms of the finite lifetimes of the excitons, i.e. the dynamic interaction between the quasi electrons and holes which form in the free-electron gas under the influence of energetic radiation. He ruled out the possibility of indirect transitions which had been proposed by Mayer. Surface roughness may also play a part in accounting for some of these phenomena.

Lithium has been investigated by Rasigni and Rasigni (1977) who measured the reflectances at the vacuum, R, and the substrate, R', sides at normal incidence for thick opaque films deposited on quartz substrates in a vacuum of 10^{-10} Torr. Figure 4.4(a) shows electron micrograms of the surfaces of lithium specimens showing regular patterns of roughness. Considerable differences were observed between the measured values of R' and those calculated from Drude's equation. The inclusion of the effects of

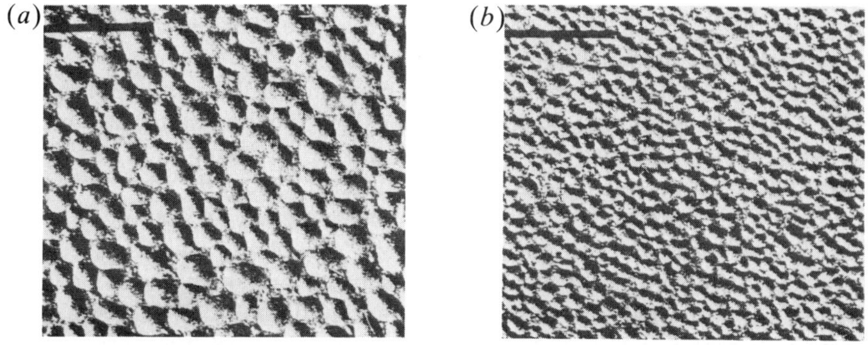

Figure 4.4 (a, b) Surfaces of lithium specimens. (From Rasigni M and Rasigni G 1977 *J. Opt. Soc. Am.* **67** 513, 514.)

interband transitions gave better, but still not perfect, agreement for wavelengths down to 0.4 μm.

A Kramers–Kronig type of calculation was carried out in order to obtain the phase change on reflection, which, together with the reflectance, could then be used in Fresnel's equations to calculate n and k. The results obtained were as follows.

(1) The interband threshold given by theory as 3 eV was found experimentally as 2.65 eV.

(2) In the low-energy region a supplementary absorption peak was observed, apparently due to collective oscillations (plasmons).

(3) The low value of R' may be due to surface plasmons. If the roughness ~2–5 nm, photons would be able to excite surface plasmons leading to absorption over a wide range of energies below the plasmon energy. For lithium, the plasmon energy is 4.68 eV thus permitting a decrease in reflectance over the energy range 2–5 eV.

Measurements were also made on lithium specimens deposited at 6 K; the observed surface roughness was considerably reduced, but the specimen was still not perfectly smooth (figure 4.4(*b*)). The results, however, showed better correlation with those predicted by the Drude theory down to a lower wavelength, thus supporting the speculation that surface roughness was responsible for some of the extra absorption.

Another study using lithium was carried out by Mathewson and Myers (1972). They evaporated the metal at a rapid rate in an ultra-high vacuum on to quartz and sapphire substrates and, using ellipsometry, found absorption edges on the sapphire at 2.4 eV at a temperature of 125 K and on the quartz at 1.4 eV and 295 K. They concluded, rather surprisingly, that the Fermi surface of lithium was just in contact with the faces of the Brillouin zone, although Cohen and Heine (1958) had also reached the same conclusion. On the other hand Ching and Callaway (1973, 1974), who carried out band structure calculations for lithium, found that the Fermi surface departed by less than 4% from the spherical shape. The onset of interband transitions was predicted at 3.28 eV (0.38 μm).

Hodgson (1966) measured the optical constants of lithium between 0.22 and 2.0 μm using internal reflection from a thick film on a silica glass prism. A graph of ε_1 versus λ^2 was linear down to about 1.5 μm giving $m_0 = 1.57$ m. The electron relaxation time decreased with photon energy and was much smaller for films than for bulk material, presumably because of the large number of imperfections in the films. The conductance, $\sigma = \varepsilon_2/2\lambda$, plotted against photon energy showed a weak absorption edge at 3.2 eV (the expected energy for direct interband transitions is 3.6 eV). A strong absorption band at an energy greater than 5.5 eV (the experimental limit) was also indicated.

Sievers (1980) using a Kramers–Kronig process carried out a reanalysis of the infrared properties of sodium, potassium and rubidium within the framework of a model using both nearly free electrons and interband transitions, but found it was not consistent. Good agreement between theory and experiment was found, however, by incorporating in the model a surface-plasmon-assisted process. When investigating thick films, Sievers found another anomalous absorption peak at a frequency greater than the interband threshold; this may be associated with either surface roughness or grain boundaries in the film. These findings were in agreement with those of N V Smith and of Inagaki *et al* (1976).

Other contributions on the optical properties of the alkali metals have been made by Ferrell (1966) Callcott and Arakawa (1974) and Whang *et al* (1971).

4.2.2 Alkaline earths together with zinc and cadmium

In the alkaline earths, which include beryllium, calcium, magnesium, strontium and barium, there are two s-electrons per atom and these would be just sufficient to fill the s-bands completely. However, the s-bands overlap with those due to the p-electrons which are added in the next groups of elements so that the alkaline earths are metallic in nature. The outermost electrons are weakly bound to the atoms, but the core electrons are strongly bound and would not be expected to contribute to the optical properties except at high photon energies. Zinc and cadmium are often considered alongside the alkaline earths as they also contain full complements of s-electrons. In terms of crystal structures, beryllium, magnesium, zinc and cadmium are hexagonal close packed (HCP), calcium and strontium are face-centred cubic (FCC) whilst barium is body-centred cubic (BCC).

Fisher *et al* (1966) determined the optical constants of thick films of barium in the ultraviolet using reflectances at 17.5 and 72.5°. The results were a good fit to a Drude model for energies less than 4 eV with an electron relaxation time of 1.6×10^{-16} s. Plots of $\mathrm{Im}(1/\varepsilon)$ and $\mathrm{Im}(1/1+\varepsilon)$ gave peaks at about 5 eV (surface plasmon) and 7.4 eV (volume plasmon).

The optical constants of beryllium were determined by Shklyarevskii and Yarovaski (1961) in the infrared (0.8 to 11 μm) using Avery's method. No anomalous optical features were found and the values of n and k could be expressed in terms of a free-electron model over this wavelength range.

Measurements of the optical properties of the HCP elements magnesium, zinc and cadmium were carried out by Lenham *et al* (1966). The surfaces were polished (i) by hand and (ii) electrolytically and the optical functions measured were the reflectance ratio, R_p/R_s, and the phase difference, $\phi_p-\phi_s$. All three elements have strong absorption bands in the visible or infrared regions, the lower energy edges of which are 0.4 eV (3.0 μm) for magnesium, 1.0 eV (1.25 μm) for zinc and 0.8 eV (1.5 μm) for cadmium. For all these metals an additional absorption band was observed at very long wavelengths (~20 μm) which was presumed to be due to transitions amongst components of split bands; this evidence clearly throws doubt on the use of free-electron theory in this region. An alternative explanation of this phenomenon, which occurs with several metals in the mid-infrared region, is the presence of slight surface roughness.

Rasigni *et al* (1975) also studied magnesium in the form of thick deposits. The surfaces obtained were not perfectly smooth, but magnesium could still be treated as an almost free-electron metal. The optical absorption spectrum has a peak due to interband transitions at 1.8 μm and this is well below the plasmon energy of 7.42 eV (0.15 μm).

Lettington (1966) has reported the results of work on the optical properties and Fermi surface of zinc which, having an HCP structure, is anisotropic. A single crystal was prepared from pure zinc powder and a surface containing the optic axis was cut and electropolished. The reflectance was measured in air, with the optic axis perpendicular to the plane of incidence, the measurement then being repeated with the optic axis parallel to the plane of incidence. By this means it was possible to calculate the complex refractive index for both the ordinary and extraordinary rays. The results show peaks in the function $2nk/\lambda$ at 1.6 eV (0.78 μm) (for the electric field parallel to the basal plane) and at 1.0 and 1.6 eV (1.24 and 0.78 μm) (for the field parallel to the c-axis). An energy band calculation for zinc was also carried out and showed that there were several situations where nearly parallel *U–K* curves were present. Transitions between these bands would thus be possible across large parts of *U–K* space and would produce strong optical absorption. The energy differences between these bands indicated absorption at exactly the same energies where these had been found experimentally.

Graves and Lenham (1968a, b) used single-crystal specimens of magnesium, zinc and cadmium at temperatures of 82 and 295 K, when measuring the absorption in the visible and near-infrared regions. Peaks in the spectra were identified with features in the band structures calculated by Harrison (1966). Other results for zinc have been published by Rubloff (1971) and Weaver (1972).

It is expected that in its optical properties, cadmium would behave similarly to zinc. Bartlett *et al* (1971) found an absorption peak at 1.4 eV (0.89 μm) for waves with electric fields parallel to the basal plane and at 0.3 and 1.2 eV for those vibrating parallel to the c-axis. Olson and Lynch (1974) measured the reflectance of cadmium between 35 and 250 nm (4.1 and 1.0 μm). At 150 nm the onset of transitions from the top of the 4d band to the Fermi level was observed in the electron-energy-loss function; these transitions persisted up to 30 eV although only 30% of their oscillator strength was used by this point. Volume and surface plasmons were found at 8.80 and 7.30 eV respectively.

Beryllium does not behave like a free-electron metal and shows only weak structures in its absorption spectrum. Similarly, the spectrum for calcium is featureless up to 2 eV.

4.2.3 Noble metals

The noble metals, gold, silver and copper are monovalent and have s- and d-bands containing a total of eleven electrons. In the periodic table they follow immediately after the transition elements to which they have many similarities; sometimes they are included in that group. The core electrons are strongly bound but there is only weak binding of the outer electrons which lie in six overlapping bands. Transitions between these bands are thus easily produced leading to broad optical absorption bands. The crystal structures are face-centred cubic and the Brillouin zones are the fourteen-sided figures known as tetrakaidecahedra. Although the Fermi surfaces might appear at first sight to depart considerably from the spherical shape and, in fact, are in contact with all the eight hexagonal faces of the Brillouin zone, they are still recognisable as being derived from a distorted free-electron sphere.

The noble metals have been the subject of very extensive optical studies, in particular gold which does not readily contaminate in air. A considerable literature has been accumulated on the optical properties of these materials. Typical results for the noble metals are illustrated in figure 4.5 which shows the variation of n and k with wavelength for copper.

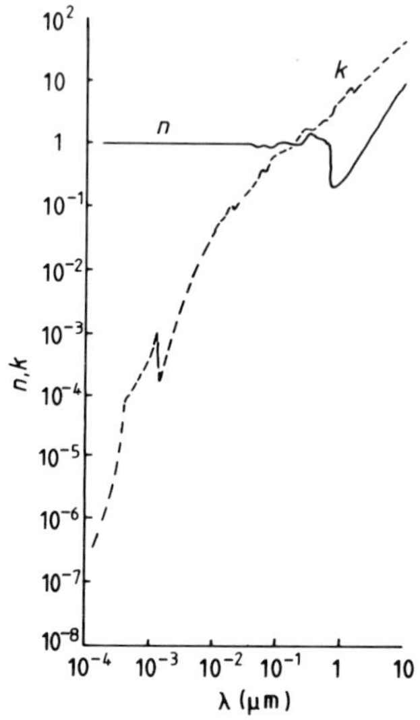

Figure 4.5 Log-log plot of n and k versus wavelength for copper. (From *Handbook of Optical Constants of Solids*, ed E D Palik (1985) courtesy Academic Press Inc.)

The validity of Drude's theory for the noble metals has been tested by Bennett and Bennett (1966) and Bennett *et al* (1968). The infrared reflectance of these metals measured on thick specimens deposited in ultra-high vacuum were significantly higher than those using less stringent preparation techniques and the values obtained were in excellent agreement with Drude's theory in the wavelength regions 3–30 μm for gold and 2–30 μm for silver. In applying the anomalous skin effect, it was assumed that the electrons were reflected specularly from the surface (but see Lenham and Treherne (1966a))—diffuse reflection giving slightly lower values of the reflectance. Near the interband transition thresholds the measured values of the reflectance fell below the calculated ones and it was speculated that this might be due to the electron relaxation time becoming wavelength dependent.

Lenham and Treherne (1966a) also applied the anomalous skin effect to the noble metals in the infrared region, 3–19 μm, measuring the functions $k^2 - n^2$ and $2nk/\lambda$ at oblique incidence at room temperature. Three types of specimen were used; hand-polished bulk, electropolished (copper) and thick, slowly deposited films. The results for the three metals were similar, showing that both the above optical functions increased with wavelength without any significant peaks or troughs. The best fit with theory was obtained using classical free-electron theory together with diffuse reflection of the electrons at the metal boundaries.

Beattie (1957) adapted Dingle's equations for the anomalous skin effect and showed that they fitted his results for silver assuming that 77% of the electrons were diffusely reflected and 23% specularly reflected at the surface.

Leveque (1983), using chemically polished single crystals, investigated silver at wavelengths near the interband transition point by means of a Kramers–Kronig analysis of the reflectance data between 3.5 and 30.0 eV. He found that, in the Drude region, the grain size affected the electron scattering, whilst, near the interband transition, errors were introduced by both the grain boundaries and voids.

Motulevich and Shubin (1965) investigated the effect of the shape of the Fermi surface of gold on its optical properties between 1 and 12 μm, i.e. mainly in the free-electron region. They concluded that small deviations from the spherical shape had little influence on the optical constants in this part of the spectrum.

Beaglehole (1966) measured the optical constants of copper and gold between 2 and 26 eV (0.05 to 0.63 μm). The optical surfaces were prepared on discs of the materials of 99.999% purity by first annealing, then electropolishing and finally removing the surface layer by reduction in the experimental chamber. A surprising feature of his results was the low value obtained for the number of electrons per atom, 2.2 for copper and 4.6 for gold. It would be expected that, for noble metals, these values should saturate at $10 + m/m^*$ electrons/atom; m/m^* being about 0.3 for copper and 0.15 for gold. The implication is that the d-electron excitations are far from saturated even at photon energies at 26 eV.

Berglund (1966) considered the effects of direct and indirect electron transitions in copper using data on photoemission as a function of U', defined as electron energy minus photon energy. Two peaks were obtained, the position of that at high energy being unaffected by the photon energy but vanishing for photon energies greater than 8.6 eV. The peak at the lower value of U' moved to lower energy with increasing photon energy. Berglund claimed that these results were consistent with the assumption that indirect transitions were dominant in copper.

Parkins *et al* (1981) investigated the noble metals at temperatures of 77, 295 and 425 K. They were studying the effects of various electron scattering mechanisms such as electron–phonon, electron–electron and electron-defect scattering, it being assumed that the scattering rate, $1/\tau$, was frequency dependent. The observed values for electron–electron scattering were two or three times larger than the theoretical estimates; it was also expected that this type of scattering would be related to both the electrical and thermal resistivities, but the incl' sion of these relationships only made the discrepancy greater. Thus, although the frequency dependence of the Drude scattering rates is qualitatively suggestive of electron–electron scattering, the quantitative evidence is uncertain.

Cooper *et al* (1965) discussed the noble metals in detail making use of the *U–K* diagrams due to Seagall (1962); that for copper is shown in figure 4.6. The reflectance of electropolished specimens were measured up to energies of 25 eV and the values analysed in terms of free electrons, interband transitions and collective oscillations. The noble metals contain partially filled conduction bands, lower, completely filled d-bands and empty excited bands. The free-electron contributions could easily be separated out as ε_2 fell rapidly to near zero as the wavelength decreased (figure 4.8). Using values of the DC conductance, the optical masses of the electrons could be calculated, giving 1.03 m (m is mass of the electron) for silver, 1.42 m for copper and 1.04 m for gold, these being in reasonable agreement with other published data.

A simplified picture of the electron energy bands in copper is given in figure 4.7 which shows that the 4s/3p (6) band is almost full whilst the 3d band (3, 4, 5) is completely filled and lies below the Fermi level; the empty 4p band (7) is above the Fermi level. Interband transitions would be expected between the 3d and U_F or between the 3d and 4p or between the 4s/3p and the 4p, these transitions being most likely at symmetry points on the Brillouin zone. Transitions from L_2' to L_{12} at the Fermi energy occur at 4.25 eV in copper; similarly, direct transitions between X_5 and X_4' occur at an energy of 2.1 eV. In addition, the transition between K_2 and K_1 requires an energy of about 3.7 eV. These transitions are marked on figure

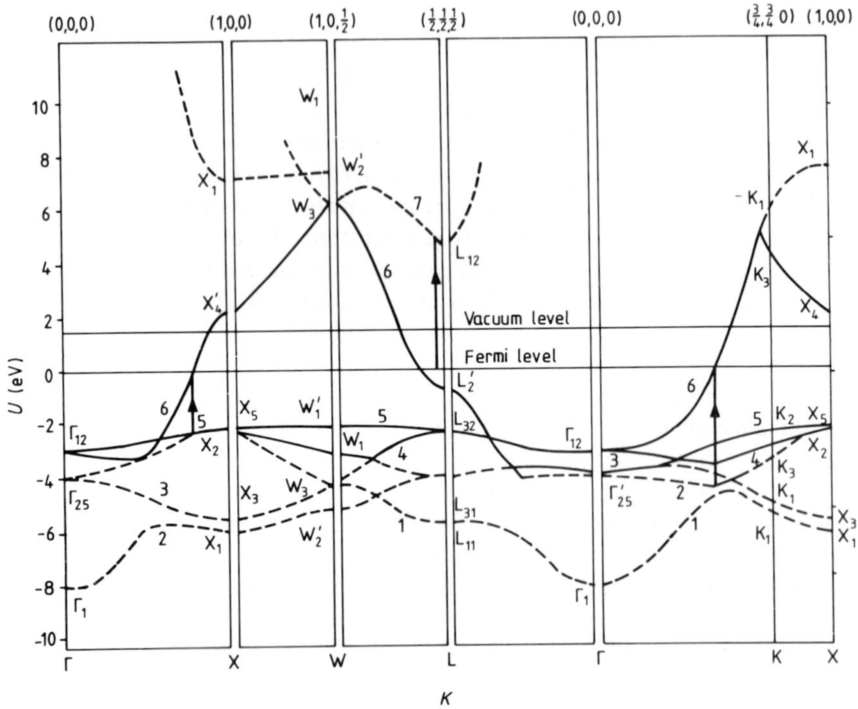

Figure 4.6 Band structure for copper. (From Sgall B 1962 *Phys. Rev.* **125** 114.)

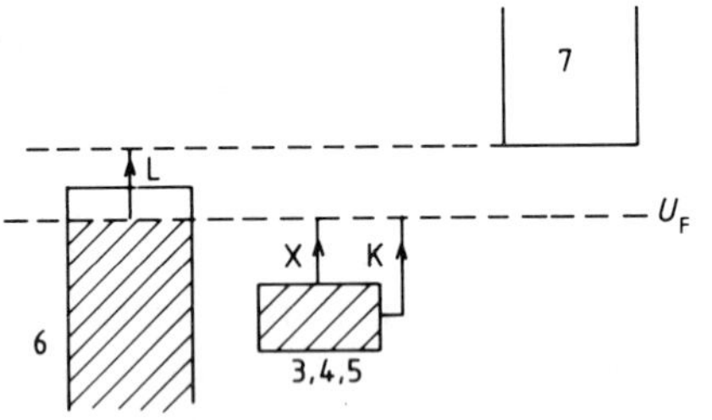

Figure 4.7 Upper energy bands for copper showing possible transitions.

4.6 and the absorption bands for copper are shown in figure 4.8. The corresponding values for silver are 3.9 eV (X-point) and 5.5 eV (L-point), while those for gold are 2.3 and 3.9 eV.

Above 10 eV the reflectances of all three metals were high and the identification of the individual transitions became difficult because of the large number of possible transitions. For gold, small peaks were observed near 7 and 15 eV due to further interband transitions and there was also a

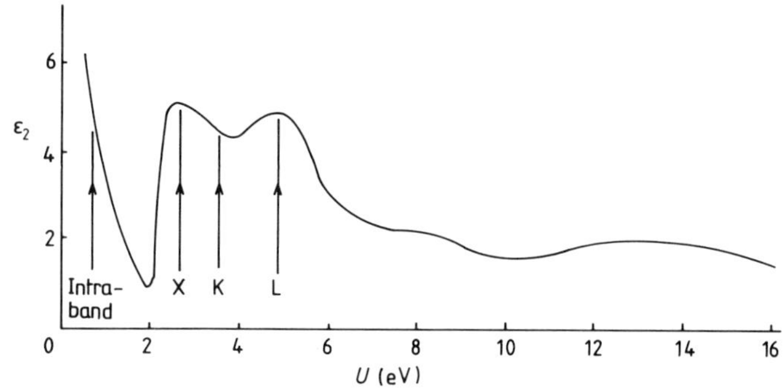

Figure 4.8 ε_2 versus U for copper.

resonance at 33 eV which might be due to collective oscillations in the s- and d-electrons; no peak was observed for the free-electron plasma.

These conclusions are in contradiction to those of Berglund discussed earlier and also of Spicer (1965) where indirect transitions were postulated. Cooper *et al* (1965) claimed that Berglund's high-energy peak could be explained in terms of direct transitions between bands 5 and 6; band 5 is almost flat right across the Brillouin zone whilst 6 shows a large change in U with zone position. However, the lower d-bands are not sufficiently flat to treat the low-energy peak in the same way and so there may be some indirect transitions here but these would not be expected to predominate and direct transitions between bands 3 or 4 and 6 would still furnish the major contribution to the optical behaviour. Spicer (1972) reexamined this controversy and, whilst agreeing that the use of direct transitions predicted the correct frequencies in the absorption structures, he noted that the theoretical curves were less sharp and also weaker than those obtained experimentally. He therefore suggested that the features due to the direct transitions were superimposed on a strong contribution from the indirect ones. Support for the dominant role of the direct transitions in copper has been provided by the photoelectric measurements of Stohr *et al* (1978).

Sets of values for the optical constants of silver and copper in the soft X-ray region (100–1200 eV, 1.0–12.5 nm) have been reported by Yanagahara *et al* (1988). These were derived from reflectance measurements on very thick films and showed satisfactory agreement with values calculated by Henke's method.

Taft and Philipp (1961) calculated the optical constants of silver from measurements of the normal incidence reflectance in the range 2–10 eV. There was a sharp minimum in k at 3.8 eV (plasmon) and a second minimum at 9.2 eV. Steinmann (1968) has given the characteristic electron-energy-loss spectrum of silver which shows a broad band at 7.4 eV in addition to a sharp peak due to plasmons broadly in agreement with Taft and Philipp.

Smith (1977) measured the optical constants of copper at 546 and 633 nm by ellipsometry, great care being taken in polishing the specimens which were then annealed and cleaned by ionic bombardment. Smith noted that, historically, the values of n have tended to increase whilst those of k have decreased as the surface quality has improved.

Dold and Mecke (1965) have reported results for the optical constants of the noble metals over the extensive range 0.1 to 20 μm. For copper and gold k showed a steady increase with wavelength but for silver there was a minimum at about 0.5 μm (2.5 eV). The values of n for all three metals exhibit minima as follows: copper 1.0 μm, silver 0.7 μm and gold 1.0 μm.

Another investigation on copper was that by Donovan *et al* (1965) where the effects of surface preparation on the reflectance were studied between 0.2 and 30.0 μm using both single-crystal and polycrystalline specimens. For electropolished specimens the reflectances were in good agreement with those from the best thick evaporated films and with Drude's theory. On the other hand, mechanically polished specimens showed conclusively that the optical properties at all wavelengths were affected by surface damage.

Liebert and Thomas (1984) studied the effects of grain boundaries using thick films of the noble metals and selecting only those surface areas which contained the grooves due to the grain boundaries. An absorption due to these boundaries was found at 0.7 eV and may have been caused by the build-up of lattice defects close to the boundaries.

4.2.4 Transition elements

The transition metals run from titanium to platinum in the periodic table and comprise titanium, vanadium, chromium, manganese, iron, cobalt, nickel, zirconium, niobium, molybdenum, technetium, ruthenium, rhodium, palladium, hafnium, tantalum, tungsten, rhenium, osmium, iridium and platinum. Their common characteristic is the progressive filling of the d-bands of electron energy levels. There are three groups of transition elements corresponding to the filling of the 3d, 4d and 5d electron shells; each group can hold ten d-electrons and hence five energy bands are required for each group. There is considerable overlapping of the s- and d-bands; both s- and d-electrons may thus contribute to electrical conduction and hence to the optical properties. Mattheiss (1974) noted some trends in the electronic structures of the transition elements. For a given crystal structure, as the total number of s- and d-electrons increases, the d-bands become narrower and, at the same time, the Fermi level rises within the d-band. These effects are illustrated in figure 4.9 which is for the first transition series and shows the BCC elements and the FCC elements. In copper, the first element beyond the transition elements, as already discussed, the d-band falls entirely below the Fermi level. In addition it has been found that the width of the d-band increases on moving vertically in the periodic table from 3d to 5d. In many of the transition elements absorption due to interband

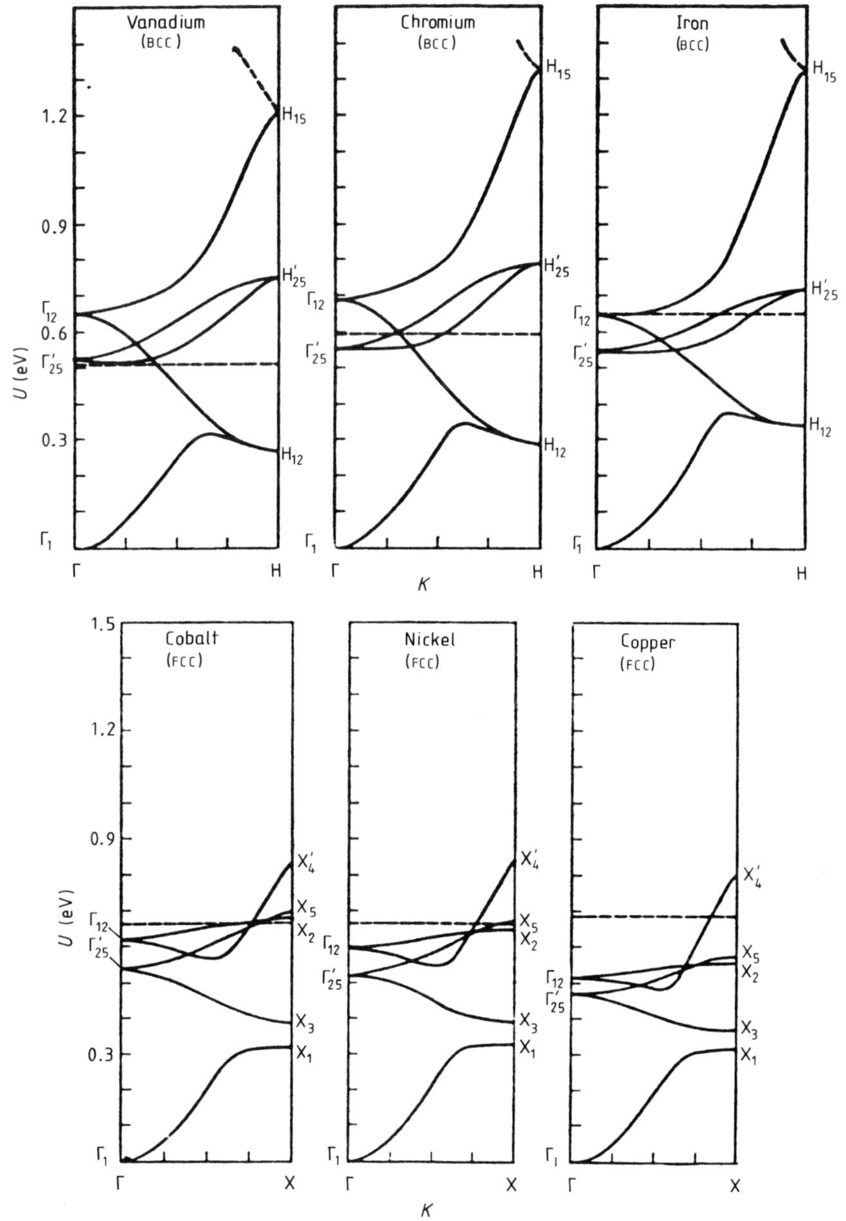

Figure 4.9 Electron energy bands for some transition metals ---, U_F. (From Mattheiss L F 1974 *Phys. Rev.* **134** A970.)

transitions from the Fermi level to the next p-band starts in the near-infrared spectrum. Table 4.2 shows the transition elements, their atomic numbers, crystal structures and electron configurations.

Several wide-ranging studies of the optical properties of this group of elements have been made, notably by Lenham and Treherne, Christy and colleagues and Weaver and co-workers. Lenham and Treherne (1966 c, d) measured ε_1 and ε_2 between 0.35 and 19.0 μm for a number of transition elements deriving the free-electron parameters and also the optical conductivity compared with the DC value. The absorption peaks of the different transition metals showed little correlation with position in the periodic

Table 4.2 Transition elements.

Element	Crystal structure	Atomic number	Number of electrons
			(in 3d shell)†
Sc	HCP	21	1
Ti	HCP	22	2
V	BCC	23	3
Cr	BCC	24	5(1)
Mn	Cubic	25	5
Fe	BCC	26	6
Co	HCP	27	7
Ni	FCC	28	8
			(in 4d shell)†
Y	HCP	39	1
Zr	HCP	40	2
Nb	BCC	41	4(1)
Mo	BCC	42	5(1)
Tc	HCP	43	6(1)
Ru	HCP	44	7(1)
Rh	FCC	45	8(1)
Pd	FCC	46	10(0)
			(in 5d shell)†
La	HCP	57	1
Hf	HCP	72	2
Ta	BCC	73	3
W	BCC	74	4
Re	HCP	75	5
Os	HCP	76	6
Ir	FCC	77	9(0)
Pt	FCC	78	9(1)

† Number of electrons in corresponding 4s, 5s and 6s shells is normally two except where shown in brackets.

table, but certain 'favoured' energies frequently occurred amongst the various elements. i.e. 0.15, 0.25, 0.6, 0.8, 1.3, 2.0, 2.3 and 2.5 eV.

Johnson and Christy (1974) calculated the optical constants of titanium. vanadium, chromium, manganese, iron, cobalt, nickel and palladium from photometric measurements of reflectance and transmittance, and ellipsometric measurements between 200 nm and 2.50 μm. They found that it was necessary to invoke interband transitions to explain the values of n, k, R and T at energies larger than 2 eV, the contribution from the free electrons being negligible in this region. They had some success in explaining their results, particularly the positions of the absorption bands, in terms of electron transitions between the various d- and s-bands numbered from 1 to 6 in order of increasing energy (figure 4.6).

Thus nickel, chromium and vanadium all have a high-energy absorption peak (5.3 eV in nickel, 6.7 eV in chromium and vanadium) which could be identified as being due to transitions between bands 1 and 6, i.e. from the lowest d-band to the s-band. Peaks at 1.8 eV in nickel and 3.4 eV in chromium and vanadium were mainly due to transitions between bands 3 and 5 whilst the peaks near 4.0 eV correspond to transitions between bands 1 and 4 and those near 2.0 eV to bands 3 and 4.

Ingram *et al* (1990) have published values for the complex dielectric properties of the first row of the transition elements (Ti, V, Fe, Co and Ni) in the energy range 4–30 eV (42–300 nm). These were calculated from electron energy loss measurements on specimens up to 0.5 mm thick and compare favourably with those of other recent workers using these elements.

Weaver *et al* (1981) give very detailed results for all the transition metals and this is a useful source book for information on these materials. Weaver and Olson (1976) studied the 4d transition elements by synchrotron radiation where transitions of the 4p–4d type were observed. They found a linear relationship between the absorption edges of the 4p band and the atomic number.

However, within the transition metals it is more usual to consider separately a number of sub-groups which have similar optical and electrical properties. Thus, the ferromagnetic elements of iron, nickel and cobalt form a convenient sub-group; another consists of the platinum-type elements of ruthenium, rhodium, palladium, osmium, iridium and platinum. Similarly, titanium, zirconium and hafnium, vanadium, niobium and tantalum, and chromium, molybdenum and tungsten are frequently considered together.

4.2.5 Ferromagnetic elements

Bueche (1948) measured the reflectances of nickel, iron and cobalt at five angles of incidence in the visible and near infrared. He recognised the difficulty of the reproducibility of the specimen surfaces and so chose those with the highest reflectances. The method of Collins and Bock (1943) was used to calculate the dielectric constants and the conductivities. Nickel showed a peak in ε_1 at 2.1 μm and also in σ at 0.8 μm.

Another study of the ferromagnetic elements was made by Vonsovskii and Sokolov (1949) using a model in which the interacting 'internal' 3d and the 'external' 4s electrons were considered responsible for the ferromagnetism, the dielectric constants and the conductivity. The optical constants were calculated for the limiting cases of high and low frequencies; those for low frequencies agreed with the experimental values near the Curie point, but the correlation was not so good at the high frequencies.

Bakulin *et al* (1972) measured the plasma resonances in iron, nickel and cobalt with the following results:

Fe:	8.0, 17.8 and 24.2 eV	(51, 70 and 155 nm)
Ni:	8.1, 20.3 and 26.4 eV	(47, 61 and 153 nm)
Co:	8.0, 19.5 and 25.5 eV	(49, 64 and 155 nm)

4.2.5(a) Nickel

Nickel has been investigated by a large number of authors. The results of Moravec *et al* (1976) based on a Kramers–Kronig analysis of normal incidence reflectances, are those shown in figure 4.10(*a*); the peak at 4.6 eV was equated with a transition at the L-point in the Brillouin zone. Vehse and Arakawa (1969) measured the reflectances of opaque films, finding inter-

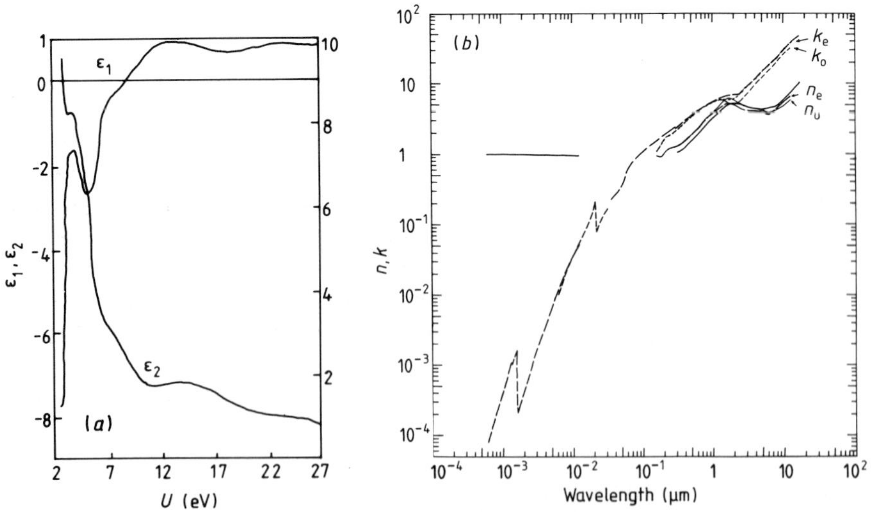

Figure 4.10 (*a*) ε_1 and ε_2 versus U for nickel. (From Moravec T J *et al* 1976 *Phys. Rev.* B **13** 3297. (*b*) log–log plots of n (solid line) and k (dashed line) for cobalt from various sources. (From *Handbook of Optical Constants of Solids II* ed E D Palik (1991), courtesy of Academic Press Inc.)

band transitions at 4.5 and 14 eV (88 and 276 nm). Ehrenreich *et al* (1963) also measured the reflectance but on an electropolished bulk specimen which gave good agreement with Johnson and Christy's results. However, the results of their Kramers–Kronig analysis to determine the dielectric constant were not so satisfactory possibly because of the arbitrary extrapolation of their values of reflectance into the short-wavelength region.

Nickel was also studied by Shiga and Pells (1969) close to its Curie point using ellipsometry, the following conclusions being made: a small peak near 1.4 eV in the absorption spectrum was identified with transitions near the W and K points of the Brillouin zone; a large peak near 5 eV was considered to be due to transitions between the L_2 and upper L_1 bands—this peak moved towards lower energies and increased in width as the temperature increased. Their values of ε_2 were very much larger than those of Johnson and Christy, perhaps because of the different methods of preparing the surfaces.

Roberts (1959) measured the optical properties of nickel over the wavelength range 0.36 to 2.65 μm, the temperature being varied between 88 and 473 K. The results were interpreted in terms of Drude's equation, good qualitative agreement being obtained. Separate terms were introduced into Drude's equation corresponding to different classes of free and bound electrons; each class, which had characteristic values of τ and mean free path, could account for a given wavelength region of the DC conductance. Although most of the DC conductance could be accounted for by just one class of electrons, the other classes also made significant contributions.

Good agreement with Johnson and Christy's results was also found by Seits and Spicer (1970b) who used a mechanically polished specimen. Lynch *et al* (1971) measured the absorptance of nickel at 4 K using a calorimetric method; the derived values of reflectance were lower than those of Johnson and Christy. They found the following absorption peaks in nickel: 1.4 eV (small), 4.7 eV (main) and 2.0 eV (shoulder). The peak near to 4.5 eV has been confirmed by Moritani *et al* (1976) and that near 2.0 eV by Studna (1975), who attributed it to the transition X_5–X_4'.

Figure 4.10(*a*) shows ε_1 and ε_2 plotted against U for nickel, which has optical properties typical of the ferromagnetic elements. ε_1 exhibits a peak at 4 eV, followed by a trough and then rises steadily with energy to become zero at about 10 eV, the plasmon energy; another peak in ε_1 also occurs at about 14 eV. ε_2 decreases rapidly with energy, showing subsidiary peaks at 4 and 14 eV corresponding to those in ε_1; that at 4 eV may be due to transitions between the d-band and the Fermi level.

4.2.5(b) Iron

In its normal state, iron has a BCC structure so its optical properties can be expected to be isotropic.

In the X-ray region (4–60 nm), Sonntag *et al* (1969) measured the absorption coefficient μ ($4\pi k/\lambda$) using synchrotron radiation; k may easily be calculated from the published graphs. A similar spectral region was covered by Wehenkel and Gauthe (1974) who used electron energy loss measurements together with Kramers–Kronig analysis. Both of these sets of results were in satisfactory agreement with one another and also with calculations using Henke's method.

Moravac *et al* (1976) made reflectance measurements in the vacuum ultra violet region analysing their results by Kramers–Kronig to give e_1 and e_2. Other results for iron have been reported by Blodgett and Spicer (1967) who found absorption peaks at 6 and 9 eV (0.21 and 0.14 μm).

The visible and near infra-red spectra have been studied by Weaver *et al* (1979), their technique being a Kramers–Kronig analysis of calorimetric absorption measurements. In the infra-red up to about 10 μm, contributions to the optical properties were observed due to electron transitions within the partially filled 3d-band, (see Jones *et al* 1972). Johnson and Christie (1974), who measured the reflectance of thick films, obtained results in good agreement with these of Jones *et al.*

The optical constants of iron in single-crystal form were measured ellipsometrically by Volken and Kruger (1965) in the visible part of the spectrum. They discovered a maximum in the derived electrical conductance at 400 nm which was interpreted as due to an interband transition. Their reflectance values were higher than those of Johnson and Christy which probably indicates a better packing density for their specimens.

Other contributors in this spectral region have been Ordal *et al* (1985, 1988), Lenham and Treherne (1966) and Siddiqui and Treherne (1977).

4.2.5(c) Cobalt

As cobalt has a hexagonal structure, the optical properties of single crystal material show anisotropy; the *e*-ray corresponds to the electric vector of the radiation vibrating parallel to the *c*-axis i.e. at right angles to the base of the hexagon. Single crystal cobalt has been investigated between 0.15 and 12.4 μm by Weaver *et al* (1979), using their calorimeter technique at wavelengths above 0.3 μm. For wavelengths greater than 1.0 μm, n_o is larger than n_e whilst k_o is smaller than k_e; at shorter wagelengths, the two sets of values converge. Both n_e and n_o show maxima at 1.7 μm whilst k_e and k_o peak at 1.4 μm; as in iron, this effect is caused by electron transitions within the unfilled 3d band.

Turning to polycrystalline cobalt which should be optically isotropic, Kirillova *et al* (1969) have published values of e_1 and e_2 between 0.1 and 0.25 μm derived by a Kramers–Kronig analysis of reflectance measurements. Kirillova and Charikov (1964) worked in the infra-red region between 2.5 and 20.0 μm; they, also, detected a contribution from interband electron transitions reaching up to about 7 μm. In addition, they obtained a value for the plasma frequency of cobalt of 18.6×10^{14} s^{-1}. A similar range of wavelengths was covered in the work of Bolotin *et al* (1973) who used Beattie's method (see section 2.8.7) with a mechancially polished specimen; they found a peak in electrical conductivity at 1.5 μm which was also due to the interband transitions. Using an ellipsometer technique, Lenham and Treherne (1966) have given results in the range 0.3–12.5 μm.

A set of values of n and k for cobalt at different temperatures up to 1900 K has been given by Gushchin *et al* (1978); this showed both n and k increasing with temperature up to about 400 °C, after which both remained approximately constant up to the melting point when each dropped sharply by 5%.

Figure 4.10(b) shows log–log plots of n and k versus wavelength for cobalt in bulk and thin film forms, covering wavelengths between 0.5 nm and 20 μm.

4.2.6 Mo, W, Zr, Ti, Nb, Ta, Cr, Mn, V, Tc, Re, Hf

This sub-group of transition metals has been extensively studied between 0.1 and 36.4 eV by Weaver and his associates. The specimens were very carefully prepared by firstly melting the metals using an electron beam, spark cutting out a sample and then hand polishing. This was followed by electropolishing in a bath of 6% perchloric acid and methanol, annealing at 1150 °C and then re-electropolishing to remove any surface damaged layer. Examination by x-rays showed that the surfaces were free of damage and consisted of large (~ 1 mm) crystallites. Measurements between 0.1 and 4.4 eV were made by an absorption calorimeter and those from 4.4 to 36 eV by using synchrotron radiation, a Kramers–Kronig analysis being carried out to evaluate n, k and e.

Titanium, zirconium and hafnium (Lynch *et al* 1975) (the HCP group) all showed five features in the spectrum of ε_2 between 0.2 and 30.0 eV these being interpreted as transitions within the d-bands; at higher energies, the core electrons became involved in transitions. Two volume and two surface plasmons for each of these metals were deduced from the electron-energy-loss function.

Weaver *et al* (1974) used the same technique to study the optical properties of vanadium, tantalum and molybdenum, these elements having the BCC structure. Typical optical characteristics of one of these metals are shown in figure 4.11 which is for molybdenum. ε_1 increases rapidly with energy, briefly going positive before plunging to a trough near 5 eV; further peaks and troughs are exhibited before ε_1 finally goes positive at the plasmon energy above 20 eV. ε_2 also shows several subsidary maxima as it decreases from an initial high value at low energy. It was found that the simple Drude model was inadequate to explain the optical behaviour of these metals because of the direct interband transitions which commenced in the infrared and extended up to about 18 eV. Features below 6 eV were attributed to transitions in the Σ and G bands of the Brillouin zone in all three metals, whilst for molybdenum an additional transition in the Δ band was noted. High-lying energy bands were identified as producing the structure at higher energies. The electron-energy-loss functions showed that two surface and two volume plasmons were required, this being confirmed for molybdenum by Manzka (1980).

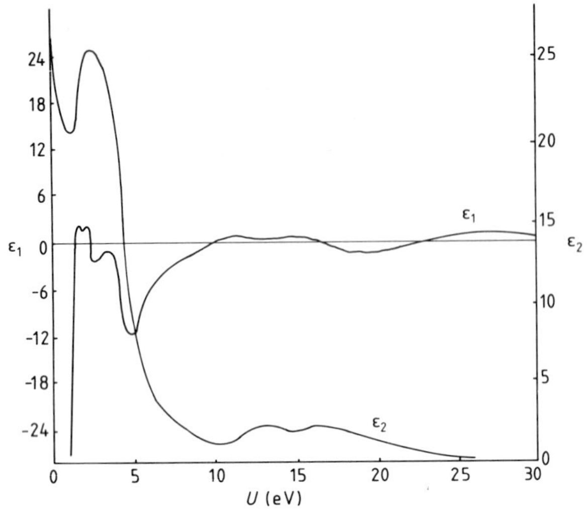

Figure 4.11 ε_1 and ε_2 versus U for molybdenum. (From Weaver J H *et al* 1981 *Physik Daten*, courtesy Fachinformationszentrum, Karlsruhe.)

In a similar study on niobium (Weaver *et al* 1973) and using the energy band calculations of Mattheiss (1970), it was noted that there were interband transitions at 2.5, 4.3 and 5.1 eV. The electron-energy-loss function showed maxima at 9.9 and 20.8 eV due to volume plasmons and at 17.7 and 9.0 eV from surface plasmons. Above 28 eV, transitions from the core electrons began to take place. For tungsten (Weaver *et al* 1975b), the Drude parameters were used to separate the contributions to ε_2 of the interband and intraband transitions; features in the spectrum of ε_2 were observed at 0.42, 0.97, 1.82, 2.35, 3.42, 5.25, 8.8, 11.3, 16.5, 22.5 and 31.5 eV, and these could he related both to the band calculations of Christensen and Feuerbacher (1974) and also to the spectra of other transition elements and thus identified with various electron transitions. From the electron-energy-loss function it was deduced that there were three volume plasmons (25.3, 15.2 and 10.0 eV) and three surface plasmons (20.8, 14.8 and 9.7 eV).

Other sets of data for some of these metals have been given by Kirillova and Charikov (1965) and by Kirillova *et al* (1968). In titanium and zirconium, the intersection of the Fermi level by a large number of energy bands gave rise to a complex pattern of interband transitions between 0.3 and 5 eV. Vanadium and niobium had interband transitions at 6.2 and 7.5 eV respectively. Plasmons were also found in this upper energy range.

Molybdenum, tungsten and tantalum were investigated by Leblanc *et al* (1964) using a novel technique. The radiation was reflected from one specimen on to a second ostensibly identical sample and the photoelectric currents which were generated were measured for both specimens, their ratio being equal to the reflectance. The specimens were cut from standard rolled stock with little surface preparation; however, they were given periodic heat treatment in vacuum to evaporate a little of the metal and so improve the surface quality. Absorption peaks were observed at 50, 80 and 110 nm; these correspond with those of Weaver at 11.3, 16.5 and 22.5 eV.

Kress and Lapeyre (1970a) carried out a Kramers–Kronig analysis on the normal incidence reflectance of molybdenum between 0.5 and 11.9 eV (in the form of films) and between 11.9 and 14.0 eV (for bulk specimens). They observed that ε_2 had shoulders at 2.2 and 4.0 eV whilst the electron-energy-loss functions showed peaks at 10.8 and 9.9 eV. The interband transitions were dominant in the region between 0.5 and 10.0 eV.

Bennett *et al* (1980) examined the relation between the optical and metallurgical properties of polished molybdenum mirrors. Well polished surfaces have surface topographies directly related to their microstructure and hence to the processing they have received. In sheet metal having a well developed texture, the polishability appeared to be independent of texture; larger grain size did not result in a lower scatter surface.

Carroll and Melmed (1974) used ellipsometry to study the optical properties of titanium, which being optically anisotropic must be investigated in several crystallographic directions. One set of measurements was made with titanium deposited on the (011 planes) of tungsten and annealed at 800 K; a second was on a mechanically polished single crystal of titanium using the (0001 plane). Smith (1972) measured the optical constants of single-crystal titanium (0001) and tungsten (110) and also the polycrystalline forms of both metals at 546 and 633 nm. He claimed to be able to detect in the optical properties the effects of contamination by argon, chlorine and sulphur.

Girouard and Truong (1984) measured the optical constants of thick films of titanium and manganese between 6 and 20 eV. They determined the reflectance at five angles of incidence, using an iteration procedure to extract the optical constants. For titanium there was an interband transition near 8.3 eV, a surface plasmon at 11.8 eV and a volume plasmon at 15.9 eV. In the case of manganese, the surface and volume plasmons were at 10 and 20 eV respectively.

Rhenium was investigated by Nomerovannya *et al* (1973) using Beattie's method. The main interband peak was found at 4.3 eV but there were other peaks at 2.5. 8, 11, 12.5. 14 and 17 eV.

Results for chromium have been given by Bos and Lynch (1970) using calorimetric means to determine the absorptance between 0.1 and 5.0 eV; their values were between 7 and 15% higher than those obtained by Johnson and Christy. Another set of results for this metal was reported by Graves and Lenham (1968b). Jones *et al* (1972) have also given values of the optical constants for chromium as well as iron and nickel up to a wavelength of 50 μm; they took into account the modification of the electron relaxation time by photon-electron-phonon interactions.

4.2.7 Ru, Rh, Pt, Pd, Os and Ir (the platinum sub-group)

The optical behaviour of the platinum sub-group is illustrated in the results for iridium which are shown in figure 4.12. ε_1 rises rapidly with energy and crosses into positive values at about 10 eV; a series of minor peaks is then produced following which ε_1 briefly goes negative again until the plasma energy at about 25 eV. ε_2 exhibits a similar series of subsidiary maxima.

Weaver *et al* (1975a) measured the optical properties of rhodium, palladium, iridium and platinum, which all have FCC structures, over the energy range 0.1 to 30 eV, features in the plot of ε_2 versus U being interpreted in terms of the band calculations of Smith (1974). At low energies, the dielectric functions were sensitive to energy bands near the L symmetry point in the Brillouin zone, particularly the spin–orbit split d-like states. For rhodium, structures in the absorptance near 2.8 eV were due to transitions between the electron energy bands 3 and 5. All these metals show minima in ε_2 near 10 eV and also structures at higher energies due to transitions to high-lying energy bands; they also need two surface and two volume plasmons. Palladium was studied by Weaver and Benbow (1975) at 4.2 K; the specimens were produced by mechanical polishing followed by boiling in aqua regia, large crystal grains being observed. A strong maximum was observed in the conductance at 0.46 eV followed by a broad shoulder at 1.25 eV, and both these features could be interpreted in terms of transitions near to the L-point.

Rhodium was also investigated by Pierce and Spicer (1973) who found interband structures at 1.1, 2.5 and 5.8 eV together with a volume plasmon at 8.8 eV and a surface plasmon at 8.1 eV.

Weaver *et al* (1975a) have reported work on rhodium and iridium. For each metal there were a pair of surface and volume plasmons; in rhodium these were at 7.9 and 8.7 eV and in iridium at 7.5 and 8.0 eV. Their results for iridium are illustrated in figure 4.12.

Iridium was also investigated by Kirillova *et al* (1972), again using Beattie's method. In order to explain the free-electron behaviour they used a two-band model which assumed the existence of two groups of free carriers with different relaxation times. Interband transitions were found at 1.01, 1.55 and 2.93 eV (between d-bands 4 and 5) and at 1.60 eV (between bands 4 and 6).

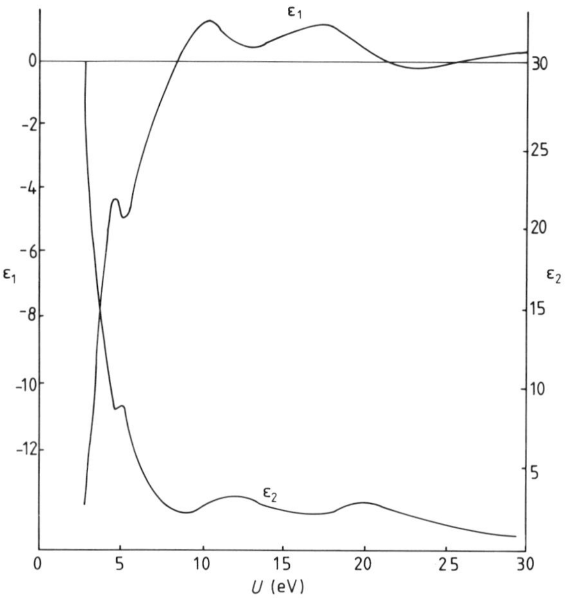

Figure 4.12 ε_1 and ε_2 versus U for iridium. (From Weaver J H *et al* 1981 *Physik Daten*, courtesy Fachinformationszentrum, Karlsruhe.)

Gluskin *et al* (1983) reported measurements of the optical properties of ruthenium and osmium in the vacuum ultraviolet region using synchrotron radiation; these two metals have HCP crystal structures. A Kramers–Kronig analysis was used to obtain the frequency dependence of the optical conductance, dielectric constant and characteristic volume and surface electron-energy-loss functions. The role played by optical excitation of the d-electrons in to the higher energy bands leading to interband conductance was also discussed. The electron-energy-loss functions in the range 5 to 30 eV were shown to be the result of the excitation of plasma oscillations of the conduction electrons.

Carroll and Melmed (1976) measured the optical constants of ruthenium (1120 plane) between 435 and 630 nm using single-crystal spark-cut specimens subsequently mechanically polished with aluminium oxide and chromic acid. There was a maximum in n at 470 nm and one in k at 520 nm. Kress and Lepeyre (1970b) also measured the optical features of ruthenium, finding an interband transition at 1.6 eV, a volume plasmon at 10.2 eV and a surface plasmon at 8.7 eV.

Vehse *et al* (1970) investigated palladium using thick films deposited rapidly at 150 Å s^{-1} and found interband peaks at 1.3, 3.6, 15.2 and 20.5 eV, a volume plasmon at 7.6 eV and a surface plasmon at 7.3 eV. Another investigation of this metal carried out by Yu and Spicer (1968) produced much lower values of reflectance, possibly because of the low evaporation rate used.

Yamagahara *et al* (1988) have reported the optical constants for platinum in the soft X-ray region using measurements of reflectance on very thick films; these agree well with values calculated by Henke's method.

4.2.8 Aluminium and indium

Aluminium is in group 3a of the periodic table, where the first p-electron states are being filled following the completion of the d-bands in the transition elements; it is trivalent and has an FCC structure. The free-electron Fermi surface is almost entirely contained within the second and third Brillouin zones; when translated back to a reduced zone model, the second zone surface consists of a series of concave structures forming a figure which contains holes. Nevertheless, aluminium behaves like a classical free-electron metal. The band structure shows two partly filled and nearly parallel energy bands and optical absorption by interband transitions is predicted at 0.5 and 1.5 eV (2.5 and 0.8 μm). The band at 1.5 eV has been observed by many investigators, but that at 0.5 eV tends to be obscured by the tail of the Drude absorption and it was only found by working at a very low temperature (4.2 K) (Bos and Lynch 1970).

On exposure to air, a thin layer of oxide rapidly forms on the surface of the aluminium and its influence on the optical properties has been the subject of a large number of investigations.

Aluminium metal, which is used for reflective coatings on optical components, has also been extensively studied, the work of Bennett being particularly well known. Bennett *et al* (1963) measured the reflectance of aluminium in ultrahigh vacuum between 0.3 and 32 μm. Using very clean and carefully prepared surfaces, they found that, between 6 and 32 μm, Drude's classical theory gave good agreement, without having to take account of the anomalous skin effect. They obtained a value of N_{eff} of 2.6 electrons/atom which agreed well with other results. However, below 6 μm their experimental results were lower than those predicted by Drude's theory and, as with the noble metals. they speculated that this might be caused by the variation of the relaxation time of the electron near a region where interband transitions become important.

Powell (1970), in an analysis between 0.04 and 72 eV, also found the Drude treatment to be inadequate, particularly for energies greater than 1 eV; however, the measured reflectance in the region 1.5 to 2.0 eV were larger than expected even taking into account the tails of the absorption bands due to electron transitions.

Beattie (1957) found that his results for aluminium fitted into Dingle's equations for the anomalous skin effect down to about 5 μm. Deviations at wavelengths lower than this were probably due to the proximity of the interband transition at 2.5 μm (0.5 eV).

Harrison (1966) studied the absorption edges associated with the nearly parallel energy bands found in aluminium. It had been found by Ehrenreich *et al* (1963) that the critical points on the joint densities of states occur away from the symmetry points on the Brillouin zone as well as at them, and that this effect was very large in aluminium where the two energy bands involved in the transitions were almost parallel. It is probable that this situation obtains for most of the polyvalent metals and may best as described as a free-electron band weakly perturbed by Bragg reflection planes. The addition of such a reflection plane permits an additional absorption mechanism in metals of this kind. They discovered an interband transition at 1.5 eV against a theoretical 1.4 eV, but there was no further structure at higher energies. Using the electron-energy-loss function, a plasma resonance was found at 15.2 eV.

Philipp and Ehrenreich (1964) also used a Kramers–Kronig method to calculate the optical constants of aluminium in the x-ray region employing as raw data values of reflectance and transmittance from a crystalline sample. It was found that the refractive index in this region could be expressed as $n = 1 - \omega_p^2/2\omega^2$ which was asymptotic to unity with increase in ω.

Shiles *et al* (1980) carried out a Kramers–Kronig analysis of data on aluminium from reflectance, transmittance, ellipsometric and electron-energy-loss measurements. This wide range of measurements was made because no single optical function can readily be measured over the whole spectral range of frequencies as is required for a full Kramers–Kronig analysis. Their results are illustrated in figure 4.13 and show the interband transition at 1.5 eV, the plasma resonance at 15 eV and a further interband transition due to core electrons at 72 eV. The analysis indicated an excess oscillator strength in the x-ray region, a possible explanation of this being a systematic error in the measurements of the absorption coefficients because of the layers of surface oxide on the thin specimens used for transmittance measurements. The data were recalculated after arbitrarily reducing the absorption at the L-edge by 14% and good agreement was then obtained with the electron density and the DC conductivity as well as with the absorption coefficients inferred from the cross sections of neighbouring atoms in the periodic table. A second conclusion was that the free-electron model was inadequate in the near infrared ($\lambda < 1$ μm), because of the interband transition at 0.5 eV (2.5 μm), and also in the ultraviolet region, due to the surface plasmon at 15 eV (80 nm) which also caused a divergence from the theoretical curve. The optical functions show evidence for both the reduction of oscillator strengths between electron energy levels and the effects on real transitions of the shielding of the conduction electrons by the core states. These results differed from those of Beattie, Lenham and Treherne and Hodgson probably because of the condition of the surface.

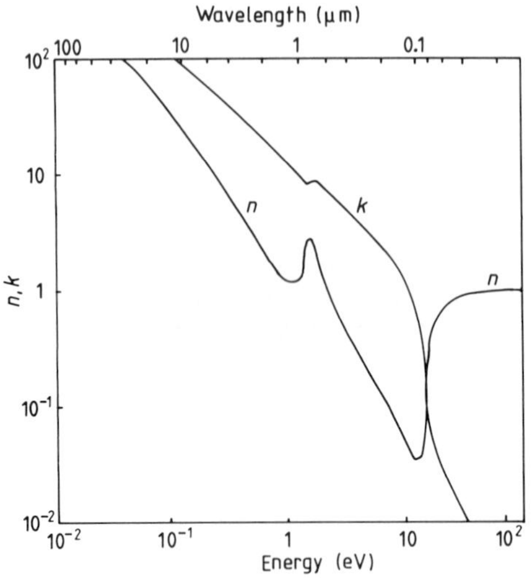

Figure 4.13 *n* and *k* versus *U* and λ for aluminium. Results of Shiles *et al.* (From *Handbook of Optical Constants of Solids*, ed E D Palik (1985), courtesy Academic Press Inc.)

Endriz and Spicer (1971) studied the effects of surface roughness on the optical properties of aluminium. Two effects could be distinguished: roughness aided coupling to the surface plasmons and roughness induced light scattering.

Hageman *et al* (1975) measured the optical constants of magnesium, aluminium, copper and silver as well as carbon and aluminium oxide between the x-ray and infrared regions. The specimens were in the form of thick (10 to 60 nm) films evaporated in a vacuum of 5×10^{-7} Torr. A film of carbon 5 nm thick was evaporated onto both sides of the specimens in order to reduce oxidisation effects. The reflectance was measured and the absorption coefficient, electron-energy-loss function and complex dielectric constant were calculated by means of a Kramers–Kronig analysis.

Other results for aluminium have been reported by Hunter (1964b), Swanson (1964) and Arakawa *et al* (1964).

Indium has been the subject of several investigations. Motulevich and Shubin (1965) were interested in mirror coatings of indium and used the normal skin effect to explain their results between 0.7 and 10.0 μm. Opaque films were made by Bor and Bartholemew (1967) who investigated several closely related metals in the range of 0.4 to 0.65 μm. Indium exhibited a sharp peak at 0.435 μm, a value which did not agree with that reported by Wesolowska (1963) who used Malé's technique. They also observed absorption bands with gallium; a sharp band was found at 590 nm with a broad band centred on 480 nm.

Lenham (1963) measured the complex refractive index for the three crystal axes of a single crystal of gallium in the range 0.4 to 8.0 μm using the reflectance of polarised light. At low energies, the results could be explained on free-electron theory whilst an absorption band due to interband transitions occurred at 570 nm for all three crystal axes.

4.2.9 Rare earths

In terms of the electron structure of the atoms, the rare earths, of which there are two groups, the lanthanides and the actinides, comprise the elements in which the 4f and 5f electron states respectively are progressively filled. There are fourteen such states for each group, thus requiring seven f-bands.

Most of the elements in the lanthanides have a hexagonal structure with the exception of europium which is BCC and ytterbium which is FCC. The first element of the series, lanthanum, has no electrons in the 4f band but one in the 5d; the next, cerium, loses the 5d electron but has two in the 4f band. After that there is a fairly straightforward build up adding one 4f electron to each successive element up to ytterbium which has the full complement of fourteen 4f electrons. The lanthanides end with lutetium which, like lanthanum, has one 5d electron. The 4f shells are near the core and do not have a great influence on the outer structure.

Ellipsometric studies on gadolinium, dysprosium and terbium were carried out by Erskine *et al* (1974) between 1.5 and 5.5 eV. The specimens were evaporated from tantalum boats and had thicknesses greater than 500 nm. The three elements contain seven, ten, and nine 4f electrons respectively and all become ferromagnetic at low temperatures; their band structures have been determined by Herbst *et al* (1972). The optical spectra of all three elements were similar, smooth curves with no sharp features to check against band models. For gadolinium the high joint density of states in the 1–3 eV region leads to a large number of allowed transitions; a broad absorption band was thus produced, centred round 1.6 eV. In dysprosium, the absorption band was broader, and it was therefore suspected that the electron energy bands were also broader.

Schuler (1966) has attempted to relate the optical properties of the rare earth elements with the interband structure and the Fermi surface. As these elements are highly reactive, the surfaces were prepared as polycrystalline thick films deposited in ultrahigh vacuum on sapphire substrates. The reflectance and transmittance were determined at normal incidence and the thicknesses, which were between 50 and 120 nm, were found independently by means of a quartz oscillator balance. k was calculated from the slope of the graph of log T versus d, whilst n was determined from the reflectance.

Gadolinium and lutetium are both trivalent and each has one 5d electron. Plots of their optical absorption against photon energy are very similar showing minima at about 0.8 eV and broad maxima between 2 and 3 eV.

At low eneries the Drude equation was approximately followed whilst the broad absorption band appears to be due to several direct interband transitions between the overlapping 5d and 6s bands. These bands are fairly flat near the Fermi energy where there is a high density of states.

Europium and ytterbium are divalent and have cubic structures; their electron configurations are similar to those of barium and strontium. Plots of the reflectance against photon energies (figure 4.14) show remarkable similarities on the one hand between those of europium and barium, both having a maximum at 1.5 eV and little structure elsewhere, and on the other between those of ytterbium and strontium. The band structures of all these elements are very similar near the Fermi level. No evidence was discovered for interband transitions due to the 4f bands, but it is possible that these were outside the energy range covered (0.3 to 5 eV).

Gadolinium, dysprosium and holmium all showed optical features associated with their magnetic properties; extra absorption bands arise because of transitions between magnetically split energy levels.

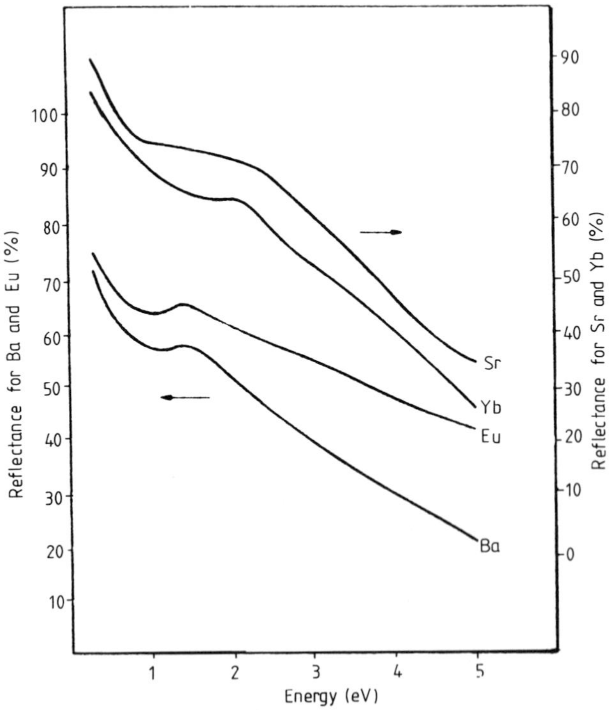

Figure 4.14 Reflectance versus energy for rare earth elements, Eu and Yb, together with Sr and Ba. (From *Optical Properties and Electronic Structure of Metals and Alloys*, ed F Abelès (1966), courtesy North-Holland Publishing.)

Dimmock *et al* (1966) concentrated on the elements gadolinium and thulium. The energy bands of these two elements near the Fermi level consist of mixed 5d and 6s states; they are relatively flat and yield high densities of states. The selection rules for interband optical transitions both with and without the inclusion of spin–orbit coupling are given for the symmetry points of the Brillouin zone and this results in the prediction of considerable polarisation-dependent structure in optical absorption and reflection in the infrared and the visible regions. An anomaly would be expected in the ultraviolet due to transitions between the 4f levels and the conduction bands above the Fermi level (7.0 eV in Gd).

Neodymium was studied by Kern (1957) in the visible and near infrared spectra. Three absorption bands were found which could be identified with transitions of the 6s and 5d electrons; the 4f electrons appeared to have no influence on the optical properties.

Work on the actinides has been much more restricted, but Weaver (1980) measured the optical absorption of polycrystalline uranium between 0.15 and 2.0 eV at a temperature of 4.2 K. Structure was observed at 0.34, 0.77 and 1.30 eV; these features were in contrast to the properties of other actinides whose reflectance at low energies have been reported as featureless. The optical properties of thorium were measured by Weaver and Olson (1977) and those of plutonium by Larson and Cash (1967).

4.2.10 Lead and tin

Lead and tin are in group 4 of the periodic table which also contains carbon, silicon and germanium, all these elements being tetravalent. Silicon and germanium are dealt with in the section on semiconductors whilst carbon (diamond) is considered under dielectrics.

Lead has been investigated in the microwave region by Brandli and Sievers (1972) who measured the absolute reflectance using a parallel plate waveguide. The metal was evaporated on to both sides of the plate, the inside surfaces being used for the measurements to avoid problems with oxidation. Good agreement was found with Drude's theory at room temperature; in fact, lead, as shown in figure 1.5, is one of the best examples to illustrate Drude's free-electron approach. Another study on lead was carried out by Golovashkin and Motulevich (1968); a strong absorption band was found centred on 2.5 eV.

Girouard and Truong (1981) studied the optical properties of thick films of white tin by measuring the reflectance at five angles of incidence. A plasma edge was observed at 13.5 eV and structure in n and k occurred at 10 eV. The imaginary part of $1/\varepsilon$ had a shoulder near 9.7 eV and a peak of 14 eV, whilst $\mathrm{Im}[1/(1+\varepsilon)]$ had a shoulder at 9 eV and a peak at 12 eV. These results indicate an interband transition near 10 eV and a plasmon at 14 eV. They may be compared with the results of Powell (1970) which gave 10.4 and 14.1 eV respectively.

4.2.11 Bismuth and antimony

These metals are group 5 elements, where the p-electron levels are continuing to be filled; they have rhombic crystal structures. They have complicated Fermi surfaces which are contained in several Brillouin zones; some of these contain electrons whilst others involve holes. There is very little overlapping of the bands and these materials are often known as semimetals as the numbers of free electrons are very low.

Hodgson (1954) used Avery's method to measure the optical properties of solid bismuth. The specimens were prepared by pouring the molten metal on to optically flat glass, removing on cooling and then polishing with metal polish. He claimed that the error in n was ± 0.3 and that in $k \pm 0.5$ and gave graphs of n and k against log l; k has a trough at a wavelength of 0.6 μm, whilst n has a peak at the same wavelength. A minimum in absorption was found at 0.3 eV, i.e. 4.0 μm, and this could be identified with an interband transition between the first filled band and the conduction band.

Lenham *et al* (1965) measured the optical properties of antimony and bismuth crystals using the method of Beattie and Conn. As these materials are anisotropic, it was necessary to study several different crystallographic directions and also to make a comparison with the results from polycrystalline specimens. It was claimed that the shapes of the dispersion curves were independent of the surface preparation and so were the best parameters to use. In addition to the free electrons, it was necessary to involve three plasmons and an interband transition in order to account for the results.

A further set of results for bismuth was published by Hagemann *et al* (1975). This was based on a Kramers–Kronig analysis using values of the transmittance of bismuth films (10–60 nm in thickness) covered on both sides by evaporated carbon to prevent oxidation. The investigation ranged from 13 to 150 eV and covered the critical region between 30 and 50 eV where normal-incidence monochromators have a low-flux output and where grazing-incidence instruments suffer from overlapping orders of interference. These authors give values of the absorption coefficient, μ, against photon energy and also a table of three values of N_{eff} calculated from ε_2, μ and Im ε^{-1}.

Potapov (1964) studied antimony and bismuth in the infrared at low temperatures. For bismuth $\varepsilon_1 = 0$ at 13 μm; at wavelengths below this the metal was transparent. In antimony the plasma wavelength was 9.3 μm; on reducing the wavelength, there was an increase in the imaginary part of corresponding to the start of an interband transition.

Shkylarevskii *et al* (1959) measured the optical constants of antimony between 1 and 12 μm using Avery's method. Both the real and imaginary parts of ε showed peaks which moved towards shorter wavelengths as the temperature was reduced. At room temperature the maximum in the absorption occurred at 6.9 μm, corresponding to a band gap of 0.18 eV.

4.2.12 Temperature dependence of optical properties of metals

Decker and Hodgkin (1980) measured the temperature-dependent changes in the reflectance of copper, silver and aluminium. They gave a theoretical interpretation of their results involving both the interband and intraband effects as well as the effects of surface roughness.

Arnold (1984) obtained empirical expressions for the temperature dependence of the absorptances of silver, aluminium, gold, copper, lead and tungsten using Drude's theory and claimed that the results were in reasonable agreement with experimental values.

Winsemius *et al* (1976) reported the temperature dependence of the optical constants of copper, silver and gold between 40 and 840 K near the interband transition points. The observed change in energy with temperature of these transitions was much larger than that predicted by band calculations. McKay and Rayne (1976) also gave data on these metals, which they interpreted in terms of photon and surface-assisted absorption.

The changes in optical constants taking place up to and beyond the melting point of copper, aluminium, gold and tin have been measured at 10.6 μm by Bruckner *et al* (1989, 1991) using a rotating analyser ellipsometer. In all four metals n tended to increase with temperature up to the melting point whereas k moved in the opposite direction; at the melting point, both n and k decreased abruptly by between 3 and 9%. For temperatures beyond the melting point, little change with temperature was observed in either n or k. Drude's free electron model fitted all the solid phase measurements adequately at this wavelength but broke down completely when applied to the liquid states.

Barnes (1966), measuring the reflectances at different polarising angles, was able to calculate the optical constants of a group of seven incandescent metals (tungsten, molybdenum, tantalum, iridium, rhenium, niobium and platinum) for temperatures ranging from 300 to 2500 K and for wavelengths between 0.47 and 2.6 μm. Rough agreement with Drude's theory was obtained with all these metals. The curves of n versus wavelength for different temperatures ran close together in the visible spectrum but spread out in the infrared; on the other hand, the corresponding curves for k tended to intersect at one particular wavelength. (This is known as the X-point and was introduced by Price (1947).)

4.3 Alloys

The optical properties of alloys can be considered in the light of a number of theoretical models which have been discussed by Abelès (1972). The first of these is the rigid band model in which it is assumed that the periodic ionic potential in the solid state model of the alloy is the same as that in the solvent metal perturbed by the difference in potentials of the solvent and the solute; it is also necessary to take into account the screening effect of the solute atoms by the conduction electrons. On alloying, the Fermi level is displaced with respect to the bottom of the conduction band by electrons from the s-p-bands in the solvent filling vacancies in the d-band of the solute. This model is thus applicable when the d-bands can be ignored, for example in alloys of the noble and polyvalent metals. The second model is that of the virtual band in which radical changes in optical behaviour are associated with the presence of the d-states due to the impurity atoms, which act as virtual bound states. The s-p-states of the impurity will merge with the conduction band of the solvent as in the rigid band model. In addition, the impurity d-states are expected to be degenerate with the host's conduction band and this will lead to an increase in the density of states at certain energies in the s-p-band but no change in the Fermi level. This model is useful for alloys of the noble and transition metals. The third model is that of the virtual crystal where the Fermi surface of an alloy can be calculated from a periodic potential given by the average periodic potential in the lattice plus a scattering potential. The latter disappears if the amplitudes of the wavefunctions of both types of atoms are nearly equal for states near the Fermi level. It is expected that the optical mass for the alloys will show a simple variation from that of one pure metal to that of the other. This model should apply to solid solutions such as alloys of the noble metals. Finally, in the coherent potential model, the movement of a conduction electron is regarded as a series of scattering processes by the alloy atoms; the theory is based on the proposition that any single scattering atom produces, on average, no further scattering. This model is useful for alloys of the transition metals.

The optical properties of alloys of silver and gold have been studied by Fukatani and Sueoko (1966) between 2.2 and 4.3 eV. The optical constants were determined from the ratio R_p/R_s of polarised light at two angles of incidence and the bulk and surface plasma oscillations were also studied. The results showed that the energy values of the bulk plasma in silver-rich alloys decrease slightly with increase in gold content; when the gold content is greater than 30% this plasma oscillation disappears. The energy values of the surface plasmons also decrease with increase in the gold concentration, but, contrary to the bulk case, this oscillation persists in alloys of any concentration.

Sterne (1967) also studied alloys of silver and gold by measuring the Faraday rotation of the plane of polarisation in reflected light. He found that the shape of the Fermi surfaces of the alloys could be calculated from the simple average potential of the constituents, thus supporting the virtual crystal model.

Alloys of gold and copper were investigated by Beaglehole *et al* (1979) using synchrotron radiation. On the weak copper side the spectrum was rather insensitive to alloying, but the situation was quite different on the rich copper side. At low energies the spectrum was consistent with a one-electron model, and that at high energies was dominated by transitions from the filled d-bands to the conduction band.

Alloys of aluminium with gold and silver were studied by Irani *et al* (1971, 1972). Aluminium has the same crystal structure as the noble metals and little distortion of the lattice would be expected on alloying. In addition, as aluminium has no d-electrons, the rigid band model ought to apply. However, the shifts with alloying of the absorption peaks in ε_2 at 3.09 and 4.54 eV respectively were only half those expected on the simple model and were better fits to a model which took account of the effects of shielding of the electrons.

An investigation of copper–nickel alloys was made by Schröder and Onengut (1967) who found that the wavelength of the main absorption edge did not change with alloying, although it disappeared completely for alloys with 20–25% nickel. The positions of the secondary absorption peaks moved slightly to lower wavelengths with increasing amounts of nickel. The optical behaviour of these alloys could best be described in terms of the virtual band model, a conclusion which was also reached by Seits and Spicer (1970), Drew and Doezema (1972) and Tokumoto *et al* (1977).

Sasovskaya and Noskov (1975) studied copper–nickel alloys between 0.3 and 17 μm. In the intraband region they fitted a two-band model to the optical properties whilst, in the interband region, the coherent potential model was the most successful.

Muldawer and Goldman (1966) made a study of a series of β-brass alloys using normal incidence reflectance measurements. The specimens, ranging from $Cu_{56}-Zn_{44}$ to $Cu_{48}-Zn_{52}$, were mechanically polished and then annealed in a vacuum of about 10^{-6} Torr. The energy range covered was between 1 and 18 eV and measurements were made at temperatures of −158, 21 and 120 °C. A Kramers–Kronig analysis was used to calculate ε_1, ε_2, the energy-loss function $[\varepsilon_2/(\varepsilon_1^2+\varepsilon_2^2)]$, n, k and the absorption coefficient μ. The main optical feature observed was an interband transition from the region of the Fermi surface to an upper level (~2.7 eV); this feature decreased as the temperature increased and also with reduced zinc concentration. A plasma resonance dependent on this transition occurs at 2.65 eV. Additional higher-energy structure was observed and this was attributed to other interband transitions, some of which were probably from an underlying d-band to the Fermi level; band structure calculations for β-brass allowed tentative identification of these transitions. The work also attempted to find the ratio N/m and the relaxation time τ, and to separate ε_1 and ε_2 into free and bound electron contributions.

Biondi and Payne (1959) also studied alloys of copper and zinc at 4.2 K by a calorimeter technique. The infrared absorption showed that the impurity relaxation time for copper was anisotropic. This increased with increasing zinc content and suggested that the Fermi surface was becoming more distorted.

A wide-ranging study of the electronic structure of IB and IIB alloys was made by Amar and Johnson (1966). This included the disordered β-phase of Cu–Zn Ag–Zn and Ag–Cd alloys and also the α-phase of Cu–Zn. In all the three alloys 'holes' were found in the corners of the first (cubic) Brillouin zones; the second (dodecahedral) zones were slightly more than half full with a significant amount of contact between the Fermi surface and the faces of the zone. The dominant feature of the frequency dependence of the optical properties of the β-phase could be explained in terms of plasma oscillations and/or interband transitions between the Fermi surface and higher levels. The chief uncertainty in making the identifications was the position of the d-bands relative to the conduction bands.

The validity of the free-electron model was investigated for alloys of gold-germanium and silver-germanium by Thèye *et al* (1983), the germanium content being varied between 4 and 20%. The Drude model with a constant electron relaxation time was found to reproduce the optical data at low energies, but, at high energies, there were deviations and also discrepancies between the average number of effective electrons per atom. The rigid band model was used, it being assumed that hybridisation took place between the s- and the p–d-bands. Rayne (1960), who also investigated copper–germanium alloys, found their optical behaviour inconsistent with the rigid band model, although they could also be reconciled with the virtual crystal model.

The alloys of the noble metals with palladium which form solid solutions have been the objects of study by Myers *et al* (1968). Additional optical absorption was found and this was attributed to the resonant bound states associated with the d-electrons in palladium, thus supporting the virtual band model. The same authors also looked at alloys of copper and silver with manganese but found little evidence for the spin splitting of the manganese bands in the alloy with copper and only a weak absorption from this effect in the silver alloys.

Black *et al* (1977) reported work on alloys of niobium with molybdenum, zirconium, vanadium and tantalum. Below 3 eV the results fitted the rigid band model but large optical structures above 4 eV could not be accounted for on any model. It was possible that there was considerable distortion of the lattice in these alloys.

4.4 Dielectrics

The optical properties of dielectrics follow closely the theoretical predictions presented in Chapter 1; these materials exhibit a series of absorption bands due to resonances of the bound electrons and of the atoms and ions, producing the characteristic dispersion curves when the refractive index is plotted against the wavelength (see figure 4.15 which is for potassium chloride). Resonances of the bound electrons occur in the ultraviolet for materials such as glass, quartz and the alkali halides, giving normal dispersion through the visible spectrum and into the near infrared; these materials have consequently been much used for spectroscopic prisms over this region. A second absorption resonance occurs in the infrared for those materials containing polar molecules; this is the reststrahlen resonance so called because of the high reflectance at this wavelength due to the large value of k. Repeated reflections of 'white' radiation by these materials leads to the isolation of these characteristic residual rays. The value of k is generally negligible except near the resonances in n.

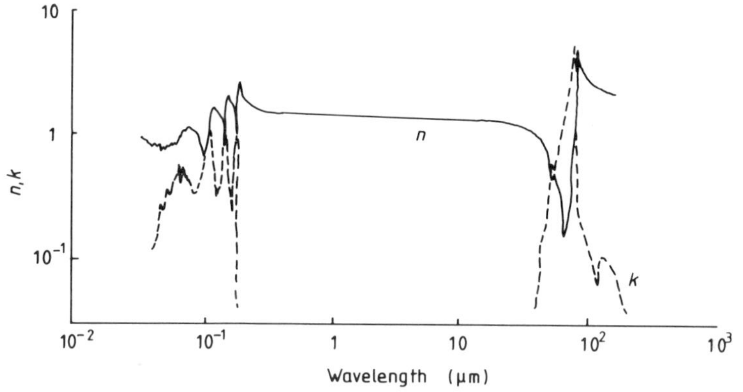

Figure 4.15 Plot of n and k versus wavelength for potassium chloride. (From *Handbook of Optical Constants of Solids*, ed E D Palik (1985), courtesy Academic Press Inc.)

The optical properties of dielectrics are of considerable importance in a number of applications, for example spectrometer components, microcircuitry (elements in capacitors), fibre optics (relative values of the refractive index for the core and the cladding) and interference filters.

Stephane *et al* (1967) measured the real and imaginary parts of the dielectric constant and the electron-energy-loss function for Pyrex, corundum, ruby, lithium fluoride and calcium fluoride in the extreme ultraviolet by three methods—Avery's, Toussey's and Kramers–Kronig. The possibility of a plasma level for these materials was also discussed.

Efimov *et al* (1980) plotted the low-frequency infrared spectra and optical constants of some borate glasses, using a Kramers–Kronig analysis. The spectra showed intense maxima due to the presence of the doubly charged metal ions relative to their oxygen coordination sphere.

Roessler and Walker (1967, 1968) measured the optical constants of magnesium oxide, lithium fluoride, sodium chloride and potassium chloride from a Kramers–Kronig analysis of the reflectance of freshly cleaved crystals. White and Straley (1968) found an anomaly in the optical properties of silver chloride close to the absorption band edge. Bauer *et al* (1974) showed that this could be explained by an artefact of the reflectance data near this wavelength causing a 10% error in the optical properties up to a photon energy of 5 eV but having no effect at higher energies. Another series of results for the alkali halide crystals has been produced by Owens (1969) and, in the infrared, Eldridge and Staal (1977); Allen and Harrington (1978) have reported measurements on these materials using a laser calorimeter.

The optical properties of diamond are covered in books by Berman (1965) and Field (1979) and also in an article by Walker (1979). Two varieties of diamond can be recognised: types I and II. Type I consists of well formed single crystals and has strong optical absorption below 300 nm and again between 6 and 13 μm. Type 11 has a plate-like structure and does not exhibit the infrared absorption. Roberts and Walker (1967) measured the absolute reflectance of diamond at nearly normal incidence at both room and liquid nitrogen temperatures, using a Kramers–Kronig analysis to calculate the phase change on reflection. Structure in ε_2 was found at 7.3, 7.8, 12.2 and 23 eV which could be related to direct interband transitions. Edwards and Ochroa (1981) measured the refractive index of diamond by a channelled spectrum method, where radiation having a broad waveband is incident on a plane parallel specimen and produces Edser–Butler-like interference fringes. The real component of the refractive index decreased from 2.70 at 0.20 μm to 2.38 at 25 μm.

Lang and Wolfe (1983) measured the optical constants of fused silica and sapphire from 0.3 to 25 μm using a Kramers–Kronig technique. Silica showed a strong absorption band at about 9 μm and a strong peak in the refractive index just below 10 μm; both these features were consistent with the reststrahlen peak of quartz at $\lambda \simeq 9$ μm. There was a second and broader absorption peak at 22 μm and one for the refractive index at 24 μm. Sapphire had absorption peaks at 18 and 24 μm with subsidiaries at 16 and 19 μm; there were corresponding peaks in the refractive index. Results for quartz have also been given by Russell and Bell (1967), Spitzer and Kleinman (1961) and by Malitson (1965) who found strong absorption due to OH ions at about 3 μm. Another study of both quartz and sapphire in the far infrared was made by Loewenstein *et al* (1973) who also used the channelled spectrum technique. Figure 4.16 shows a set of averaged values of n and k versus wavelength for quartz (ordinary ray).

Dobrowolski *et al* (1983) used a synthesis program whereby the constants in the dispersion equations were adjusted until a good fit was obtained between the measured and calculated values of reflectance and

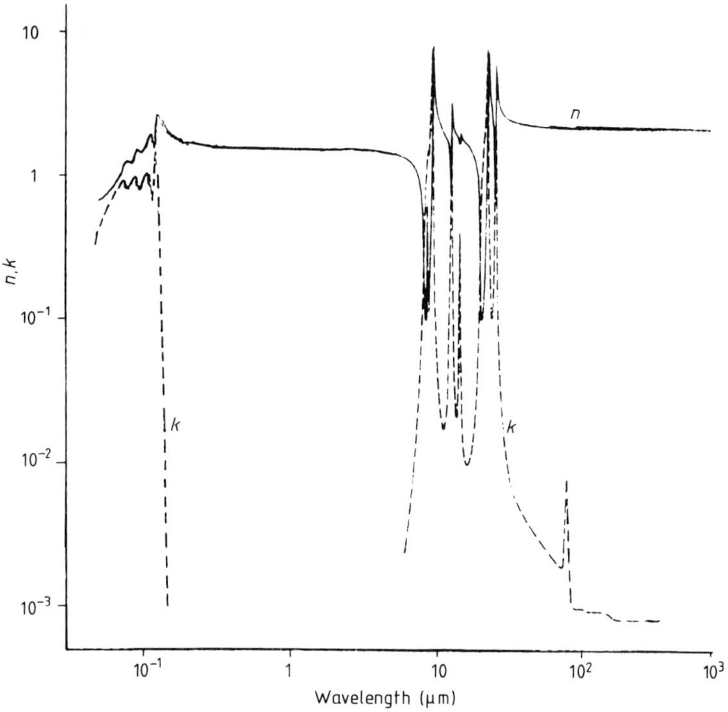

Figure 4.16 Log-log plot of *n* and *k* versus wavelength for quartz. (From *Handbook of Optical Constants of Solds*, ed E D Palik (1985), courtesy Academic Press Inc.)

transmittance. More data than the required minimum were used and this proved to be a sensitive method for obtaining the optical constants; it was applied to magnesium fluoride, zinc sulphide, magnesium oxide and silicon.

4.5 Ferroelectrics

The ferroelectrics are insulators which are characterised by very high values of the dielectric constant; *n* may reach values of over 100. The ferroelectrics also exhibit some of the properties of the ferromagnetics such as permanent dipoles, hysteresis effects and a Curie temperature above which the ferroelectric properties disappear. These materials are clearly of considerable practical interest in the manufacture of capacitors because of their very large values of ε and n.

The archetypal ferroelectric material is barium titanate, $BaTiO_3$, which is one of the perovskite lattice series of compounds having the general form ABO_3. This lattice may be regarded as a set of BO_6 octahedra arranged in a cubic structure, linked together by the shared oxygen atoms with the A atoms occupying the spaces in between. The ferroelectric properties can be largely explained in terms of the behaviour of the octahedra.

The optical properties of barium and strontium titanates have been studied by Bauerle *et al* (1978) from a Kramers–Kronig analysis of reflectance data; figure 4.17 shows ε_1 and ε_2 plotted against photon energy for barium titanate. It was possible to identify the peaks in these functions in terms of interband transitions.

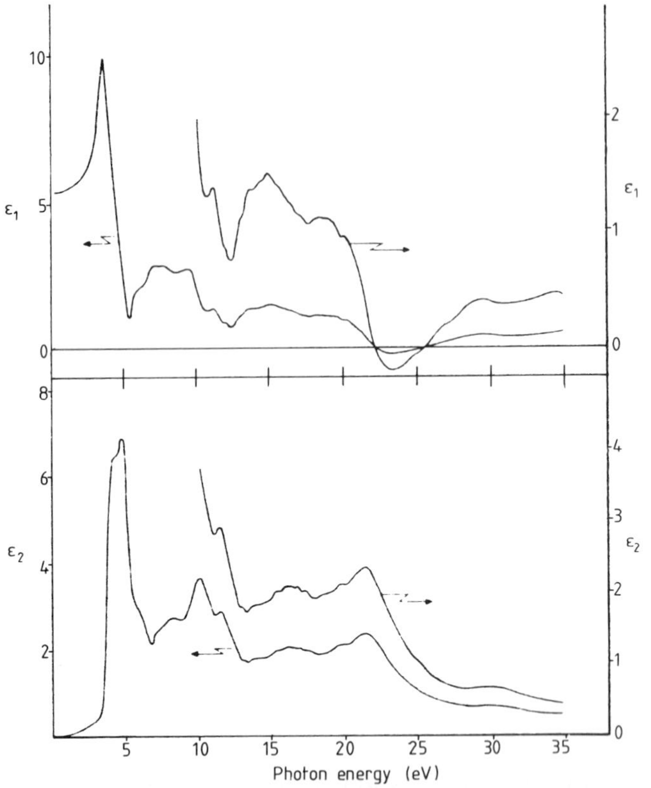

Figure 4.17 ε_1 and ε_2 versus photon energy for barium titanate. (From Bauerle D *et al* 1978 *Phys. Rev.* B **29** 181.)

Brody *et al* (1983) have measured the refractive indices of ferroelectrics in the near-microwave region by means of a Michelson interferometer. Mamedro (1982) carried out measurements on ferroelectric crystals of the type ABO_3 in the vacuum ultraviolet using a Kramers–Kronig analysis.

There have been a number of international meetings on ferroelectricity and the proceedings of a meeting held at Kobe, Japan in 1985 have been published as a supplement to the *Japanese Journal of Applied Physics*. Several of the papers deal with the optical properties of various ferroelectrics, e.g. Goudonnet *et al* (1985) and Hikachi *et al* (1985).

4.6 Semiconductors

Moss (1961) has given an exhaustive account of the optical properties of semiconductor materials including band structures for the different materials; in addition, he had edited a four-volume handbook on semiconductors (1980) of which volume 2 deals with the optical properties. The reader is referred to these books for fuller details on all the materials.

4.6.1 Silicon

This element has a diamond-type structure and the minimum energy gap is an indirect transition of 1.1 eV. Measurements of its optical constants have been made by Salzburg and Villa (1957) using a single crystal and also by Dunlop and Watters (1953) at radio frequencies, where $\varepsilon = 11.7$. Figure 4.18 shows plots of n and k versus λ for silicon from various sources.

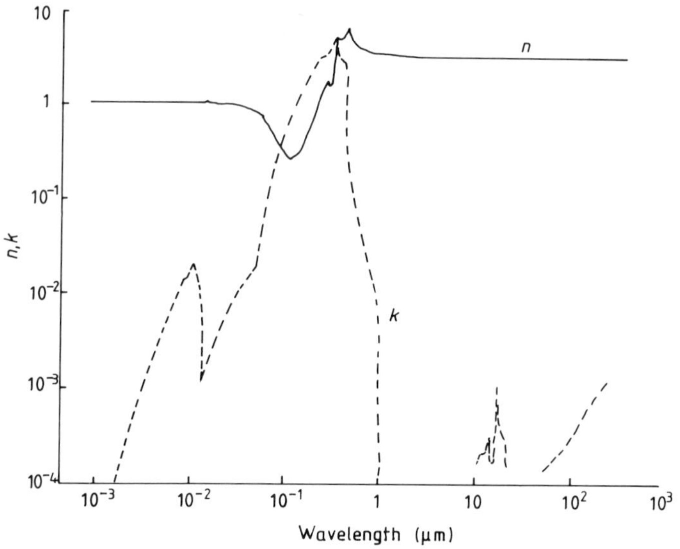

Figure 4.18 Log-log plot of n and k versus wavelength for silicon. (From *Handbook of Optical Constants of Solids* en E D Palik (1985), courtesy Academic Press Inc.)

Edwards and Ochroa (1980) measured the infrared refractive index of silicon using fringes of equal chromatic order caused by multiple reflections at the parallel surfaces. Aspnes and Studna (1983) used spectroscopic ellipsometry to determine the complex dielectric constants of silicon, germanium and several compounds. Rather than attempt to correct for the presence of overlayers of oxides, these were removed as far as possible, the real-time capability of the instrument being used to assess surface quality during cleaning. Unlike much previous work, there was good agreement between the results for different specimens when the measurements were made on smooth, clean and undamaged samples maintained in an inert atmosphere. Differences could generally be understood in terms of inadequacy of preparation, although the Kramers–Kronig analysis sometimes exhibited the effects of improper extrapolations.

Philipp (1972) studied the influence of oxide layers on the optical constants of silicon. Loewenstein (1973) and Icenogle *et al* (1976) have also reported results for silicon, whose refractive index falls steadily from 3.45 at 2 μm to 3.42 at 12 μm.

4.6.2 Germanium

Germanium also has a diamond-type lattice, its band structure showing a direct gap of about 0.8 eV and also an indirect gap of 0.66 eV. Plots of n and k against the wavelength show similar features to those for silicon. Measurements of the optical properties of germanium have been made by Salzburg and Villa (1954), Briggs (1950), Avery and Clegg (1953), Philipp and Taft (1959) and Potter (1966).

Icenogle *et al* (1976) determined the refractive indices of both germanium and silicon by the method of angle of minimum deviation between 2.5 and 12 μm. n fell steadily from 4.1 at 2.5 μm to 4.02 at 10 μm.

Donovan *et al* (1963) measured the normal-incidence reflectance of germanium after using an electropolishing technique to obtain an optically smooth surface; they presented their results in the form of graphs.

Goldschmidt (1984) measured the optical properties of single-crystal germanium up to 500 °C by means of the ratio R_p/R_s at the pseudo-Brewster angle. He claimed that the variation with time in the thickness of the surface oxide was small enough to be ignored when using this technique, although this would not be the case with ellipsometry.

Hutchinson *et al* (1982) measured the surface and bulk absorption of germanium at a wavelength of 10.6 μm using an adiabatic laser calorimeter, the specimens being polished conventionally with aluminium oxide. It was found that 50% of the absorption in a specimen 10 mm thick was caused by the surface of the specimen, and could be reduced by about 10% by repolishing with colloidal silica.

Edwin *et al* (1982) determined the refractive index of germanium between 8 and 14 μm by measuring the angle of minimum deviation of a germanium prism. n fell from 4.0055 at 8 μm to 4.0012 at 14 μm, an accuracy of ±0.0003 in refractive index being claimed.

Ilin *et al* (1987) measured the optical constants of germanium between 10 and 200 μm; anomalous dispersion was observed between 28 and 40 μm due to a two-phonon interaction with lattice vibrations. At very long wavelengths, $e = 16.03 \pm 0.04$.

Jellison and Lowndes (1987), using nanosecond time resolved ellipsometry in the region 1.96–3.17 eV (0.40–0.64 μm) have measured the optical constants of liquid silicon and germanium during the melting produced by pulsed irradiation from a laser. Their results for germanium were in good agreement with those of Hodgson (1961) but the values of n for silicon were about 30% lower than those published by Shvarov *et al* (1977). Table 4.3 gives a summary of Jellison and Lowndes results.

Table 4.3 Results of Jellison and Lowndes (1987) for n and k of liquid silicon and germanium.

Energy (eV)	λ (μm)	Silicon		Germanium	
		n	k	n	k
0.5	2.48			9.0	10.8
1.0	1.24	4.6	6.6	5.6	8.0
2.0	0.62	4.0	5.1	3.4	5.5
3.0	0.41	2.4	4.9	2.0	4.4
3.5	0.35	1.8	4.3	1.6	4.0

4.6.3 Cadmium sulphide

Cadmium sulphide is used extensively for making photoconductive cells; its fundamental, indirect, energy gap of 2.4 eV makes it a suitable material for work in the visible and ultra-violet regions. The normal crystal structure is HCP which thus leads to optical anisotropy. However, a second form of CdS can be made, FCC in structure, and this would be isotropic in its optical behaviour.

In the ultra-violet region (0.1–0.6 μm), Cardona and Harbeke (1965) measured the reflectances of platelets of CdS from which they extracted n and k by Kramers–Kronig analysis. A similar technique was used by Balkanski and Petroff (1964) between 0.06 and 1.24 μm using crystals of CdS. Both these studies detected a series of peaks in k which could be correlated with electron transitions in the energy band structure and also showed satisfactory correlation with values calculated by Henke's formula.

Over the main absorption band at about 0.48 μm, Brodin and his collaborators have made several sets of measurements. Brodin (1961) determined n_e and n_o by Kramers–Kronig analysis of reflectance measurements; he found that n_e was about 2% larger than n_o. Brodin and Strashnikova

(1966), using an interference method, also measured n in both states of polarisation; in addition to the main peak at 0.48 μm, a second broad absorption band was found centred round 0.28 μm.

Bienewski and Czyzek (1963) used the minimum deviation prism method to produce accurate values of n between 0.5 and 1.4 μm. The same technique was used at wavelengths of 1 to 14 μm by Francis and Carlson (1960) producing values of n which link satisfactorily with those of Bienewski and Czyzek. Devore (1951) was able to express his results for the e- and o-rays in the visible and near infra-red in the form of a Sellmeier type equation, as below, where λ is in Å:

$$\text{o-ray} \qquad n^2 = 5.913 + 2.441 \times 10^7/(\lambda^2 - 0.803 \times 10^7)$$

$$\text{e-ray} \qquad n^2 = 7.197 + 3.322 \times 10^7/(\lambda^2 - 0.843 \times 10^7).$$

Collins (1959) measured the reflectance for unpolarised radiation between 20 and 48 μm which covers the infra-red absorption band; he detected two phonons, a transverse optical phonon at 41.5 μm and a longitudinal optical phonon at 37.8 μm. Balkanski and Besson (1961) working in the same wavelength region, found the explanation of their reflectance results needed a multiple phonon process i.e. several phonons produced by the absorption of a single photon. Manabe *et al* (1967) measured the reflectance between 20 and 100 μm and at temperatures of 100 and 300 K; over this region, n and k could be calculated from a Lorentz-type dispersion formula. Between 100 and 1000 μm Hattori *et al* (1971) determined n from the wavelengths of interference maxima; n was constant over this wavelength range, 3.27 for n_e and 3.11 for n_o (n_e is parallel to the c-axis, perpendicular to the base of the hexagon).

The variation with temperature of n for CdS between 100 and 600 K has been reported by Vitrikhovskii *et al* (1968). Both n_e and n_o increased with temperature, the value of $\mathrm{d}n/\mathrm{d}T$ being about $8 \times 10^{-5}/°\mathrm{C}$ for both cases. Samara (1983) measured $\mathrm{d}(\ln e)/\mathrm{d}T$ at 300 K and at frequencies between 1 and 100 kHz; his values were $22.0 \times 10^{-5}/°\mathrm{C}$ for the c-axis and $18.7 \times 10^{-5}/°\mathrm{C}$ for the a-axis.

Figure 4.19 shows log–log plots of n and k versus wavelength for the o- and e-rays of cadmium sulphide from various sources.

4.6.4 Zinc sulphide, selenide and telluride

These three materials normally exist in the FCC crystal structure although HCP forms can also be made. All the materials exhibit two main absorption bands, one in the ultra-violet due to electron excitations across the main indirect energy gap between the valence and conduction bands and the second in the infra-red caused by resonance between the incident photons and lattice vibrations (phonons). Over the visible and near infra-red regions their behaviour is typically normal dispersive with k having very low values and n steadily decreasing from about 3.0 to 2.5; consequently, they are very suitable materials for making optical components for use in this spectral region.

Li (1984) has given a comprehensive review of the optical properties of all three materials, presenting tables of refractive index versus wavelength at a number of temperatures between 93 and 618 K; he also produced tables of dn/dT at different wavelengths. Several other studies have covered the optical properties of all these materials, examples being Feldman *et al* (1979), Hattori *et al* (1971) and Manabe *et al* (1967).

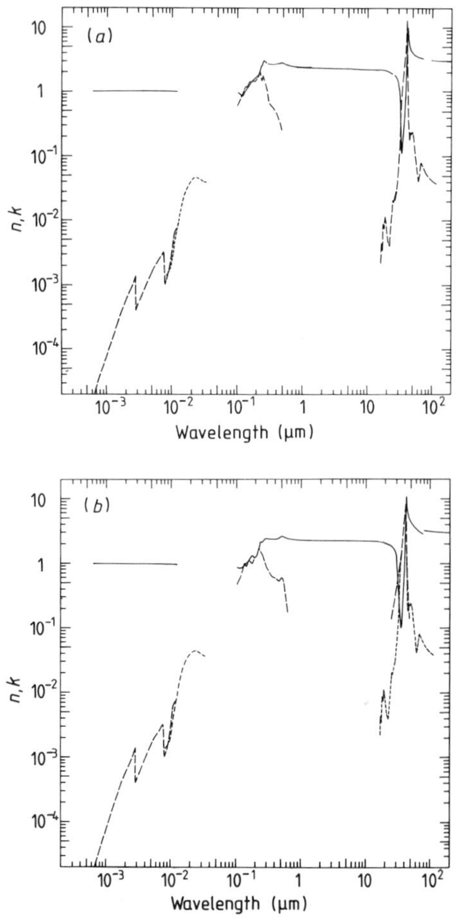

Figure 4.19 Log-log plots of n (solid line) and k (dashed line) for (*a*) o-rays and (*b*) e-rays of CdS from various sources. (From *Handbook of Optical Constants of Solids II* ed E D Palik (1991), courtesy Academic Press Inc.)

At short wavelengths (0.04–2.0 μm), Freeouf (1973) measured reflectances using synchrotron radiation and analysed his results by Kramers–Kronig analysis to get e_2 for both ZnSe and ZnTe. Over the range 0.04–1.24 μm, Gautron *et al* (1982) used this same technique and their results for ZnSe correlate well with those of Freeouf where wavelength overlap occurs. Both sets of results show good agreement with values calculated by Henke's formulae.

Over the main energy gap of ZnSe, the work of Balkanski and Petroff (1964) and Aven *et al* (1961) may be quoted. Balkanski and Petroff found n had a maximum value of 3.61 at 0.28 μm with a secondary maximum of 1.42 at 0.095 μm; n_{max} from Aven *et al* was 3.64 also at 0.28 μm.

There have been a number of accurate investigations of all three materials over the visible and near infra-red region; thus, Sliker and Jost (1966) and Shozawa *et al* (1962) both used the minimum deviation method between 0.5 and 1.5 μm. Debenham (1984) has reported a very accurate set of values of n between 0.4 and 13.0 μm; she used the prism technique on specimens grown by chemical vapour deposition and then applied a least-squares fit. In this region, where they behave in a normal dispersive manner, it is possible to fit the values of n for all three materials into Sellmeier-type dispersion equations. For example, Devore (1951) has quoted the equation below for ZnS, whilst Marple (1964) gave the formulae for ZnSe and ZnTe; in all cases, λ is measured in μm:

$$\text{ZnS:} \quad n^2 = 5.164 + 0.121\ \lambda^2/(\lambda^2 - 0.0732)$$

$$\text{ZnSe:} \quad n^2 = 4.00 + 1.90\ \lambda^2/(\lambda^2 - 0.113)$$

$$\text{ZnTe:} \quad n^2 = 4.27 + 3.01\ \lambda^2/(\lambda^2 - 0.142).$$

Very accurate values of n quoted to five decimal places have been given by Feldman *et al* (1979) for ZnS and ZnSe between 0.5 and 18.0 μm using a minimum-deviation technique.

In the region between 20 and 85 μm, which covers the infra-red absorption peak, Manabe *et al* (1967) determined n and k for ZnSe and ZnTe from the reflectance, assuming that Lorentz-type oscillator equations could be applied (see section 1.5). At 300 K they detected a transverse optical phonon at 48.3 μm, and a longitudinal phonon at 40.7 μm (ZnSe). The corresponding values for ZnTe were 56.5 and 48.5 μm. Repeating their work at 100 K they found that the phonon wavelengths had decreased by about 2% in both materials. Klein (1986) working with ZnS found the ultra-violet resonance at 0.242 μm and the infra-red resonance at 36.71 μm; he gave $e_0 = 8.393$ and $e_\infty = 5.105$. Miles (1977) found structures in the absorption coefficient between 11 and 33 μm and needed to postulate a process of multiphonon absorption to explain them. Hattori *et al* (1971) extended measurements from 150 to 1000 μm using an interference technqiue with thick film specimens; in this range, n again becomes normally dispersive. For

ZnSe, *n* decreases steadily from 3.07 to 3.01 whilst for ZnTe, the decrease is from 3.20 to 3.15; *k* is negligible for both materials over this range.

At very low frequencies (below 20 Hz) *e* can be determined from measurements of capacitance; for ZnSe, Strzalkowski *et al* (1976) found $e = 9.25$ at 296 K and when extrapolated to 0 K, $e_o = 8.80$ i.e. $n_o = 2.97$. Samara (1983) gave $n_o = 2.88$ i.e. $e_o = 8.29$ for frequencies between 1 and 100 kHz.

In the case of ZnTe, Cardona and Haensel (1970) measured the absorption coefficient at X-ray wavelengths. They found a doublet of lines at 0.0133 and 0.0139 μm due to the electronic M-transition and their results were in very good agreement with those calculated using Henke's method.

Cardona and Greenaway (1963) measured the reflectance of cleaved ZnTe between 0.05 and 1.0 μm; the reflectance exhibits structures around 0.01 μm which may be correlated with electron transitions from the d-band to the conduction band in zinc.

The values of *n* from various sources for ZnS, ZnSe and ZnTe are illustrated in figure 4.20.

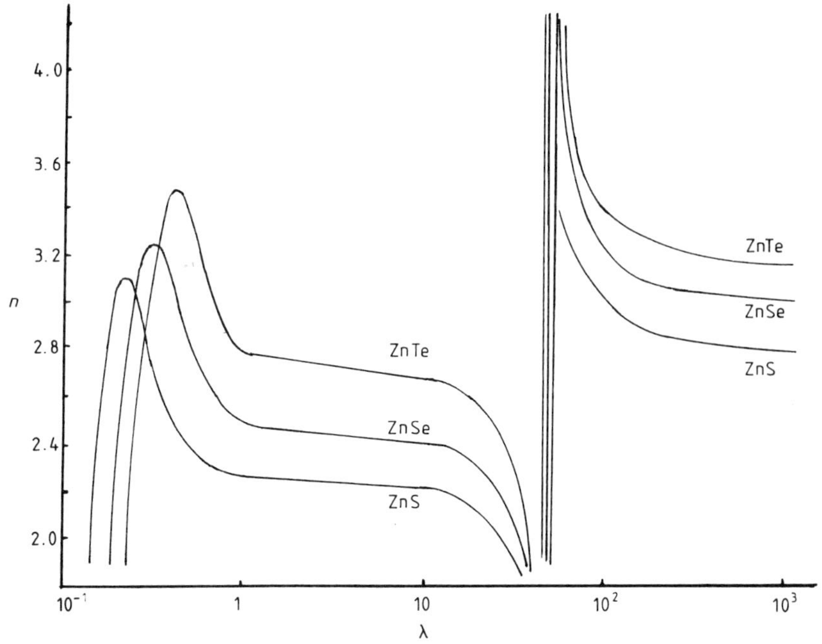

Figure 4.20 Plots of *n* versus λ for ZnS, ZnSe and ZnTe.

4.6.5 Indium antimonide, arsenide and phsophide

Cardona *et al* (1970) measured the absorption coefficient of indium antimonide between 14 and 155 eV (8–89 nm) using their synchrotron radiation method; several peaks were observed, three in the region of 20 eV (62 nm) being ascribed to interband transitions. The optical properties of InSb and InAs have been investigated by Philipp and Ehrenreich (1963) using a Kramers–Kronig analysis of reflectance measurements between 1.5 and 25 eV (0.05 and 0.83 μm). Graphs were produced of R, e_1 and e_2 versus photon energy and the results could be explained in terms of interband transitions and plasma oscillations. In the range 1.5–6.0 eV, Aspnes and Studna (1983) determined n and k for all three materials by spectroscopic ellipsometry; for InSb, n reached a maximum value of 3.525 at 0.372 μm and the corresponding values for InP and InAs are 4.415 at 0.403 μm and 3.761 at 0.291 μm.

Dealing with indium antimonide only, Moss *et al* (1957) measured n and k in the visible and near infra-red using an interference method and found n_{max} at 0.37 μm. In the far infra-red between 20 and 220 μm, Yoshinga and Oetjen (1956) measured the reflectance at 45° and the normal incidence transmittance; the reflectance reached a maximum at 54.6 μm. Sanderson (1965) used reflectance in conjunction with Kramers–Kronig analysis and a Lorentz oscillator to determine the lattice absorption parameters; he found $e_\infty = 15.68$, a transverse optical phonon at 58.87 μm and a longitudinal optical phonon at 52.52 μm. Bermudez and Ritz (1982) measured e_1 and e_2 between 250 and 900 nm by ellipsometry on crystals of InSb cleaved in UHV.

Cardona (1961) gave the temperature coefficient of n for InSb as $(11.9 \pm 0.2) \times 10^{-5}/°\text{C}$.

4.6.6 *Gallium antimonide, phosphide and arsenide*

In the case of gallium arsenide, Cardona (1961) measured k using synchrotron radiation between 15 and 155 eV. Theeten *et al* (1978) used ellipsometry to measure n and k in the range 1.5–6 eV, finding $n_{max} = 5.107$ at 2.88 eV (0.43 μm) whilst, at 0.546 μm, the optical constants could be represented by $4.025 \pm 0.01 \ - \text{i}(0.030 \pm 0.05)$. Christensen *et al* (1974) employed a calorimeter method in the infra-red region. A multiphonon approach was used by Cochran *et al* (1961) to explain their reflectance and transmittance measurements over the absorption band. Over the same region, Holm *et al* (1977) and Kachire *et al* (1974) used oscillator models to explain their results. They found a longitudinal phonon at 34.22 μm and a transverse phonon at 37.22 μm. Another set of results in this same region was obtained by Johnson *et al* (1967) from transmittance measurements, again employing a fit to an oscillator model; although their values of n fitted well with those of Holm *et al*, the k values were much larger. Sturge (1962) determined the optical properties of GaAs between 0.6 and 2.75 eV (0.45 and 2.0 μm) and at temperatures from 10 to 294 K using a transmission interference technique. Sell *et al* (1974) measured n by double beam reflectometry near the fundamental energy gap (1.4 eV), their results being in good agreement with those of Marple (1964). Casey *et al* (1975) determined the absorption coefficient over the same region obtaining satisfactory correlation with the values of Sturge.

Aspnes and Studna (1983) also studied the gallium compounds, phosphide, arsenide and antimonide between 0.2 and 0.8 μm by the method described in section 4.6.1. Their values for the maximum values of n and the wavelengths at which they occur are as follows: GaAs, 5.502 at 2.9 eV; GaP, 5.406 at 3.6 eV; GaSb, 3.808 at 3.3 eV. Burkhard *et al* (1982) used ellipsometry between 365 and 1100 nm in their work on thick films grown by liquid phase epitaxy; their results were in broad agreement with those of Aspnes and Studna.

The temperature coefficient of n for GaAs as determined by Cardona *et al* (1960) was $(4.5 \pm 0.2) \times 10^{-5}/°C$.

4.6.7 Lead sulphide, selenide and telluride

A review of the optical properties of these materials has been given by Dalven (1973). In the medium and far infra-red, the reflectance of PbS has been measured by Dixon and Reidl (1965) who obtained the following values for the phonons: transverse = 140.8 μm, longitudinal = 47.2 μm. In addition, $e_\infty = 16.8$. Schoolar and Dixon (1965) worked in the near infra-red and visible spectrum, measuring the reflectance and transmittance at a small angle of incidence. Walton and Moss (1963) have reported the optical properties of PbSe and PbTe in the range 0.2–8.0 μm. Other workers in this field include Avery (1951) and Scanlon (1958) who determined the

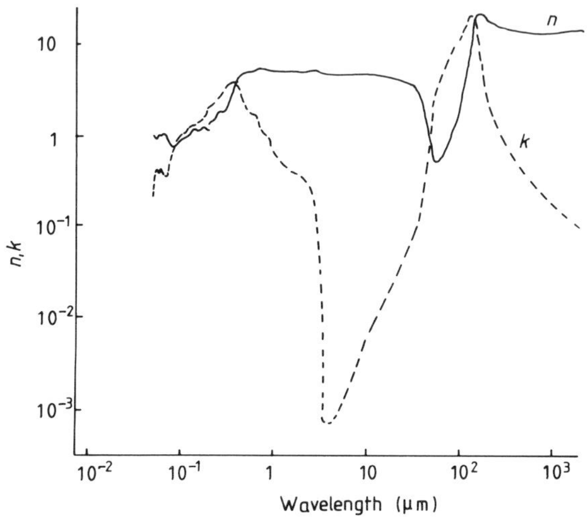

Figure 4.21 Log-log plot of n and k versus wavelength for lead sulphide. (From *Handbook of Optical Constants of Solids I*, ed E D Palik (1985), courtesy Academic Press Inc.)

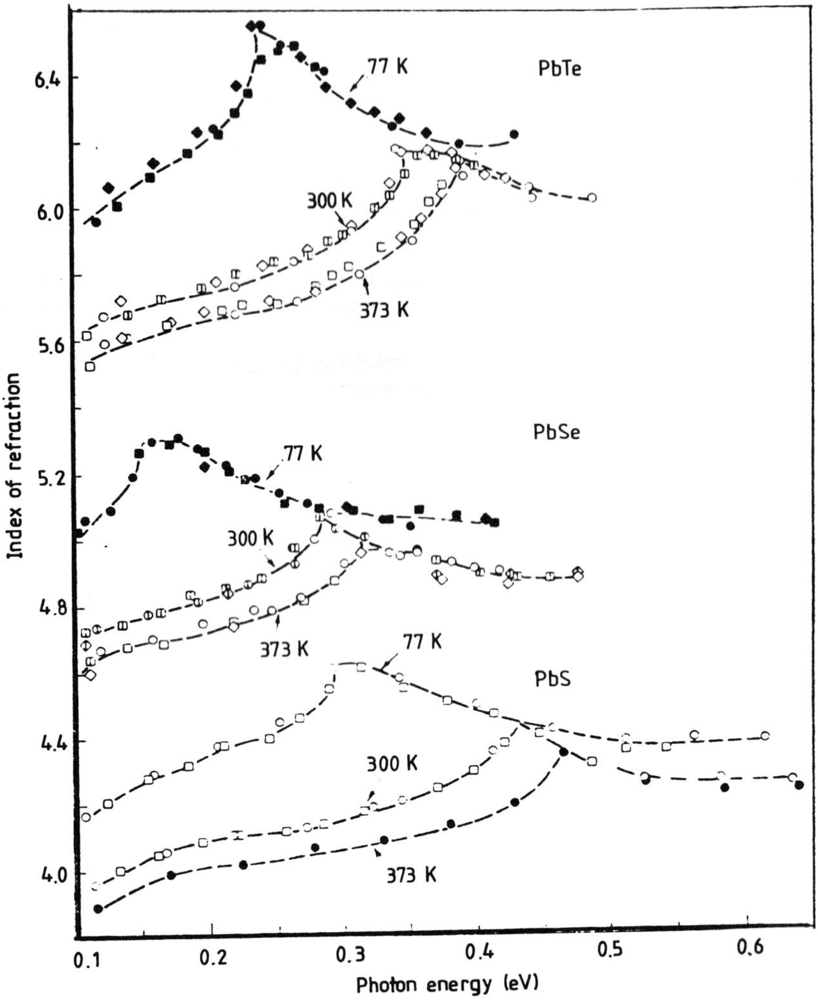

Figure 4.22 *n* versus photon energy for PbS, PbSe and PbTe at temperatures of 77, 300 and 373 LK. (From Zemel *et al.* 1965 Phys. Rev. A **140** 330.)

Table 4.4 Optical data for PbS, PbSe and PbTe.

Property	PbS	PbSe	PbTe
e_0	16.9	21.0	41.5
e_∞	17.2	22.9	32.8
Longitudinal optical phonon (μm)	48.8	75.1	87.8
Transversal optical phonon (μm)	153.2	227.1	315.7
$a \times 10^5/^\circ C$	7.2	6.0	6.0

absorption coefficient. In the ultra-violet, work has been reported by Cardona and Greenaway (1963), Rossi and Paul (1967) and Cardona and Haensel (1970). Ravindra and Srivastava (1980) determined e_0, e_∞ and the longitudinal and transverse phonons for all three materials at room temperature and their values are given in table 4.4. Zemel *et al* (1965) claimed that the optical properties of films of these materials grown on alkali halide substrates were similar to those for bulk between 2 and 15 μm. Figure 4.21 shows n and k for lead sulphide plotted against wavelength collected from a number of sources, whilst in figure 4.22 the values of n are plotted against photon energy for all three materials at temperatures of 77, 300 and 373 K taken from Zemal *et al* (1965).

References

Abelès F 1966 in *Optical Properties and Electronic Structure of Metals and Alloys* (Amsterdam: North-Holland) p 553

—— 1972 *Optical Properties of Solids* (Amsterdam: North-Holland) p 93

Allen S D and Harrington J A 1978 *Appl. Opt.* **17** 1679

Amar H and Johnson K 1966 in *Optical Properties and Electronic Structure of Metals and Alloys* (Amsterdam: North-Holland) p 586

Arakawa E T, Herickhoff R J and Birkhoff R D 1964 *Phys. Rev. Lett.* **12** 319

Arnold G S 1984 *Appl. Opt.* **23** 1434

Aspnes D E, Kinsbron E and Bacon D D 1980 *Phys. Rev.* B **21** 3290

Aspnes D E and Studna A A 1983 *Phys. Rev.* B **27** 985

Aven M, Marple D T F and Segall B 1961 *J. Appl. Phys.* **32** 2261

Avery D G 1950 *Phil. Mag.* 41 1028

—— 1951 *Proc. Phys. Soc.* B **64** 1087

Avery D G and Clegg P L 1953 *Proc. Phys. Soc.* B **66** 512

Bakulin E A, Bredov M M and Vasil'ev V A 1972 *Sov. Phys.-Solid State* **13** 3114

Balkanski M and Besson J M 1961 *J. Appl. Phys.* **32** 2292

Balkanski M and Petrov Y 1964 *Int. Conf. Physics of Semiconductors, Paris* (Paris: Dunod) 245

Barnes B T 1966 *J. Opt. Soc. Am.* **56** 1546

Bartlett R J, Lynch D W and Rosei R 1971 *Phys. Rev.* B **3** 4074

Bauer R S. Spencer W E and White J J 1974 *J. Opt. Soc. Am.* **64** 830

Bauerle D, Braun W, Saile V, Sprussel G and Koth E E 1978 *Z. Phys.* B **29** 179

Beaglehole D 1966 in *Optical Properties and Electronic Sructure of Metals and Alloys* (Amsterdam: North-Holland) p 154

Beaglehole D, de Crescenzo M, Thèye M L and Vuye G 1979 *Phys. Rev.* B **19** 6303

Beattie J R 1957 *Physica* **23** 898

Bennett H E and Bennett J M 1966 in *Optical Propertes and Electronic Structure of Metals and Alloys* (Amsterdam: North-Holland) p 175

Bennett H E, Bennett J M, Ashley E J and Motyka M J 1968 *Phys. Rev.* **165** 755

Bennett H E, Silver M and Ashley E J 1963 *J. Opt. Soc. Am.* **53** 1089

Bennett J M, Wong S M and Krauss G 1980 *Appl. Opt.* **19** 3562

Berglund C N 1966 in *Optical Properties and Electronic Structure of Metals and Alloys* (Amsterdam: North-Holland) p 285
Berman R 1965 *The Physical Properties of Diamond* (Oxford: Clarendon)
Bienewski T M and Czyzek S J 1963 *J. Opt. Soc. Am.* **53** 496
Biondi M A and Payne J A 1959 *Phys. Rev.* **115** 1522
Black E S, Lynch D W and Olson C G 1977 *Phys. Rev.* B **16** 2337
Blodgett A J and Spicer W E 1967 *Phys. Rev.* **158** 514
Bolotin G A, Noskov M M and Sasovshaya I I 1973 *Phys. Met. Metallog.* **35** 24
Bor J and Bartholemew C 1967 *Proc. Phys. Soc.* **90** 1153
Bos L W and Lynch D W 1970 *Phys. Rev.* B **2** 4567
Brandli G and Sievers A J 1972 *Phys.* Rev. B **51** 3550
Briggs H B 1950 *Phys. Rev.* **77** 287, 728
Brodin M S 1961 *Sov. Phys. Solid State* **2** 2292
Brodin M S and Strashnikova M I 1966 *Sov. Phys. Solid State* **8** 549
Brody P S, Sattler J P and Simmons G T 1983 *Ferroelectrics* **50** 645
Brown H M and Brodie D E 1972 *Can. J. Phys.* **50** 2502
Bruckner M, Schafer J H, Schiffer C and Uhlenbusch J, 1991 *J. Appl. Phys.* **70** 1642
Bruckner M, 1989 *J. Appl. Phys.* **66** 1326
Bueche F 1948 *J. Opt. Soc. Am.* **38** 806
Burkhard H, Dinges H W and Kuphel E 1982 *J. Appl. Phys* **53** 655
Burmudez V M and Ritz U H 1982 *Phys. Rev.* B **26** 3297
Callaway J and Wang C S 1977 *Phys. Rev.* B **16** 2095
Callcott T A and Arakawa E T 1974 *J. Opt. Soc. Am.* **64** 839
Cardona M 1961 *Proc. Int. Conf. on Semiconductor Physics, Prague 1960* (Prague: Czech Academcy of Science) p 349
Cardona M and Greenaway D L 1963 *Phys. Rev.* **131** 98
Cardona M and Haensel R 1970 *Phys. Rev.* B **1** 2605
Cardona M and Harbeke G 1965 *Phys. Rev.* A **137** 1467
Cardona M, Gudat W, Sonntag S and Yu P Y 1970 *Proc. Int. Conf. on Physics of Semiconductors, Cambridge 1970* (Oak Ridge, TN US Atomic Energy Commission) p 208
Carroll J J and Melmed A J 1971 *J Opt Soc. Am.* **61** 470
—— 1974 *J. Opt. Soc. Am.* **64** 514
—— 1976 *J. Opt. Soc. Am.* **66** 1050
Casey H C, Sell D D and Wecht K W 1975 *J. Appl. Phys.* **46** 250
Ching W Y and Callaway J 1973 *Phys. Rev. Lett.* **30** 44
—— 1974 *Phys. Rev.* B **9** 5115
Christensen N E and Feuerbacher B 1974 *Phys. Rev.* B **10** 2349
Christensen C P, Joiner R, Nieh S K T and Steier W H, 1974 *J. Appl. Phys* **45** 4957
Cochran W, Fray S J, Johnson F A, Quarrington T E and Williams N 1961 *J. Appl. Phys. Suppl.* **32** 2102
Cohen M H 1958 *Phil. Mag.* **3** 762
—— 1966 in *Optical Properties and Electronic Structure of Metals and Alloys* (Amsterdam: North-Holland) p 66
Cohen M H and Heine V 1958 *Adv. Phys.* **7** 395
Collins R T 1959 *J. Appl. Phys.* **30** 1135

Cooper B R, Ehrenreich H and Philipp H R 1965 Phys. Rev. **138** A494–507
Dalven R 1973 *Solid State Phys.* **28** 179
Debenham M 1984 *Appl. Opt.* **23** 2238
Decker D L and Hodgkin V A 1980 *Proc. Symp. Laser Induced Damage in Optic Materials, Boulder, Colorado* (Washington, DC: NBS, 1981) p 190
Devore J R 1951 *J. Opt. Soc. Am.* **41** 416
Dimmock J O, Freeman A J and Watson R E 1966 in *Optical Properties and Electronic Structure of Metals and Alloys* (Amsterdam: North-Holland) p 237
Dixon J R and Reidl H R 1965 *Phys. Rev.* A **140** 1283
Dobrowolski J A, Ho F C and Waldorf A 1983 *Appl. Opt.* **22** 3191
Dold B and Mecke R 1965 *Optik* **22** 435
Donovan T M and Ashley E J 1964 *J. Opt. Soc. Am.* **54** 1141
Donovan T M. Ashley E J and Bennett H E 1963 *J. Opt. Soc. Am.* **53** 1403
Donovan T M, Bennett H E and Ashley E J 1965 *J. Opt. Soc. Am.* **55** 612
Drew H D and Doezema R E 1972 *Phys. Rev. Lett.* **28** 1581
Dunlop W C and Watters P L 1953 *Phys. Rev.* **92** 1396
Edwards D F and Ochroa E 1980 *Appl. Opt.* **19** 4130
—— 1981 *J. Opt. Soc. Am.* **71** 607
Edwin R P, Dudermil M T and Lamera M 1982 *Appl. Opt.* **21** 878
Efimov A M, Polushkin Yu I, Mikhailov V A, Fomina T N and Arkatova T G 1980 *Fiz. Khim. Stekla (USSR)* **6** 706
Ehrenreich H, Philipp H R and Segall B 1963 *Phys. Rev.* **132** 1918
Eldridge J E and Staal P R 1977 *Phys. Rev.* B **16** 4608
Endriz J G and Spicer W E 1971 *Phys. Rev.* B **4** 4144
Erskine J L, Blake G A and Flaten C J 1974 *J. Opt. Soc. Am.* **64** 1332
Feldman A, Malitson I, Horwitz D, Waxler R M and Dodge M J 1979 *NBS (US) Tech. Note.* **993** 63
Ferrell R A 1966 in *Optical Properties and Electronic Structure of Metals and Alloys* (Amsterdam: North-Holland) p 78
Field J E 1979 *Properties of Diamond* (New York: Academic)
Fisher E I, Fujita S and Weissler G L 1966 *J. Opt. Soc. Am.* **56** 1560
Francis A B and Carlson A J, 1960 *J. Opt. Soc Am.* **50** 118
Freeouf J L 1973 *Phys. Rev.* B **7** 3810
Fukatani H and Sueoko O 1966 in *Optical Properties and Electronic Structure of Metals and Alloys* (Amsterdam: North-Holland) pp 564–73
Gautron J, Raisin C and Lemasson P 1982 *J. Phys. D: Appl. Phys.* **15** 153
Girouard F E and Truong V V 1981 *Appl. Opt.* **20** 4144
—— 1984 *J. Opt. Soc. Am.* **81** 76
Gluskin E S, Druzhinin A V, Kirillova M M. Kochubei V J, Numerovannaya I V and Maevskii V M 1983 *Opt. Spektrosk.* **55** 537
Goldschmidt D 1984 *Proc. SPIE Int. Opt. Eng.* **452** 110
Golovashkin A I and Motulevich G P 1968 *Sov. Phys.-JETP* **26** 881
Goudonnet J P, Godefroy C, Inagaki T and Moretti P 1985 *Jpn. J. Appl. Phys.* Suppl No. 24.2 269
Graves R H A and Lenham A P 1968a *J. Opt. Soc. Am.* **58** 126
—— 1968b *J. Opt. Soc. Am.* **58** 884
Gushchin V S, Shararev K M, Baum B A and Geld P V 1978 *Sov. Phys. Dokl.* **23** 344

Hagemann H-J, Gudet W and Kunz C 1975 *J. Opt. Soc. Am.* **65** 742
Harrison W A 1966 *Phys. Rev.* **147** 467
Hattori T, Homma Y, Mitsuishi A and Tacke M 1971 *Opt. Commun.* **7** 229
Henke B L 1981 *A I P Conf. Proc.* **No 75** 146
Henke B L, Lee P, Tanoka T J, Shimabukura R and Fujikawa B K 1982 *At. Data and Nucl. Data Tables* **27** 1
Herbst J F, Lowy D N and Watson R E 1972 *Phys. Rev.* B **6** 1913
Hikachi Y, Wiimi T and Saito H 1985 *Jpn. J. Appl. Phys.* Suppl. No. 24.2 1039
Hodgson J N 1961 *Phil. Mag.* **6** 509
—— 1963 *J. Phys. Chem. Solids* **24** 1213
—— 1964 *Proc. Phys. Soc.* B **67** 269
—— 1966 in *Optical Properties and Electronic Structure of Metals and Alloys* (Amsterdam: North-Holland) p 64
—— 1968 *J. Phys. Chem. Solids* **29** 2175
—— 1964a *J. Phys., Paris* **25** 154
—— 1964b *J. Opt. Soc. Am.* **54** 208
Holm R T, Gibson J W and Palik E D 1977 *J. Appl. Phys.* **48** 212
Hutchinson C J, Lewis C, Sawaji J A and Pitt J A 1982 *Appl. Opt.* **21** 1490
Icenogle H W, Platt B C and Wolfe W L 1976 *Appl. Opt.* **15** 2348
Ilin S Yu, Propopenko V T and Yaskov A D 1987 *Opt. Spectrosc.* **62** 781
Inagaki T, Emerson L C, Arakawa E T and Williams M W 1976 *Phys. Rev.* B **13** 2305
Ingram J C, Nebesmy K W and Pemberton J E 1990 *Appl. Sur, Sci.* **45** 247
Irani G B. Huen T and Meeten F 1971 *Phys. Rev.* B 3 2385
—— 1972 *Phys. Rev.* B **6** 2904
Ives H E and Briggs H B 1936 *J. Opt. Soc. Am.* **26** 238
—— 1937 *J. Opt. Soc. Am.* **27** 181, 395
Jellison G E and Lowndes D H 1987 *Appl. Phys. Lett.* **51** 352
Johnson C J, Sherman G H and Weil R 1969 *Appl. Opt.* **8** 1667
Johnson P B and Christy R W 1974 *Phys. Rev.* B **9** 5056
Jones M C, Palmer D C and Tien C L 1972 *J. Opt. Soc. Am.* **62** 353
Kachare A H, Spitzer W G, Euler P K and Kahan A 1974 *J. Appl. Phys.* **45** 2938
Karlsson A, Myers H P and Wallden L 1967 *Solid State Commun.* **5** 971
Kern E 1957 *Z. Phys.* **148** 38
Kirillova M M and Charikov B A 1965 *Opt. Spectrosc.* **17** 134
Kirillova M M, Nomerovannaya L V, Bolotin G A, Mayavskiy V M, Noskov M M and Bolotina M S 1968 *Phys. Met. Metall.* **25** 81
Kirillova M M, Nomerovannaya L V and Noskov M M 1972 *Phys. Met. Metall.* **34** 61
Klein G A 1986 *Appl. Opt.* **25** 1873
Koelling D D 1981 *Rep. Prog. Phys.* **44** 139
Kress K A and Lapeyre G J 1970a *J. Opt. Soc. Am.* **60** 1681
—— 1970b *Phys. Rev* B **2** 2532
Landolt-Börnstein 1985 *New Series III* Vol 15b (Berlin: Springer)
Lang M I and Wolfe W I 1983 *Appl. Opt.* **22** 1267
Larson D T and Cash D L 1967 *J. Nuclear Mat.* **24** 232
Leblanc L J, Ferrell J S and Juenker D W 1964 *J. Opt. Soc. Am.* **54** 956

Lee *et al* 1988 *Phys. Rev.* B **37** 2918
Lenham A P 1963 *Proc. Phys. Soc.* **82** 933
Lenham A P and Treherne D M 1966 *J. Opt. Soc. Am.* **56** 1076
—— 1966a *J. Opt. Soc. Am.* **56** 683
—— 1966b *J. Opt. Soc. Am.* **56** 752
—— 1966c *J. Opt. Soc. Am.* **56** 1137
—— 1966d in *Optical Properties and Electronic Structure of Metals and Alloys* (Amsterdam: North-Holland) p 196
—— 1967 *J. Opt. Soc. Am.* **57** 476
Lenham A P. Treherne D M and Metcalfe R J 1965 *J. Opt. Soc. Am.* **55** 1072
Lenham A P, Treherne D M and Woodall A J 1966 in *Optical Properties and Electronic Structure of Metals and Alloys* (Amsterdam: North-Holland) p 40
Lettington A H 1966 in *Optical Properties and Electronic Structure of Metals and Alloys* (Amsterdam: North-Holland) pp 147–53
Leveque G 1983 *J. Opt. Soc. Am.* **57** 473
Li H H 1984 *J. Phys. Chem. Ref. Data* **13** 103
Liebert M and Thomas L K 1984 *Z. Metallkd.* **75** 273
Loewenstein E V, Smith D R and Morgan R L 1973 *Appl. Opt.* **12** 398
Lynch D W, Olsen G G and Weaver J H 1971 *Solid State Commun.* **9** 2195
—— 1975 *J. Opt. Soc. Am.* B **11** 3617
McKay J A and Rayne J A 1976 *Phys. Rev.* B **13** 673
Malitson J H 1965 *J. Opt. Soc. Am.* **55** 1205
Mamedro A M 1982 *Ferroelectrics* **45** 55
Manabe A, Mitsuishi A and Yoshinga H 1967 *Japan J. Appl. Phys.* **6** 593
Manzka R 1980 *Phys. Status Solidi* **97** 157
Marple D T F 1964 *J. Appl. Phys.* **25** 539
Mathewson A G and Myers H P 1972 *Phil. Mag.* **25** 853
Mattheiss L F 1970 *Phys. Rev.* B **1** 373
—— 1974 *Phys. Rev.* **134** A970
Mayer H and El Naby M H 1963a *Z. Phys.* **174** 269
—— 1963b *Z. Phys.* **174** 280
—— 1963c *Z. Phys.* **174** 289
Mayer H and Heitel B 1966 in *Optical Properties and Electronic Structure of Metals and Alloys* (Amsterdam: North-Holland) p 47
Meeson A 1972 *J. Phys., Paris* **33** 371
Miles P H 1977 *Appl. Opt.* **16** 2891
Monin J 1973 *Acta Electron.* **16** 139
Monin J and Boutry E A 1974 *Phys. Rev.* B **9** 1309
Moravec T J, Rife J C and Dexter R N 1976 *Phys. Rev.* B **13** 3297
Moravec T J, Rife T C and Dexter R W 1976 *Phys. Rev.* B **13** 3297
Moritani A, Kundo K and Nakai J 1976 *Jpn. J. Appl. Phys.* **15** 1549
Moss T S 1961 *Optical Properties of Semiconductors* (London: Butterworth)
—— (ed) 1980 *Handbook on Semiconductors* (Amsterdam: North-Holland)
Moss T S, Smith S D and Hawkins T D F 1957 *Proc. R. Soc.* B **70** 776
Motulevich G F and Shubin A A 1965 *Sov. Phys.-JETP* **19** 1549
Muldawer L and Goldman H J 1966 in *Optical Properties and Electronic Structure of Metals and Alloys* (Amsterdam: North-Holland) p 574

Myers H P, Wallden I and Karlsson A 1968 *Phil. Mag.* **18** 725
Nilsson P O 1974 *Solid State Phys.* **29** 139
Nomerovannya L V, Kirillova M M. Noskov M M and Bolotin G A 1973 *Phys. Met. Metall.* **36** 66
Olson C G and Lynch D W 1974 *Phys. Rev.* B **9** 3159
Ordal M A, Long L L, Bell R J, Bell S E, Bell R R, Alexander R W and Ward C A 1983 *Appl. Opt.* **22** 1099
Ordal M A, Bell R J, Alexander R W, Long L L and Querry M R 1985 *Appl. Opt.* **24** 4493
—— 1987 *Appl. Opt.* **26** 744
Ordal M A, Bell R J, Alexander R W, Newquist L A and Querry M R 1988 *Appl. Opt.* **27** 1203
Overhauser A W and Butler R N 1976 *Phys. Rev.* B **14** 3371
Owens J C 1969 *Phys. Rev.* **181** 1228
Palik E D 1985 *Handbook of Optical Constants of Solids* (New York: Academic)
Palik E D 1991 *Handbook of Optical Constants of Solids II* (New York: Academic)
Palmer R E and Schnatterly S E 1971 *Phys. Rev.* B **4** 2329
Parkins G R, Lawrence W E and Christy R W 1981 *Phys. Rev.* B **23** 6408
Payan R and Roux D 1969 *Opt. Commun.* **1** 37
Phillipp H R 1972 *J. Appl. Phys.* **43** 2835
Phillipp H R and Ehrenreich H 1963 *Phys. Rev.* **129** 1530
Philipp H R and Ehrenreich H 1964 *J. Appl. Phys.* **35** 1007
Philipp H R and Taft E A 1959 *Phys. Rev.* **113** 1002
Pierce D T and Spicer W E 1973 *Phys. Status Solidi* **60** 689
Potapov E V 1964 *Sov. Phys.-JETP* **20** 307
Potter R F 1966 *Phys. Rev.* **150** 562
Powell C J 1970 *J. Opt. Soc. Am.* **60** 78
Price D J 1947 *Proc. Phys. Soc.* **59** 131
Rasigni G, Palmari G P and Rasigni M 1975 *Phys. Rev.* B **12** 1121
Rasigni M and Rasigni G 1977 *J. Opt. Soc. Am.* **67** 54
Rasigni M, Rasigni G and Palmari G P 1976 *Phys. Rev.* B **14** 4235
Ravindra N M and Srivastava V K 1980 *Phys. Status Solidi* **58** 311
Reichmann J and Feldman C 1969 *J. Opt. Soc. Am.* **59** 1404
Roberts R A and Walker W C 1967 *Phys. Rev.* **161** 720
Roberts S 1959 *Phys. Rev.* **114** 104
Roessler D M and Walker W C 1967 *J. Opt. Soc. Am.* **57** 835
—— 1968 *J. Opt. Soc. Am.* **58** 279
Rossi C E and Paul W 1967 *J. Appl. Phys.* **38** 1803
Rubloff G W 1971 *Phys. Rev.* B **3** 285
Russell E E and Bell E E 1967 *J. Opt. Soc Am.* **57** 241
Saga S, Inona K. Taniguchi T, Shu S and Sahi M 1983 *J. Phys. Soc. Jpn.* **52** 1848
Salzburg C D and Villa J J 1954 *J. Opt. Soc. Am.* **43** 579
—— 1957 *J. Opt. Soc. Am.* **47** 244
Samara G A 1983 *Phys. Rev.* B **27** 3494
Sanderson R B, 1965 *J. Phys. Chem. Solids* **26** 803
Sasovskaya I T and Noskov M M 1975 *Opt. Spectrosc.* **39** 64
Scanlon W W 1958 *Phys. Rev.* A **109** 47

Scanlon W W 1959 *J. Phys. Chem. Solids* **8** 423/442
Schoolar R H and Dixon J R 1965 *Phys. Rev.* **137** A667
Schröder K and Onengut D 1967 *Phys. Rev.* **162** 628
Schuler C 1966 in *Optical Properties and Electronic Structure of Metals and Alloys* (Amsterdam: North-Holland) p 221
Segall B 1962 *Phys. Rev.* **125** 109
Seits D H and Spicer W E 1970a *Phys. Rev.* B **2** 1676
—— 1970b *Phys. Rev.* B **2** 1694
Sell D D, Casey H C and Wecht K N 1974 *J. Appl. Phys.* **45** 2650
Shiga M and Pells G P 1969 *J. Phys. C: Solid State Phys.* **2** 1847
Shiles E, Sasuki T, Inokuti M and Smith D Y 1980 *Phys. Rev.* B **22** 1612
Shiozawa L R, Barrett, Jost J M, Chotkevys C P and Devlin S S 1962 *Research on II–Vi Compound Semiconductors* Clevite Corp Cleveland Ohio, Final Report on the Office of Aerospace Research, US Air Force Contract AF33 (616)–6865
Shklyarevskii I N, Adveenko A A and Padulka V G 1959 *Opt. Spectrosc.* **6** 326
Shklyarevskii I N and Yaroveski P G 1961 *Opt. Spectrosc.* **11** 355
Shvarov K M, Baum B A and Gel'd V 1977 *High. Temp.* **15** 548
Siddiqui and Treherne 1977 *Infrared Phys.* **17** 33
Sievers A J 1980 *Phys. Rev.* B **22** 1600
Sliker T and Jost J M 1966 *J. Opt. Soc. Am.* **56** 130
Smith N V 1969 *Phys. Rev.* **183** 634
—— 1970 *Phys. Rev.* B **2** 2840
—— 1974 *Phys. Rev.* B **91** 1365
Smith T 1972 *J. Opt. Soc. Am.* **62** 774
—— 1977 *J. Opt. Soc. Am.* **67** 48
Sonntag B, Haensel R and Kunz C 1969 *Solid State Commun.* **7** 597
Spicer W E 1965 *Phys. Rev.* **139** A565
—— 1972 *Optical Properties of Solids* (Amsterdam: North-Holland) p 755
Spitzer G W and Kleinman D A 1961 *Phys. Rev.* **121** 1324
Steinmann W 1968 *Phys. Stat. Solidi* **28** 437
Stephane G. Lammomie J-C and Robin S 1967 *J. Opt. Soc. Am.* **57** 486
Sterne E A 1967 *Phys. Rev.* **157** 544
Stohr J, Wehner P S, Williams R S, Apai G and Shirley D A 1978 *Phys. Rev.* B **17** 587
Strzalkowski I, Joshi S and Crowell C R 1976 *Appl. Phys. Lett.* **28** 350
Studna A A 1975 *Solid State Commun.* **16** 1063
Sturge M D, 1962 *Phys. Rev.* **127** 768
Swanson N 1964 *J. Opt. Soc. Am.* **54** 1130
Taft E A and Philipp H R 1961 *Phys. Rev.* **121** 1100
Theeten J P, Aspnes D E and Chang R P H, 1978 *J. Appl. Phys.* **49** 6097
Thèye M-L, Nguyen Van V and Fisson S 1983 *Phil. Mag.* B **47** 31
Tokumoto M, Drew H D and Bagchi A 1977 *Phys. Rev.* B **14** 3497
Vehse R E and Arakawa E T 1969 *Phys. Rev.* **180** 695
Vehse R E, Arakawa E T and Williams M W 1970 *Phys. Rev.* B **1** 517
Vitrikhovskii N I, Gudymenko L F and Maznichenko A F 1968 *Sov. Phys. Semicond.* **2** 732
Volken H T and Kruger J 1965 *J. Opt. Soc. Am.* **55** 842

Vonsovskii S V and Sokolov A V 1949 *J. Exp. Theor. Phys.* USSR **19** 615
Walker J 1979 *Rep. Prog. Phys.* **42** 1605
Walton A K and Moss T S 1963 *Proc. Phys. Soc.* **81** 509
Weaver J H 1972 *Phys. Rev.* B **5** 2829
—— 1980 *J. Opt. Soc. Am.* **70** 1030
Weaver J H and Benbow R L 1975 *Phys. Rev.* B **12** 3509
Weaver J H, Colevita E, Lynch D W and Rosei R 1979 *Phys. Rev.* B **19** 3850
Weaver J H, Krafta C, Lynch D W and Koch E E 1981 *Physik Daten* (Karlsruhe: Fachinformationszentrum)
Weaver J H, Lynch D W and Olson G G 1973 *Phys. Rev.* B **7** 4311
—— 1974 *Phys. Rev.* B **10** 501
—— 1975a *Phys. Rev.* B **11** 1416
Weaver J H and Olson G G 1976 *Phys. Rev.* B **14** 3251
—— 1977 *Phys. Rev.* B **15** 4602
Weaver J H, Olson G G and Lynch D W 1975b *Phys. Rev.* B **12** 1293
Wehenkel C and Gauthe B 1974 *Phys. Status Solidi* B **64** 515
Wesolowska C and Dobierzewska-Mozrzymas E 1963 *Acta Phys. Pol.* **24** 65
Whang U S, Arakawa E T and Callcott T A 1971 *J. Opt. Soc. Am.* **61** 740
White J J and Straley J W 1968 *J. Opt. Soc. Am.* **58** 759
Winsemius P, van Kampen F F, Langkiek H P and van West C G 1976 *J. Phys. F: Metal Phys.* **6** 1583
Wood R W 1933 *Phys. Rev.* **44** 353
Yanagahara M, Niwaro M, Konde T and Sato S 1988 *Japan J. Appl. Phys.* I **27** 666
Yoshinga H and Oetjen R A 1956 *Phys. Rev.* **101** 526
Yu A J-C and Spicer W E 1968 *Phys. Rev.* **169** 497
Zemel J N, Jensen J D and Schoolar R H 1965 *Phys. Rev.* **140** 330

5 Thin Films (Preparation and Theory)

5.1 The importance of thin films

The thin films discussed in this chapter are defined as continuous laminar structures having thicknesses ranging from a few nanometres up to about 1000 nm and which transmit a measurable amount of radiation. This definition distinguishes them from, on the one hand, the thick opaque films treated as bulk materials in Chapter 4 and, on the other, the discontinuous films dealt with in Chapter 8. In recent years such films have acquired considerable importance in view of their increasing usage in a wide range of technical applications. For example, thin films of dielectrics are used as coatings on the components of optical systems bringing about considerable reductions in the losses due to reflection. A stack of thin films of suitable materials and thicknesses may be used as a filter for either transmitting or rejecting a required band of wavelengths. Thin films of metals are used for high-reflectance coatings of mirrors and also for decorative finishes. Microcircuits have a layer structure and, with the progressive development of ever smaller units, the layers are approaching the thin-film category. Another area of interest is that of thin oxide layers on the surface of bulk metals.

In addition to these practical considerations, thin films are interesting in their own right as their properties may be different from those of bulk solids, thus presenting a number of intriguing problems to the theoretician.

5.2 Production of thin films

It is possible to produce unsupported thin films by the etching or chemical thinning of bulk materials, but such films are not in common use for optical measurements at the present time because of the lack of control on their final thickness and the non-uniformity of the film produced. One of the few uses of this technique is in the production of some specimens for electron microscopy. Another method of producing thin films, which can be used with some polymeric materials, is to allow a drop of the parent liquid to spread over a water surface, but again such films are rarely uniform and they are unsuitable for optical purposes.

Almost all the thin films used in either research or industry are made by deposition on to a suitable substrate following evaporation in a high vacuum or by sputtering in an inert gas atmosphere. For many years evaporation *in vacuo* was virtually the only satisfactory means of making thin films, but the RF sputtering process has now been developed as a viable alternative.

The evaporation technique may also be used to produce the non-transparent, thick metal films (thicknesses greater than 1000 nm) which have been used instead of bulk specimens in many optical measurements as described in Chapter 4.

5.2.1 Preparation of substrates

In order to prepare high-quality films it is necessary to start with a good clean substrate. Fused quartz and some glasses are frequently used as their surfaces can be brought to a high polish. Their major disadvantage is that they may introduce large strains into the films due to differential thermal expansion, so it is best to use a substrate of similar expansion coefficient to that of the film material. If crystalline or epitaxial films are required, a crystalline substrate must be used; electropolished or cleaved crystals of germanium, silicon, rock-salt or other alkali halides have been used for this purpose as well as split mica. Best results are obtained if the crystals can be cleaved *in vacuo*; however, the working areas of such crystals may be limited by cleavage steps. It is possible to prepare large-area single-crystal surfaces which are optically flat, smooth and have no surface damage or lattice distortion at the surface (Seraphin 1962).

The most commonly used glass substrates are microscope slides. These are fire polished in manufacture and are fairly smooth (RMS roughness about 1 nm), but they are not optically flat so are not suitable for work of the highest precision. Optical flats which are flat to a few hundredths of a fringe are commercially available but these are generally less smooth than slides (RMS roughness about 2–3 nm). They can be further polished in a slurry of polishing compound in water, constantly stirred and covering the surface to a depth of about 15 mm, the final polish being obtained using pure water as the lubricant.

After polishing, it is vital that the substrates are thoroughly cleaned to remove all traces of grease or oil which can have a disastrous effect on thin-film surfaces. The procedure recommended by Bennett and Bennett (1967) is as follows.

(1) Any existing film should be removed using a suitable solvent, for example nitric acid (silver), caustic soda (aluminium), potassium cyanide (gold), hydrogen peroxide (germanium). This process must not be allowed to continue for longer than absolutely necessary otherwise the surface of the substrate may be chemically etched and will then need to be repolished.

(2) The substrate is washed by hand using a cotton wool pad soaked in a detergent solution.

(3) A rinse is then given in distilled water.

(4) An ultrasonic cleaning bath containing detergent in distilled water is used for at least 15 minutes.

(5) The ultrasonic bath is repeated but now using only distilled water, also for 15 minutes.

(6) A final wash is given in conductivity water; this is continuously circulated and monitored for the presence of contaminants.

(7) The substrate is dried in vacuum.

Bennett and Bennett did not recommend cleaning the substrate in a glow discharge, as has been the practice by some workers, because of the danger that some of the electrode material might be sputtered on to the surface. After all the above precautions have been taken, the main problem is caused by dust particles falling on the surface during storage and in transfer to the deposition chamber. One method of avoiding this is to coat the cleaned surface with collodion and then cover with sticky tape; the whole covering can then be stripped off under the vacuum.

With crystalline substrates, if it is not required to produce epitaxial films, the same polishing technique as that described above may be used. However, the production of epitaxial films requires further preparation. As discussed in Chapter 2, mechanical polishing introduces surface damage, the damage extending roughly to a depth of twice the diameter of the polishing particles, and it is necessary to remove the damaged layer caused by the previous grade of polishing medium before proceeding. It is possible in some cases to complete the polishing by electropolishing; if care is not taken an 'orange-peel' effect is produced but this can be avoided by simultaneously using a lapping cloth and dripping the electrolyte on to it. Finally, the surface should be examined in the electron microscope to check that it is crystalline.

5.2.2 Evaporation techniques

The evaporation process makes use of the fact that, in a low-pressure environment, a material will evaporate at a much lower temperature than its normal boiling point. In addition, under vacuum conditions, the molecules of the evaporated material will have long mean free paths and thus travel in a straight line from the source to the substrate producing a sharp-edged film. Most important of all, the absence of collisions with the residual gas produces a film which is, initially, free from contamination; prolonged exposure of the surface, even in the best vacuum conditions, will, however, ultimately result in it becoming contaminated.

The early film evaporation plant (see figure 5.1) consisted of a glass bell-jar maintained at a pressure of about 10^{-6} Torr by means of oil or mercury diffusion pumps backed by an oil-lubricated rotary pump. The film materials were evaporated from electrically heated tungsten wires or boats and allowed to settle on glass or other substrates placed about 10 cm above the sources. Although it was possible to make films of uniform thickness by this process, they were subject to contamination by oil or mercury vapour backstreaming from the pumps as well as to oxidation by the residual air in the system. In addition, in many experiments the films were taken out of the evaporation unit for subsequent optical measurements in the open air where the oxidation and contamination problems would increase. It is not surprising therefore that many of the earlier results with films were not consistent.

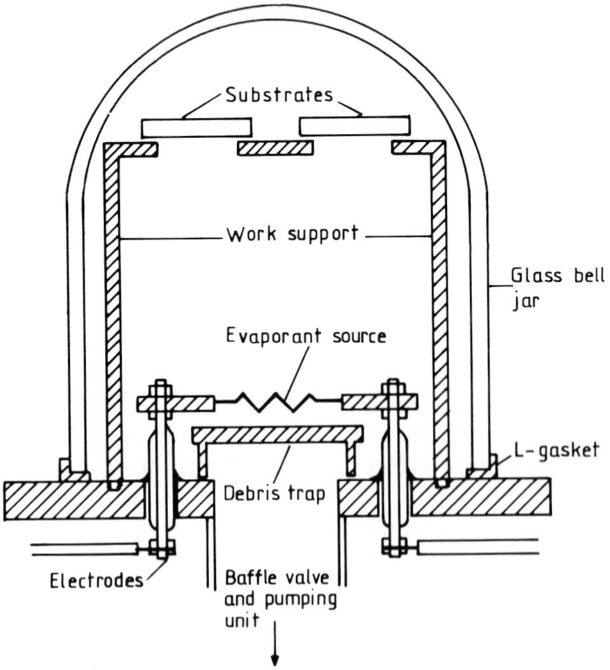

Figure 5.1 An evaporation unit. (From *Introduction to the Theory and Practice of High Vacuum Technology*, by L Ward and J P Bunn (1967), courtesy Butterworth Scientific Ltd.)

The technique has been continually improved in a number of ways. A liquid nitrogen trap placed between the diffusion pump and the evaporation chamber effected a large reduction in the vapour pressure in the chamber due to the pump fluid. It also became possible to make measurements on the film *in situ*, although this introduced some new problems such as the influence of the window materials on the optical measurements, and the limitations imposed on the angles of incidence by the geometry of the fixed windows.

True ultrahigh-vacuum systems constructed entirely from stainless steel together with the use of vapour-free pumps, such as ion or evaporation pumps, represented a further step in the technique of producing uncontaminated films. Optical measuring systems were able to take advantage of the technology being developed for the production of microcircuits where ultraclean conditions were required. Using stainless steel systems vacua down to 10^{-10} Torr may be achieved by (i) baking the evaporation chamber at about 400 °C for at least 24 hours to remove the gases and vapours adsorbed on all the internal surfaces including the substrate, (ii) using a liquid nitrogen trap to act as a cryo-pump for any residual vapours and (iii) using vapour-free pumps for both the main and backing pumps.

The evaporation source has also undergone considerable development from the simple tungsten wires which were originally used to evaporate materials such as aluminium; when molten, aluminium wets the wire forming a liquid bead which acts virtually as a point evaporation source. Tungsten, molybdenum or tantalum boats are available for those materials which do not wet these metals and would, on melting, fall off a wire. These boats are particularly useful for dielectrics which are usually in powder form and are often very refractory: the partial pressures of these high-melting-point boat materials are very low, and negligible quantities are evaporated alongside the evaporants. Ceramic crucibles heated by tungsten spirals or tailor-made fitted foil heaters are also available for refractory substances, although the outgassing process for ceramics can be very protracted and it is difficult for it to be completed satisfactorily.

A different method of evaporation is by electron beam heating, a suitable electron gun being described by Heavens (1959). An electrically heated tungsten loop acts as the electron source and the material to be evaporated is placed at the centre of the loop and raised to a positive potential of several kilovolts; the electrons may also be focused on to the specimen by a negatively charged cylinder surrounding the anode. The bomardment of the electrons on the anode raises its temperature to melting point and a pendant drop is formed from which the evaporation takes place. This system has the considerable advantage of reducing the amount of 'foreign' material in the system but can only be used for conductors of electricity. Cox *et al* (1964) describe a technique for the evaporation of refractory materials such as silicon, aluminium oxide and silicon dioxide by means of an electron gun.

Thin films may also be produced by evaporation by laser beams (Smith and Turner 1965).

Thin films of alloys may also be produced by the evaporation process. Direct evaporation of the alloy is generally unsatisfactory as dissociation is liable to take place and the composition of the resultant film is then difficult to determine unless an electron probe analyser is incorporated in the apparatus. It is therefore considered to be better practice to evaporate the constituents of the alloy separately and allow the alloying process to take place on the substrate which may have to be heated to improve the mobility of the deposited materials.

A review of the various techniques for producing thin films by evaporation, covering a wide range of materials, has been given by Olsen *et al* (1945); Holland (1961) has also provided useful information on this topic. More recent reviews of this field have been given by Caswell (1963) and Campbell (1976).

5.2.3 Sputtering techniques

The technique of cathodic sputtering consists of bomarding a negatively charged sample of the required material with heavy positive ions which have sufficient kinetic energy to dislodge clusters of molecules from the surface. These molecules then fall on to a substrate and eventually build up into a film. The technique can be applied to any metallic material, element or alloy, and as the process takes place at a low temperature (if required the substrate can be cooled with a liquid gas) the problems of dissociation encountered in the evaporation methods are avoided. The serious disadvantage of sputtering is that the positive ions needed to initiate the process can only be produced in a gas discharge necessitating the presence of a gas at a pressure of about 10 Torr. One of the inert gases such as argon, xenon or krypton is used in this technique thereby minimising chemical reactions with the film; however, the inert gas may become trapped within the film or adsorbed on its surface.

With a DC voltage applied to the anode, sputtering is a very slow process and a considerable time is needed to form a thickness of tens of nanometres. A positive charge builds up on the cathode, especially with semiconductors and insulators, and this inhibits the arrival of the ions and hence reduces the efficiency of the sputtering process. If a radiofrequency electric field is applied, the charge is neutralised every half cycle and the sputtering rate is greatly increased (see Anderson *et al* 1962). In addition, the high-frequency electric field causes the production of extra ionisation of the inert gas and hence a greater sputtering rate. Radiofrequency sputtering is particularly useful for the production of films of alloys and cermats which are conductors of electricity but are also very refractory. Reviews of sputtering techniques and their problems have been given by Maissel (1966) and Patten *et al* (1981).

A variation of the sputtering technique, ion beam sputtering, uses an external source of ions to initiate the sputtering process. The advantages are: (i) the deposited film is not exposed to high ion fluxes and (ii) the deposition can be carried out in clean ultra-high vacuum conditions. Franks (1981) and Allen (1985) have discussed this technique and have also considered the effects of ion energy and current density on the properties of the growing film. It was claimed that films deposited by this technique showed no grain boundary cracks because of the high mobility of the islands; such films should thus be a close approximation to bulk materials.

Little work has been reported on the production and use of sputtered layers in the measurement of optical constants, although McNeil *et al* (1985) have discussed the ion beam deposition of optical materials such as TiO_2 SiO_2 Al_2O_3, Cu, Ag, Al and Mo. However, sputtered films have been used in the measurement of the magneto-optical properties of films, a topic which is largely outside the scope of this book (see Carey *et al* 1985, Hunt 1967).

5.2.4 Ion assisted evaporation techniques

There are a number of methods of forming thin films in which ions are used to supplement the evaporation process; it has been shown by McNeil *et al* (1984) and Martin *et al* (1984) that films produced by such techniques have high packing densities giving bulk-like properties. These processes have been reviewed by Martin and Netterfield (1986) and by Teer (1988) and two of them, ion plating and ion clustered beam deposition, are described below.

Ion plating was first described by Mattox (1964) and involves a combination of thermal evaporation with ionisation in a glow discharge; a more modern arrangement has been given by Tagaki (1982). Evaporation takes place from a wire or crucible into a glow discharge in argon gas at a pressure of between 10^{-1} and 10^{-2} Torr which is maintained by applying a potential difference between the source (anode) and the substrate (cathode). Some of the evaporated atoms become ionised in the plasma and the positive metal ions are then accelerated across the cathode dark space to the substrate. They suffer many collisions on the way and, when they finally reach the cathode, they have energies of about 30 eV which is sufficient to cause sputter cleaning, so allowing the film subsequently formed by the neutral atoms to adhere strongly. The technique is particularly useful for coating plastics and for depositing hard, wear-resistant or anticorrosion layers on components to be used as moving parts in machinery; all sides of the substrate become coated in this process. One problem is that, if the accelerating voltage is too high, the high energy ions will tend to sputter away any deposit already formed so the deposition conditions need to be controlled carefully. Silver, gold, aluminium, chromium and nickel have all been deposited successfully by ion plating. If the inert gas is replaced by a reactive gas, this technique can be used to apply films of very hard compounds such as carbides or nitrides.

The process of ion clustered beam deposition, which has been described by Younger (1985), uses an evaporation source such as electron beam heating to vaporise the metal inside a crucible at a pressure of between 1 and 10 Torr. As the vapour expands adiabatically out of the crucible it cools until it is supersaturated when the atoms coalesce to produce aggregates of up to 1000 atoms. The clusters with the highest thermal energy rise into the ionisation region where electron bombardment produces ionisation in about 20% of them. A potential difference of up to 10 kV is applied between the source and the substrate and, as in ion plating, this accelerates the ions which then bombard the substrate cleaning it by sputtering before the arrival of the neutral particles which form a strongly adhesive layer.

5.2.5 Multi-layer films

The production of multi-layer films can be effected by providing several evaporation sources within the same vacuum chamber and operating each in succession. The substrate may either be moved mechanically under each source in turn or, more usually, the sources are arranged so that each can produce a uniform layer at the substrate. In the majority of cases it is necessary to monitor the thickness of each layer carefully and ways of doing this are described in the next section. Films containing up to a hundred separate layers have been produced by the evaporation process.

5.3 Measuring and monitoring film thickness

A number of techniques are available for the measurement of the thickness of a film and also for monitoring the thickness and rate of growth during deposition. Behrndt (1966) and Piegari and Masetti (1985) have reviewed several of these methods.

One of the simplest methods is to use a Talystep machine which gives a direct reading of the thickness; a review of this group of instruments has been given by Eschbach and Verhoyen (1974). The thickness is measured by a stylus attached to an extremely light arm which is pivoted so that a small vertical movement of the stylus causes a large movement of the arm inside a search coil and so generates a small EMF; the amplified signal can be fed eventually to a pen recorder which is used to draw out the step at the edge of the film. The technique can be used with continuous films of thickness greater than about 100 nm, and can also be used to scan the film surface to investigate changes in thickness. It is claimed that differences of a few ångstrom units can be detected by this type of equipment (Bennett and Dancy 1981). If this method is to be used regularly, it is necessary to check the calibration of the Talystep against films of known thickness. There is a danger in using this machine that the surface of the film may be scratched, and so it should only be used on films after all the optical measurements have been completed.

A technique in which the film thickness is determined in terms of the mass per unit area is x-ray scattering and this is particularly useful for metallic specimens. Nag and Ward (1971) used this method with nickel films. The intensity of the x-rays scattered by the film is measured and, for a given element, this is proportional to the mass per unit area of that element. The characteristic x-ray wavelengths for the film material are used for these measurements and so it is possible to determine the areal masses of each component in the film, information which could be invaluable when dealing with alloy films. The x-ray unit is calibrated for each element by scattering measurements on samples of known thicknesses. This technique can be used on both continuous and island films, the calculation of the actual thickness of the film being dependent on the assumed value of the density of the deposited film; as this may be different from the density of the bulk material the technique could lead to error. One further advantage of this method is that thickness measurements may be made, without damage, on the same films on which the optical experiments are made. In addition only small areas of the films are needed.

Another method in which the mass per unit area is measured is the quartz crystal oscillator whose resonant frequency in the X or Y direction is changed by the presence of a deposited layer. The frequency change produced may be calibrated against the deposited mass (or be provided by the manufacturer). This technique is especially useful for monitoring the progress of film deposition as the resonant frequency can be read continuously. However, it has to be remembered that the monitored film is not the actual film being produced for investigation and the assumption has to be made that both films have the same properties.

The optical properties of the films themselves may be used for measuring and monitoring film thickness. Thus the reflectance and/or transmittance may be continuously measured so that films of the required reflection or transmission properties can be produced. This monitoring technique is particularly useful when low-reflectance coatings are required, such as for the blooming of optical components, so that the deposition can be stopped as soon as the reflectance has reached its minimum value.

The optical monitoring of films is vital when making multi-layer films for optical filters when the thicknesses of the various layers are critical if the filter is to have exactly the desired properties. Sophisticated information feedback systems have been devised to indicate when the exact required optical properties have been achieved with a given layer so that the next layer can be started.

The thicknesses of conducting films can also be deduced from their electrical resistance. Turner *et al* (1963) describe a method for continuously monitoring the thicknesses of films by this technique.

5.4 Film growth and structure

This topic has been dealt with in detail in a large number of publications, for example Thun (1963), Lewis and Anderson (1978) and Venables *et al* (1984).

A thin film is produced by molecules firstly adhering to the substrate and subsequently redistributing themselves by thermal diffusion on the surface. There is evidence that molecules may make up to a million such readjustments before taking up their final positions on the substrate. When the quantity of deposited material is small, and especially if the surface is contaminated, surface tension forces predominate over the thermal dispersion forces and the deposit collects in the form of droplets or islands each of which is initially roughly spherical in shape. Such discontinuous films and their properties are discussed in more detail in Chapter 8. As more material continues to arrive at the substrate, the film islands grow in size and decrease in number by amalgamation; their outlines become irregular and their profiles depart considerably from the original spheres. They will eventually join together to form a layer which, although it is electrically continuous, may not be homogeneous as it may contain a large number of vacancies. In addition, as growth starts from a number of randomly orientated islands, the final film will generally be polycrystalline in structure.

The growth of gold deposits on sodium chloride substrates has been discussed by Schmeisser (1974) in terms of the mean times for re-evaporation t_e, and re-orientation, t_r of the gold atoms. Both these times are related exponentially to the respective binding energies of the atoms to the surface, E_b and E_d. Thus, $t_e = t_o \exp(E_b/kT)$ and $t_r = t_o \exp(E_d/kT)$, where t_o is approximately equal to the reciprocal Debye frequency. The diffusive flux of atoms into an already existing hemispherical cluster of radius r is given by:

$$\mathrm{d}N/\mathrm{d}t = 2\pi r N_1/\sqrt{Dt_e}\{K_1(r/\sqrt{DT_e})\big/K_o(r/\sqrt{Dt_e})\}$$

where K_1 and K_o are Bessel functions of the first and zeroth order and D is the diffusion constant and is equal to $a_o^2 \exp(-E_d/kT)/4\,t_o$, a_o being the mean distance between neighbouring sites.

Velfe *et al* (1982) also working with gold deposited on rocksalt, quoted a formula for N_c, the number of visible clusters formed in terms of the deposition rate, R_1, the time, t and the substrate temperature, T, namely

$$N_c = QR_1^m t^s \exp(E_v/kT)$$

where $E_v \doteqdot 0.1$ eV, $m = s = 1/3$ and $Q = 3450$ (cm$^{-4/3}$).

Gates and Robins (1987) have also reported studies of the nucleation rate of gold clusters deposited on sodium chloride substrates. They noted that, as the clusters passed through a certain range of sizes during growth, they showed exceptionallly high mobility on the surface and were thus able to reorientate themselves easily. They devised a model for nucleation based on the following assumptions: (1) only those clusters nucleating at defect sites are able to survive long enough to grow to a size, *a*, where this high mobility starts and continues up to another size, *b*, where it ends; (2) except for brief transit periods, the density of visible clusters $\gg$ the density of sub-visible clusters; (3) provided the temperature is high enough, a cluster reaching size *a* is able to break away from a defect site and then become attached to the free surface of the material; (4) the remaining defect sites are occupied by sub-visible clusters; (5) an expression for the probability that the highly mobile clusters of size *a* will survive to grow to size *b* without colliding with another large cluster. They were able to show that these assumptions led to a formula for the density of clusters which had the same powers for R_1, t and T as that of Velfe *et al.*

When a thick film is examined in profile by electron microscopy it is found that, once a continuous layer has been formed, growth has proceeded in a columnar manner. Figure 5.2 shows a typical electron micrograph of a thin-film cross section. During growth, spaces may be produced between the columns and, on exposure to the atmosphere, these can become filled with air or with water vapour thus affecting the properties of the film.

Dirks and Leamy (1977) have dealt with the columnar microstructure of thin films. The structure consists of a low-density or void network surrounding an array of parallel uniform-sized rods of high density. The column orientation in films deposited at oblique incidence is always more nearly perpendicular to the substrate than the vapour beam direction. The structure is only observed in materials which exhibit a limited atomic mobility. A simple geometric argument based on shadowing of the vapour beam by atoms within the growing film is sufficient to explain the gross features of the structure. The shadowing mechanism can be demonstrated by means of atomistic simulations of the deposition process, which also yield columnar structure. In such films the columns are oriented at an angle b, such that $\tan b = \frac{1}{2} \tan a$, where a is the angle between the normal and the vapour beam. The microstructure of thin films has also been discussed by Macleod (1982); this structure has great influence on the optical and mechanical properties of the films and is responsible for many of the difficulties experienced with the fabrication and subsequent performance of optical coatings. Macleod gives an account of some of the effects of this structure on the film properties, and on the origins of the structure.

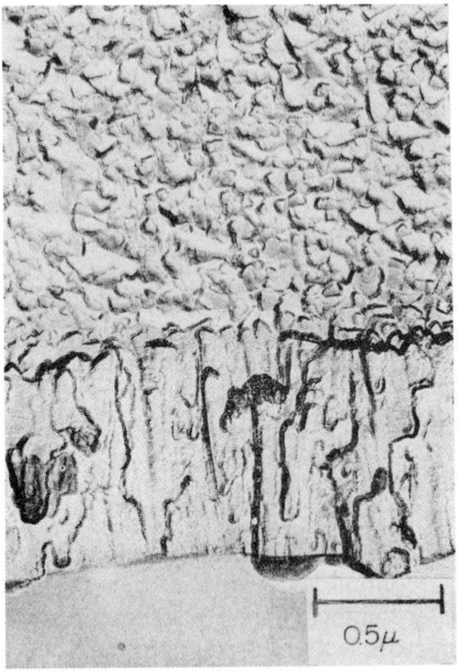

Figure 5.2 Columnar structure of a film. (From Okamoto K *et al* 1971 *J. Phys. Soc. Jpn.* **31** 1374.)

5.4.1(a) Epitaxial growth

A development which started in about 1960 is that of the epitaxial growth of films and this has now reached the stage where reliable single-crystal films can readily be produced for many materials; Matthews (1967) has published a review article on this subject. The basis of this technique is that the films are grown on a substrate whose crystal structure is a close match with what of the film material. If the substrate is also heated during deposition, the molecules of the film material will absorb sufficient thermal energy to keep reorientating themselves until they match their own crystal structure with that of the substrate, which is the condition of minimum potential energy. The result will he a film which is a single crystal and whose interface with the substrate is relatively smooth even on an atomic scale. Both these conditions will produce a reduction in electron scattering, enhance electron mobility and so affect the optical properties of the film.

In the earliest experiments epitaxy was attempted on substrate materials such as freshly cleaved rock-salt or mica mounted on an electric heating element. Temperatures of about 400 to 500 °C were needed to generate the necessary thermal mobility, and the quality of the film depended on the cleanliness of the substrate and the lack of contamination in the 'vacuum'.

A variation on the above procedure was to raise the temperature of the substrate to the melting point of the deposited material; this process being known as 'epitaxial melting'.

Modern techniques for growing epitaxial films, known as molecular beam epitaxy, make use of ultrahigh-vacuum systems capable of producing and maintaining vacua down to pressures of 10^{-10} Torr. Systems of this kind were described earlier in §5.2.2 and must be constructed of stainless steel and pumped by devices which do not contaminate the deposition chamber. Before deposition starts the system must be 'baked' at a temperature of 400 °C for up to 96 hours in order to outgas the walls of the vessel and its contents. Once this outgassing process has been carried out, the chamber is not allowed up to atmospheric pressure except on rare occasions for maintenance as the surfaces would immediately become contaminated with adsorbed gases and the whole of the outgassing procedure would have to be repeated. Consequently, the loading of the specimens is carried out in a second chamber and they are transferred mechanically under vacuum to the operational chamber by remote control.

Evaporation of the film materials takes place from separate ovens each provided with a remotely controlled shutter so that the stream of molecules can be rapidly cut off and the film thickness accurately controlled without having to shut down the oven. Mixtures or alloys can be deposited by opening the shutters of two or more ovens at the same time; if the components of the mixture are reactive, they can be deposited together in the correct proportions when the compound will be formed on the substrate.

Reviews covering the field of molecular beam epitaxy have been published by Wood (1980) and Joyce (1985).

5.4.1(b) Low dimensional structures—nanostructures

These devices are the outcome of a further development in thin film technology, in which the films produced are perfectly crystalline but have thicknesses of less than about 200 nm; they have been designated low dimensional structures or nanostructures. Devices made in this way can be expected to play an increasing role in the microminiaturisation of integrated circuits and hence in the improvement of response times.

A simple planar device such as that described above, in which the motion of the electrons will be severely constrained in the thickness direction, has been called a quantum well. It is also possible to produce quantum wires of very small diameter in which the electrons are constrained in two dimensions and quantum dots or small islands where confinement of motion takes place in all three dimensions.

The effect of confining the electrons in such physically small regions is to cause a reduction in the widths of their available energy bands (see section 1.6); these bands must still be able to accommodate all the electrons in the solid, so the density of states will be much greater than in solids of larger dimensions and this will lead to enhanced electrical and optical properties. In a review article, Perrenboom *et al* (1981) and Arada (1986) have discussed the electronic properties of metallic particles whose dimensions are only a few nm. The quasi-continuous energy bands are split into several narrow sub-bands and this leads to novel effects, the so-called 'quantum size effects'. One effect which has already been exploited is the change in band gap which is produced in quantum wires; thus the band gap becomes increasingly large as the wires get smaller in diameter.

Three main techniques are available for the production of nanostructures. The first is molecular beam expitaxy (MBE) which was described in the previous section; this technique has also been reviewed by Panish (1980) and described in detail by Parker (1985). In this process, the film materials are evaporated from specially designed ovens in a very good vacuum of about 10^{-9} Torr and then fall onto carefully cleaned and heated crystal substrates. The second technique is called liquid phase epitaxy (LPE) where a hot solution saturated with the material to be deposited is slowly cooled below the saturation temperature, crystallisation taking place on the heated seed crystalline substrate. In the third technique, chemical vapour deposition (CVD) or vapour phase epitaxy (VPE) the crystals are grown directly from the vapour phase; chemical reactions are produced in the vapour phase and the resulting compounds are deposited on a heated substrate. The reacting materials are carried in a stream of hydrogen and heated to the reaction temperature by a radio frequency unit; one variation of this method is MOCVD where the materials used are metal-organics such as metal alkyls. Dupuis (1984) has reviewed these VPE techniques.

In nanotechnology, the substrates used for deposition must be crystalline and atomically as near perfect as possible. They also must be contamination free and, having been formed and prior to use, they are kept in a separate vacuum chamber from which they can be transferred to the deposition chamber without exposure to air. During deposition, the substrates are heated to allow the deposited particles to gain sufficient kinetic energy so that they can move around freely until they find the position of least potential energy, which is generally in the crystal lattice pattern of the substrate.

Using one of the above techniques, atoms can be built up in a controlled manner layer by layer on the substrate; continuous monolayers can be produced rather than the island films which are the subject of chapter 8. It is also common practice to observe the film growth continuously by electron or ion microscopy and, in this way, much information on the patterns of film growth has been obtained. In addition, it is now possible to grow films which are virtually free from dislocations and these are available commercially.

At the present time, most interest is being directed towards nanostructures of semiconductor materials because of their importance in integrated microcircuits. One of the main problems encountered is that of heteroepitaxy where there is a mismatch between the basic crystallographic dimensions of the substrate and of the film; thus silicon and germanium have a 4% mismatch which is large enough to cause difficulties. It is then useful to put down one or more intermediate layers of an alloy of the two materials which have crystal lattice sizes between the two extremes.

Dingle *et al* (1974) have published absorption spectra of GaAs as a function of layer thickness as this was varied between 4000 and 50 Å; as the film became thinner, the first absorption maximum, due to excitation of electrons from the valence to the conduction band, shifted from 1.52 eV to 1.63 eV and, in addition, several subsidiary peaks appeared. These features may be explained in terms of the electrons forming narrow sub-bands in the thinner material, the main energy gap becoming wider as the thickness decreased. The electron-in-a-box theory described in chapter 1 gives a good account of these phenomena.

Quantum wires may be made by deposition through submicron gratings produced by lithography (Kojami *et al* 1990). Another technique for quantum wires is to make porous silicon from bulk material by partial electrochemical dissolution in electrolytes containing HF. The residual material consists of a forest of very fine silicon rods (the quantum wires) which have dimensions of microns in length and nanometers in width but having still retained the crystal structure of silicon (Bomchil *et al* 1989. It has been possible to stimulate red photoluminescence from material of this kind having wires about 4 nm in diameter so demonstrating the considerable change in bandgap which has taken place. Further reduction of the sizes of the wires leads progressively to orange, yellow and green photoluminescence. Review articles have been written by Sakoki (1992) and Weisbuch (1993).

5.4.2 Electrical properties of thin films

The electrical resistance of a metal is generated by the collisions encountered by the free conduction electrons as they proceed through the material under the influence of an externally applied electric field, such as that due to an electromagnetic wave. These collisions may occur between the conduction electrons and other electrons, lattice atoms, crystal boundaries or the boundaries of the metal. In terms of the free-electron gas theory a mean free-electron path may be postulated, and at temperatures of 300 K this is of the order of 10^{-8} m, i.e. it is much greater than the atomic spacing. If the metal consists of a large number of randomly oriented microcrystals the numerous crystal boundaries will make a considerable contribution to the electrical resistance. Similarly, a thin film constrained between the two interfaces with the atmosphere and the substrate will also have an increased resistance because of the high proportion of electron collisions with the film boundaries. It would therefore be expected that the electrical properties of thin metallic films would be significantly different from those of the bulk material and investigation of this phenomenon has been a fruitful field of research.

Another phenomenon connected with electrical resistance is the anomalous skin effect discussed in Chapter 1, and this clearly has considerable implications for thin films in which the 'skin' represents a major portion of the structure. For very high-frequency radiation ($\sim 10^{14}$ Hz) the absorption coefficients of metals reach very large values and the applied electric field is attenuated to a negligible value in a distance much less than the mean free path of an electron in the metal. Consequently, the electric field operates only on those electrons which have been deflected by collisions into the regions close to the interfaces and the resistance of the metal becomes very large as only a small proportion of all the electrons in the film will be deflected in this way. The anomalous skin effect is enhanced at low temperatures when the mean free paths of the electrons become very long and an even smaller proportion of them occupy the field penetration depth.

5.5 Optics of thin films

There are a number of alternative formulations for expressing the optical functions of a thin film on a substrate in terms of the optical constants of the film material and the substrate, the thickness of the film and the angle of incidence of the applied radiation. Amongst the better known formulae are those due to Meyer (1950), Schopper (1952), Boehme (1982), Born and Wolf (1975) and Abelès (1957). The optics of thin films have also received attention from Azzam (1985), who gives explicit equations for the polarising angles for a highly reflective substrate covered by a transparent thin film, and from Girard and Delisle (1982), concerning the propagation of light through a material of finite thickness separating two semi-infinite media, in particular the case where $d \sim \lambda$ is studied.

In the first instance the equations given by Born and Wolf will be derived using the formulae developed in Chapter 1. Brief accounts will then be given of the equations due to Schopper, Boehme and Abelès.

The equations for the optical functions at an interface derived in Chapter 1 will be applied to the general case of a thin absorbing film on an absorbing substrate; this can then readily be adapted to the more usual case involving a non-absorbing substrate. The optical situation for the case of the absorbing substrate is illustrated in figure 5.3. Radiation originating in a medium of refractive index n_1 is incident at an angle θ_1 on the film which is of thickness d_2 and complex refractive index $n_2 - \mathrm{i}k_2$. The radiation will be refracted at an angle θ_2 and, on meeting the substrate which is of complex refractive index $n_3 - k_3$ and thickness d_3, will be further refracted at an angle θ_3. Finally the beam re-emerges into the first medium, n_1, at the

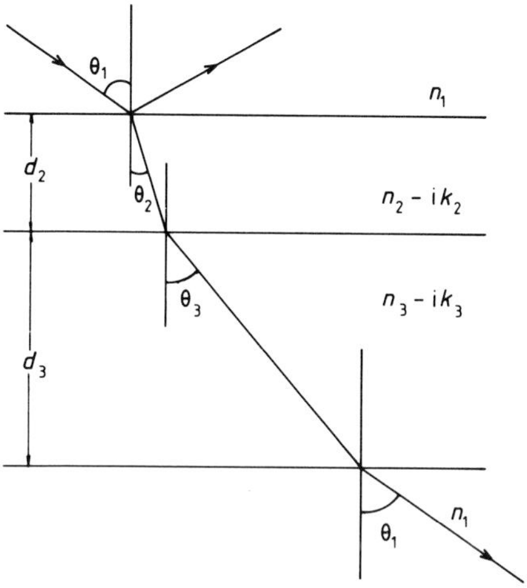

Figure 5.3 Thin absorbing film on an absorbing substrate.

angle θ_1. Both the film and the substrate will be considered to be isotropic, planar and of uniform thicknesses.

In §1.4.1, expressions were derived for the reflection and transmission coefficients at the interface between two media. Adopting the notation used in figure 5.3, these expressions are as follows for s-polarised radiation:

$$r_{1s} = \rho_{1s} \exp(i\phi_{1s}) = \rho_{1s}(\cos\phi_{1s} + i\sin\phi_{1s}) \tag{5.1a}$$

$$= \frac{n_1\cos\theta_1 - (u_2 - iv_2)}{n_1\cos\theta_1 + (u_2 - iv_2)} \tag{5.1b}$$

$$t_{1s} = \tau_{1s} \exp(i\chi_{1s}) = \tau_{1s}(\cos\chi_{1s} + i\sin\chi_{1s}) \tag{5.2a}$$

$$= \frac{2n_1\cos\theta_1}{n_1\cos\theta_1 + (u_2 - iv_2)} \tag{5.2b}$$

$$r_{2s} = \rho_{2s} \exp(i\phi_{2s}) = \rho_{2s}(\cos\phi_{2s} + i\sin\phi_{2s}) \tag{5.3a}$$

$$= \frac{u_2 - iv_2 - (u_3 - iv_3)}{u_2 - iv_2 + (u_3 - iv_3)} \tag{5.3b}$$

$$t_{2s} = \tau_{2s} \exp(i\chi_{2s}) = \tau_{2s}(\cos\chi_{2s} + i\sin\chi_{2s}) \tag{5.4a}$$

$$= \frac{2(u_2 - iv_2)}{(u_2 - iv_2) + (u_3 - iv_3)}. \tag{5.4b}$$

In the case of p-polarised radiation, the corresponding equations are

$$r_{1p} = \rho_{1p} \exp(i\phi_{1p}) = \rho_{1p}(\cos\phi_{1p} + i\sin\phi_{1p}) \tag{5.5a}$$

$$= \frac{(n_2^2 - k_2^2)\cos\theta_1 - n_1u_2 - i(2n_2k_2\cos\theta_1 - n_1v_2)}{(n_2^2 - k_2^2)\cos\theta_1 + n_1u_2 - i(2n_2k_2\cos\theta_1 + n_1v_2)} \tag{5.5b}$$

$$t_{1p} = \tau_{1p} \exp(i\chi_{1p}) = \tau_{1p}(\cos \chi_{1p} + i \sin \chi_{1p}) \tag{5.6a}$$

$$= \frac{2n_1(n_2 - ik_2)}{(u_2 - iv_2)n_1/\cos \theta_1 + (n_2 - ik_2)^2} \tag{5.6b}$$

$$r_{2p} = \rho_{2p} \exp(i\phi_{2p}) = \rho_{2p}(\cos \phi_{2p} + i \sin \phi_{sp}) \tag{5.7a}$$

$$= \frac{(u_2 - iv_2)(n_3 - ik_3)^2 - (u_3 - iv_3)(n_2 - ik_2)^2}{(u_2 - iv_2)(n_3 - ik_3)^2 + (u_3 - iv_3)(n_2 - ik_2)^2} \tag{5.7b}$$

$$t_{2p} = \tau_{2p} \exp(i\chi_{2p}) = \tau_{2p}(\cos \chi_{2p} + i \sin \chi_{2p}) \tag{5.8a}$$

$$= \frac{2(n_2 - ik_2)(n_3 - ik_3)}{(n_3 - ik_3)^2 + (n_2 - ik_2)^2(u_3 - iv_3)/(u_2 - iv_2)}. \tag{5.8b}$$

Heavens' (1965) treatment for the optics of thin films is illustrated in figure 5.4. The reflected and transmitted amplitudes at the various interfaces after a beam of unit amplitude is incident on the film at angle θ_1 are shown in the figure.

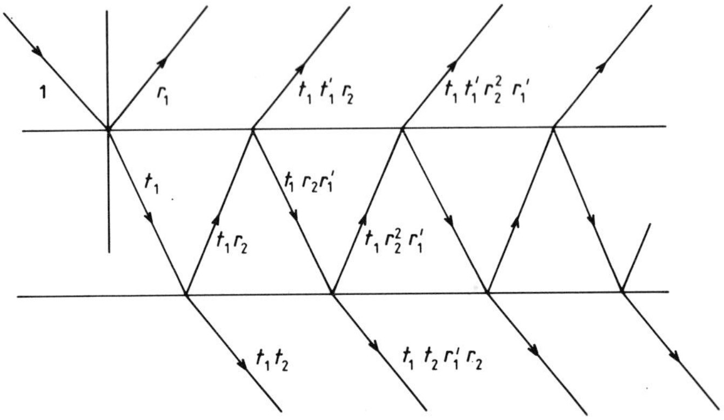

Figure 5.4 Multiple reflections and transmissions by a thin film.

The amplitudes of the successive reflected waves are

$$r_1, t_1t_1'r_2, t_1t_1'r_2^2, t_1t_1'r_1'^2r_2^2, \ldots \tag{5.9}$$

where r_1 is the reflection coefficient for s-polarised or p-polarised radiation appropriate to the first interface, r_2 that for the second interface and r_1' is that for reflection at the first interface from the substrate side; t_1, t_1' and t_2 are defined similarly for the transmission terms. Expressions for these functions are given in equations (5.1) to (5.8).

Now, $r_1' = -r_1$ so that these amplitudes transform to

$$r_1, t_1t_1'r_2, -t_1t_1'r_2^2, t_1t_1'r_1^2r_2^2, \ldots. \tag{5.10}$$

In addition, the phase change across the film is equal to

$$\delta_2 = n_2 d_2 \cos\theta_2\, 2\pi/\lambda = \mathrm{d}_2(u_2 - \mathrm{i}v_2)2\pi/\lambda \tag{5.11a}$$

$$= x_2(u_2 - \mathrm{i}v_2). \tag{5.11b}$$

Thus, the sum of the reflected amplitudes is given by

$$r = r_1 + t_1 t_1' r_2 \exp(-2\mathrm{i}\delta_2) - t_1 t_1' r_1 r_2^2 \exp(-4\mathrm{i}\delta_2)$$

$$+ t_1 t_1' r_1^2 r_2^3 \exp(-6\mathrm{i}\delta_2) - \cdots \tag{5.12a}$$

$$= \frac{r_1 + t_1 t_1' r_2 \exp(-2\mathrm{i}\delta_2)}{1 + r_1 r_2 \exp(-2\mathrm{i}\delta_2)} \tag{5.12b}$$

$$= \frac{r_1 + r_2 \exp(-2\mathrm{i}\delta_2)(t_1 t_1' + r_1^2)}{1 + r_1 r_2 \exp(-2\mathrm{i}\delta_2)}. \tag{5.12c}$$

Applying the principle of conservation of the energy of the wave,

$$t_1 t_1' = 1 - r_1^2. \tag{5.13}$$

Hence,

$$r = \frac{r_1 + r_2 \exp(-2\mathrm{i}\delta_2)}{1 + r_1 r_2 \exp(-2\mathrm{i}\delta_2)} \tag{5.14a}$$

$$= \frac{r_1 \exp(\mathrm{i}\delta_2) + r_2 \exp(-\mathrm{i}\delta_2)}{\exp(\mathrm{i}\delta_2) + r_1 r_2 \exp(-\mathrm{i}\delta_2)} \tag{5.14b}$$

$$= \frac{\rho_1 \exp(\mathrm{i}\phi_1) + \rho_2 \exp[\mathrm{i}(\phi_2 - 2x_2u_2)] \exp(-2x_2v_2)}{1 + \rho_1\rho_2 \exp[\mathrm{i}(\phi_1 + \phi_2 - 2x_2u_2)] \exp(-2x_2v_2)}. \tag{5.14c}$$

Similarly, the sum of the transmitted amplitudes is given by

$$t = t_1 t_2 \exp(-\mathrm{i}\delta_2) + t_1 t_2 r_1' r_2 \exp(-3\mathrm{i}\delta_2) + t_1 t_2 r_1'^2 r_2^2 \exp(-5\mathrm{i}\delta_2) + \cdots \tag{5.15a}$$

$$= t_1 t_2 \exp(-\mathrm{i}\delta_2)\,[1 - r_1 r_2 \exp(-2\mathrm{i}\delta_2) + (r_1 r_2)^2 \exp(-4\mathrm{i}\delta)] \tag{5.15b}$$

$$= \frac{t_1 t_2 \exp(-\mathrm{i}\delta_2)}{1 + r_1 r_2 \exp(-2\mathrm{i}\delta_2)} \tag{5.15c}$$

$$= \frac{\tau_1\tau_2 \exp[\mathrm{i}(\chi_1 + \chi_2 - u_2x_2)] \exp(-x_2v_2)}{1 + \rho_1\rho_2 \exp[\mathrm{i}(\phi_1 + \phi_2)] \exp[-\mathrm{i}(2x_2u_2)] \exp(-2x_2v_2)}. \tag{5.15d}$$

The reflected intensity (reflectance) is given by

$$R = rr^*$$

$$= \frac{\rho_1^2 \exp(2x_2v_2) + \rho_2^2 \exp(-2x_2v_2) + 2\rho_1\rho_2 \cos(\phi_1 - \phi_2 + 2x_2u_2)}{\exp(2x_2v_2) + \rho_1^2\rho_2^2 \exp(-2x_2v_2) + 2\rho_1\rho_2 \cos(\phi_2 + \phi_1 - 2x_2u_2)} \tag{5.16b}$$

and for the transmitted intensity (transmittance),

$$T_s = tt^* \bar{n}_3 \cos\theta_3 / n_1 \cos\theta_1 \tag{5.17a}$$

$$= \frac{n_3 \cos\theta_3 \tau_1^2 \tau_2^2}{n_1 \cos\theta_1 [\exp(2x_2 v_2) + \rho_1^2 \rho_2^2 \exp(-2x_2 v_2) + 2\rho_1 \rho_2 \cos(\phi_2 + \phi_1 - 2x_2 u_2)]} . \tag{5.17b}$$

In the expression for T_p, $n_3 \cos\theta_3 / n_1 \cos\theta_1$ is replaced by $(n_3/\cos\theta_3)/(n_1/\cos\theta_1)$.

Expressions for the phase changes on reflection and transmission at a thin film may also be obtained using equations (5.14b) and (5.15c) respectively. In the case of reflected radiation, the appropriate expressions for r_1 and r_2 from equations (5.1) and (5.3) for s-polarised waves and equations (5.5) and (5.7) for p-polarised radiation are fed into equation (5.14b) which is then manipulated to make the denominator become real. The ratio of the imaginary to the real parts in the numerator then gives the tangent of the phase change. Thus

$$\tan\phi_r = \mathscr{A}/\mathscr{B} \tag{5.18}$$

where

$$\begin{aligned} \mathscr{A} = {} & \rho_1 \sin\phi_1 [\exp(2v_2 x_2) - \rho_2^2 \exp(-2v_2 x_2)] \\ & + \rho_2 (1 - \rho_1^2) [\sin(\phi_2 - 2u_2 x_2)] \end{aligned} \tag{5.19}$$

and

$$\begin{aligned} \mathscr{B} = {} & \rho_1 \cos\phi_1 [\exp(2v_2 x_2) + \rho_2^2 \exp(-2v_2 x_2)] \\ & + \rho_2 (1 - \rho_1^2) [\cos(\phi_2 + 2u_2 x_2)] . \end{aligned} \tag{5.20}$$

The case for the phase change on transmission is somewhat more complicated and will be worked out in more detail. Starting with equation (5.15d), the amplitude of the transmitted wave may be written in the form

$$t = \frac{\tau_1 \tau_2 \exp(x_2 v_2) \exp[i(\chi_1 + \chi_2 - x_2 u_2)] \exp[i(2x_2 u_2)]}{\exp(2x_2 v_2) \exp[i(2x_2 u_2)] + \rho_1 \rho_2 \exp[i(\phi_1 + \phi_2)]} . \tag{5.21}$$

On rationalising this expression by multiplying top and bottom by the conjugate of the denominator, the numerator, N, becomes

$$\begin{aligned} N = {} & \tau_1 \tau_2 \exp(v_2 x_2) \exp[i(\chi_1 + \chi_2 + u_2 x_2)] \exp(2v_2 x_2) \exp(-2iu_2 x_2) \\ & + \rho_1 \rho_2 \exp[-i(\phi_1 + \phi_2)] . \end{aligned} \tag{5.22}$$

Now, the imaginary and real parts are extracted from equation (5.22), their quotient being equal to the tangent of the phase change on transmission, χ_t. Thus

$$\tan\chi_t = \frac{A \sin C - B \cos C}{B \sin C + A \cos C} \tag{5.23}$$

where

$$\begin{aligned} A &= \exp(2x_2v_2)\cos(2x_2u_2) + \rho_1\rho_2\cos(\phi_1+\phi_2) \\ B &= \exp(2x_2v_2)\sin(2x_2u_2) + \rho_1\rho_2\sin(\phi_1+\phi_2) \\ C &= \chi_1 + \chi_2 + x_2u_2. \end{aligned} \tag{5.24}$$

In equation (5.23), tan χ_t is of the form

$$\tan(C-D) = \tan(\chi_1+\chi_2+x_2u_2-D) \tag{5.25}$$

where

$$\tan D = \frac{B}{A} = \frac{\exp(2x_2v_2)\sin(2x_2u_2) + \rho_1\rho_2\sin(\phi_1+\phi_2)}{\exp(2v_2x_2)\cos(2x_2u_2) + \rho_1\rho_2\cos(\phi_1+\phi_2)}. \tag{5.26}$$

It should be noted that this phase change is measured from the first interface as in the case of reflection.

The phase changes on transmission for the s- and p-states of polarisation are obtained by inserting the appropriate forms of $\phi_1, \phi_2, \chi_1, \chi_2, \tau_1, \tau_2, \rho_1$ and ρ_2.

5.5.1 Non-absorbing film on absorbing substrate

This situation occurs with a thin oxide layer on a thick metal base; the waves will be rapidly attenuated to zero in the metal and the equations above for the various optical functions may be used by putting in a real refractive index for the film and a complex index for the substrate.

5.5.2 Absorbing film on non-absorbing substrate

This situation may also be dealt with by replacing the complex refractive index of the substrate by a real refractive index in the previous equations.

5.5.3 Substrate of finite thickness

In the treatment given in §5.5, no account has been taken of reflections at the lower surface of the substrate. Expressions for the optical functions of a thin film on a substrate of finite thickness may readily be obtained by treating the system as a double film. Effective optical functions for the system involving the substrate and lower interface of the film can be evaluated using the same analysis as that above; these are then used in place of the functions of the second interface for radiation incident from air on to the film. Thus, the reflected amplitude, using the notation in figure 5.3, is given by

$$r = \frac{r_1 + r_2\exp(-2\mathrm{i}\delta_2) + r_3\exp[-2\mathrm{i}(\delta_2+\delta_3)] + r_1r_2r_3\exp(-2\mathrm{i}\delta_3)]}{1 + r_1r_2\exp(-2\mathrm{i}\delta_3) + r_1r_3\exp[-2\mathrm{i}(\delta_2+\delta_3)] + r_2r_3\exp(-2\mathrm{i}\delta_3)} \tag{5.27}$$

whilst the transmitted amplitude is

$$t = \frac{t_1 t_2 t_3 \exp[-i(\delta_2 + \delta_3)]}{1 + r_1 r_2 \exp(-2i\delta_3) + r_1 r_3 \exp[-2i(\delta_2 + \delta_3)] + r_2 r_3 \exp(-2i\delta_3)} \quad (5.28)$$

The reflectance is given by the product rr^*. Similarly, the transmittance is equal to tt^* as both the incident and emergent beams are in the same medium. In the above equations r_3 and t_3 are the reflection/transmission coefficients at the final substrate/air interface.

5.5.4 Approximate formulae for optical functions

Abelès (1953, 1963) has derived expressions for the optical functions of a very thin film on a non-absorbing substrate in the form of power series in d/λ up to $(d/\lambda)^3$. These expressions were obtained by expanding the exponential and trigonometrical functions in equations (5.12) to (5.15) and collecting together the terms in each of the powers of d/λ. In general, these approximate equations are easier to handle numerically than the full equations and, as will be discussed later, they can lead to some interesting relations for ε_1 and ε_2. The ranges of d/λ over which the equations can be applied are rather limited (up to about $d/\lambda = 0.04$), but may be increased by considering terms in higher powers. However, these terms become increasingly complicated and soon offer no advantages over the full equations; in any case, the availability of computing facilities means that the full expressions can usually be readily evaluated. Abelès' equations are as follows:

$$R(R') = (A_0 \pm A_1 x \pm A_2 x^2 \pm A_3 x^3)/(B_0 + B_1 x + B_2 x^2 + B_3 x^3) \quad (5.29a)$$

$$T = 4C_0/(B_0 + B_1 x + B_2 x^2 + B_3 x^3) \quad (5.29b)$$

where, for s-polarisation, the values of the A and B coefficients are

$$A_0(B_0) = (Y_1 \mp Y_3)^2 \quad (5.30a)$$

$$A_1(B_1) = 2(Y_3 \mp Y_1)\varepsilon_2 \quad (5.30b)$$

$$A_2(B_2) = (\varepsilon_1 - \varepsilon_3)^2 + \varepsilon_2^2 - (Y_1^2 + Y_3^2)(\varepsilon_1 - \varepsilon_3) + Y_1^2 Y_3^2 \quad (5.30c)$$

$$A_3(B_3) = -\tfrac{2}{3}\varepsilon_2(Y_3 \mp Y_1)(\varepsilon_1 - \varepsilon_3 - Y_1 Y_3). \quad (5.30d)$$

The corresponding expressions for p-polarisation are;

$$A_0(B_0) = (Z_1 \mp Z_3)^2 \quad (5.31a)$$

$$A_1(B_1) = 2(Z_3 \mp Z_1)\varepsilon_2[1 \mp Z_1 Z_3 n_1^2 \sin^2\theta_1/(\varepsilon_1^2 + \varepsilon_2^2)] \quad (5.31b)$$

$$\begin{aligned} A_2(B_2) = {} & (Z_1^2 + Z_3^2)(\varepsilon_3 - \varepsilon_1) + \varepsilon_1^2 + \varepsilon_2^2 \\ & + \{Z_1^2 Z_3^2[(\varepsilon_1 - \varepsilon_3)^2 + \varepsilon_2^2]/(\varepsilon_1^2 + \varepsilon_2^2)\} \\ & \mp [4Z_1 Z_3 \varepsilon_3 \varepsilon \varepsilon_2^2/(\varepsilon_1^2 + \varepsilon_2^2)] \end{aligned} \quad (5.31c)$$

$$A_3(B_3) = \tfrac{2}{3}\varepsilon_2\{(Z_3 \mp Z_1)(2\varepsilon_3 - \varepsilon_1) \mp Z_1 Z_3[1 + (2\varepsilon_2^2 - 3\varepsilon_1\varepsilon_2)/(\varepsilon_1^2 + \varepsilon_2^2)]\}. \tag{5.31d}$$

$C_0 = Y_1 Y_3$ for s- and $Z_1 Z_3$ for p-polarisation. It may be noted that, when dealing with the case of s-polarisation, and considering only the terms up to A_2 and B_2, the quantity $(1 - R' - T)/(1 - R - T)$ becomes equal to Y_s/Y_1, i.e. a constant. There is thus an additional relationship between the three photometric quantities R, R' and T, known as Wolter s relationship (Wolter 1937), and it implies that n, k and d are not independent. It is therefore not possible to determine all three parameters from measurements of R, R' and T in the s-state of polarisation when the films are sufficiently thin for Abelès' equations to hold. This restriction does not apply in the p-state as the ratio above does not reduce to a constant and, theoretically, measurements in the p-state allow the determination of all three parameters. However, such determinations will still not in general be well conditioned and this is a common situation when attempts are made to calculate the optical constants and thickness from photometric measurements.

Abelès (1957) also provided a set of equations for the phase changes on reflection and transmission at thin films and these are given below

$$\tan\phi_s = \frac{2Y_1 x(n_3^2 - \varepsilon_1 + Y_3\varepsilon_2 x)}{(n_1^2 - n_3^2) - 2Y_3\varepsilon_2 x + x^2[Y_1^2 Y_3^2 + (n_3^2 - n_1^2)(\varepsilon_1 - S^2) - (\varepsilon_1 - S^2)^2 - \varepsilon_2^2]} \tag{5.32}$$

$$\tan\chi_s = \frac{-x[Y_1 Y_3 - S^2 + \varepsilon_1 + \tfrac{1}{2}(Y_1 + Y_3)\varepsilon_2 x]}{Y_1 + Y_3 + \varepsilon_2 x - \tfrac{1}{2}(Y_1 + Y_3)(\varepsilon_1 - S^2)x^2)x^2} \tag{5.33}$$

$$\tan\phi_p = \frac{2Z_1 x\{Z_3^2(\varepsilon_1^2 + \varepsilon_2^2) - \varepsilon_1[\varepsilon_1^2 + \varepsilon_2^2 + (Z_3 S)^2] + Z_1\varepsilon_2 x(\varepsilon_1^2 + \varepsilon_2^2 - 2\varepsilon_1 S^2)\}}{\{(Z_1^2 - Z_3^2)(\varepsilon_1^2 + \varepsilon_2^2) + 2Z_3\varepsilon_2 x[(Z_1 S)^2 - (\varepsilon_1^2 + \varepsilon_2^2)] + x^2\{(Z_1^2 - Z_3^2)(S^2 - \varepsilon_1)(\varepsilon_1^2 + \varepsilon_2^2) + Z_1^2 Z_3^2[(\varepsilon_1 - S^2)^2 + \varepsilon_2^2] - (\varepsilon_1^2 + \varepsilon_2^2)^2\}\}} \tag{5.34}$$

$$\tan\chi_p = \frac{-x[(Z_1 Z_3 + \varepsilon_1)(\varepsilon_1^2 + \varepsilon_2^2) - Z_1 Z_3\varepsilon_1 S^2 + \tfrac{1}{2}\varepsilon_2 x(Z_1 + Z_3)(\varepsilon_1^2 + \varepsilon_2^2)]}{(Z_1 + Z_3)(\varepsilon_1^2 + \varepsilon_2^2) + x\varepsilon_2(\varepsilon_1^2 + \varepsilon_2^2 + Z_1 Z_3 S^2) + \tfrac{1}{2}x^2(Z_1 + Z_3)(\varepsilon_1^2 + \varepsilon_2^2)(S^2 - \varepsilon_1^2)} \tag{5.35}$$

where

$$S = n_1 \sin\theta_1 = \varepsilon_3^{1/2}. \tag{5.36}$$

Another set of equations from which the reflectances and transmittances of thin films may be obtained has been given by Schopper (1952) and subsequently developed by Boehme (1982). In these equations R, R' and T are expressed in terms of trigonometrical and hyperbolic quantities. Boehme's formulae for the optical functions of a thin film are as follows:

$$R = \bar{A}/\bar{N} \quad (5.37)$$

$$R' = \bar{B}/\bar{N} \quad (5.38)$$

$$T = \bar{C}/\bar{N} \quad (5.39)$$

where

$$\bar{A} = \cosh(2\beta + I_1 - I_3) - \cos(2\alpha + J_1 - J_3) \quad (5.40a)$$

$$\bar{B} = \cosh(2\beta - I_1 + I_3) - \cos(2\alpha - J_1 + J_3) \quad (5.40b)$$

$$\bar{C} = \cosh(I_1 + I_3) - \cos(J_1 + J_3) - \cosh(I_1 - I_3) + \cos(J_1 - J_3) \quad (5.40c)$$

$$\bar{N} = \cosh(2\beta + I_1 + I_3) - \cos(2a + J_1 + J_3) \quad (5.40d)$$

$$2\alpha = 2a2\pi d/\lambda \qquad 2\beta = 2b2\pi d/\lambda \quad (5.41a)$$

$$= 2u_2x_2 \qquad = 2v_2x_2 \quad (5.41a)$$

$$a^2 - b^2 = n_2^2 - k_2^2 - n_3^2 \sin^2\theta_3 \quad (5.42)$$

$$2ab = 2n_2k_2 \quad (5.43)$$

$$I_1 = \operatorname{arctanh}[2an_1\cos\theta_1/(a^2 + b^2 + n_1^2\cos^2\theta_1)] \quad (5.44a)$$

$$I_3 = \operatorname{arctanh}[2an_3\cos\theta_3/(a^2 + b^2 + n_3^2\cos^2\theta_3)] \quad (5.44b)$$

$$J_1 = \arctan[2bn_1\cos\theta_1/(a^2 + b^2 - n_1^2\cos^2\theta_1)] \quad (5.44c)$$

$$J_3 = \arctan[2bn_3\cos\theta_3/(a^2 + b^2 - n_3^2\cos^2\theta_3)]\,. \quad (5.44d)$$

The corresponding expressions for the phase changes are

$$\tan\phi = \frac{-[\sinh(2b + I_3)\sin J_1 - \sin(2a + J_3)\sinh I_1]}{\cosh(2b + I_1)\cos J_3 - \cos(2a + J_3)\cosh I_1} \quad (5.45)$$

$$\tan(\phi' - 2g) = \frac{-[\sinh(2b + I_1)\sin J_3 - \sin(2a + J_1)\sinh I_3]}{\cosh(2b + I_1)\cos J_3 - \cos(2a + J_1\cosh I_3} \quad (5.46)$$

$$\tan(\chi - g) = [\sinh(b + I_1 + I_3)\sin a - \sin(a + J_1 + J_3)\sinh b + \sinh(b + I_1)$$
$$\times \sin(a + J_3) - \sin(a + J_1)\sinh(b + I_3)]/$$
$$[\cosh(b + I_1 + I_3)\cos a - \cos(a + J_1 + J_3)\cosh b + \cosh(b + I_1)$$
$$\times \cos(a + J_3) - \cos(a + J_1)\cosh(b + I_3)] \quad (5.47)$$

here

$$g = n_3\cos\theta_3\, 2\pi d/\lambda \quad (5.48)$$

When expanded, and the terms in pure numbers, x, x^2, x^3 etc are collected, these expressions lead to the same formulae as those given by Abelès.

5.6 Matrix methods for calculating *R* and *T* for thin films

Another powerful technique for calculating the optical properties of films, which is particularly useful when dealing with multi-layers, is to employ matrices.

The relationship between the magnitudes of the electric vectors on either side of a discontinuity may be expressed in matrix form as follows:

$$\begin{bmatrix} E_1^+ \\ E_1^- \end{bmatrix} = \begin{bmatrix} P & Q \\ S & W \end{bmatrix} \begin{bmatrix} E_2^+ \\ E_2^- \end{bmatrix} \tag{5.49}$$

where the positive signs represent the transmitted wave and the negative signs the reflected wave; P, Q, S and W are the coefficients of the matrix. Figure 5.5 shows the E and H components of a plane wave passing through a thin film which has boundaries a, with medium of refractive index n_0 and b with medium of refractive index n_2; the film itself of thickness d has a refractive index of n_1. All the refractive indices may be complex. Across the boundary between any two media the tangential components of E and H must be continuous (section 1.3). Transmitted waves are treated as positive whilst reflected waves are negative.

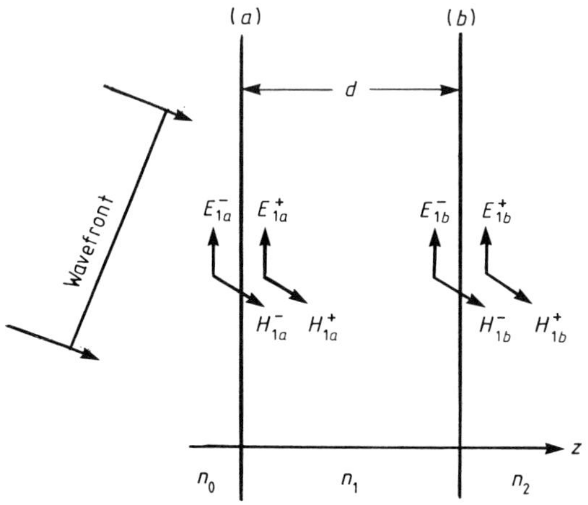

Figure 5.5 *E* and *H* components of a plane wave passing through a thin film.

Consider the situation at boundary *b*; it is assumed that the substrate is infinite so there is no negative wave in n_2.

Thus:

$$E_b = E_{2b}^+ + E_{1b}^- \quad (5.50a)$$

and

$$H_b = n_1 E_{1b}^+ - n_1 E_{1b}^- \quad (5.50b)$$

where $n_1 = n_1/\cos\theta$ for p-waves and $n_1 \cos\theta$ for s-waves.

Then;

$$E_{1b}^+ = \tfrac{1}{2}(E_b - H_b/n_1) \quad (5.51a)$$

$$E_{1b}^- = \tfrac{1}{2}(E_b - H_b/n_1) \quad (5.51b)$$

$$H_{1b}^+ = \tfrac{1}{2}(n_1 E_b + H_b) \quad (5.51c)$$

$$H_{1b}^- = \tfrac{1}{2}(H_b - n_1 E_b) \quad (5.51d)$$

The situation at the a boundary may now be obtained by multiplying the expressions above by the appropriate phase factor: exp(id) for the positive moving waves and exp($-id$) for the negative moving waves where $d = 2\pi n_1 d \cos\theta_1/\lambda$. If n_1 is complex, $d = 2\pi(u - iv)d/\lambda$.

Thus:

$$E_{1a}^+ = \tfrac{1}{2}(E_b + H_b/n_1)\exp(id) \quad (5.52a)$$

$$E_{1a}^- = \tfrac{1}{2}(E_b - H_b/n_1)\exp(-id) \quad (5.52b)$$

$$H_{1a}^+ = \tfrac{1}{2}(n_1 E_b + H_b)\exp(id) \quad (5.52c)$$

$$H_{ia}^- = \tfrac{1}{2}(H_b - n_1 E_b)\exp(-id). \quad (5.52d)$$

Then:

$$E_a = E_{1a}^+ + E_{1a}^-$$
$$= E_b((\exp(id) + \exp(-id))/2 + H_b((\exp(id) - \exp(-id))/2 \quad (5.53)$$

and

$$H_a = H_{1a}^+ + H_{1a}^-$$
$$= E_b n_1(\exp(id) - \exp(-id))/2 + H_b(\exp(id) - \exp(-id))/2 \quad (5.54)$$

So

$$E_a = E_b \cos d + H_b/n_1 \mathrm{i} \sin d \quad (5.55a)$$

$$H_a = \mathrm{i} n_1 E_b \sin d + H_b \cos d. \quad (5.55b)$$

These two equations may be expressed in matrix form as follows

$$\begin{vmatrix} E_a \\ H_a \end{vmatrix} = \begin{vmatrix} \cos d & \mathrm{i}/n_1 \sin d \\ \mathrm{i}\, n_1 \sin d & \cos d \end{vmatrix} \begin{vmatrix} E_b \\ H_b \end{vmatrix} . \tag{5.56}$$

The matrix is known as the characteristic matrix of the film; its determinant is equal to unity. The optical admittance of the film is given by the ratio $Y = H_a/E_a$.

Another important result can be obtained from the above mathematics. From equations (5.50) and (5.52) the following relations apply:

$$E_{1a}^{+}/E_{1b}^{+} = H_{1a}^{+}/H_{1b}^{+} = \exp(\mathrm{i}d) \tag{5.57a}$$

$$E_{1a}^{-}/E_{1b}^{-} = H_{1a}^{-}/H_{1b}^{-} = \exp(-\mathrm{i}d). \tag{5.57b}$$

These equations may be expressed in matrix form to give:

$$\begin{vmatrix} E_{1a}^{+} \\ E_{1a}^{-} \end{vmatrix} = \begin{vmatrix} \exp(\mathrm{i}d) & 0 \\ 0 & \exp(-\mathrm{i}d) \end{vmatrix} \begin{vmatrix} E_{1b}^{+} \\ E_{1b}^{-} \end{vmatrix} \tag{5.58a}$$

$$\begin{vmatrix} H_{1a}^{+} \\ H_{1a}^{-} \end{vmatrix} = \begin{vmatrix} \exp(\mathrm{i}d) & 0 \\ 0 & \exp(-\mathrm{i}d) \end{vmatrix} \begin{vmatrix} H_{1b}^{+} \\ H_{1b}^{-} \end{vmatrix} . \tag{5.58b}$$

The matrix

$$\begin{vmatrix} \exp(\mathrm{i}d) & 0 \\ 0 & \exp(-\mathrm{i}d) \end{vmatrix}$$

represents the film itself.

It may be noted that the determinant of this matrix is again equal to 1.

Similarly, it may be shown (Hayfield and White 1964) that the characteristic matrix for a plane interface is:

$$\begin{vmatrix} 1/t & r/t \\ r/t & 1/t \end{vmatrix} = \frac{1}{t} \begin{vmatrix} 1 & r \\ \mathrm{r} & 1 \end{vmatrix} \tag{5.59}$$

where r and t are the reflectance and transmittance of the interface; once again the determinant is equal to 1.

Making use of the multiplicative properties of matrices, the matrix representing a thin film is the product of three matrices; that for the air-film interface, that for the slab of film and that for the film substrate interface. Thus, using the previous notation, this matrix becomes

$$\frac{\exp(\mathrm{i}\delta_2)}{t_1 t_2} \begin{vmatrix} 1 & r_1 \\ r_1 & 1 \end{vmatrix} \begin{vmatrix} 1 & 0 \\ 0 & \exp(-2\mathrm{i}\delta_2) \end{vmatrix} \begin{vmatrix} 1 & r_2 \\ r_2 & 1 \end{vmatrix} \tag{5.60}$$

$$= \frac{\exp(\mathrm{i}\delta_2)}{t_1 t_2} \begin{vmatrix} 1 + r_1 r_2 \exp(-2\mathrm{i}\delta_2) & r_2 + r_1 \exp(-2\mathrm{i}\delta_2) \\ r_1 + r_2 \exp(-2\mathrm{i}\delta_2) & r_1 r_2 + \exp(-2\mathrm{i}\delta_2) \end{vmatrix} . \tag{5.61}$$

The ratio of reflected to incident amplitudes for radiation incident on the film is given by the ratio S/P in this matrix, i.e. by

$$[r_1 + r_2 \exp(-2i\delta_2)] / [1 + r_1 r_2 \exp(-2i\delta_2)]$$

and the ratio of transmitted to incident amplitudes by $1/(P \times$ matrix multiplier), namely

$$t_1 t_2 \exp(-i\delta_2) / [1 + r_1 r_2 \exp(-2i\delta_2)] .$$

These relationships are the same as those obtained in equations (5.14a) and (5.15a) respectively.

In order to calculate the effect of a substrate of finite thickness d_3, it is only necessary to multiply equation (5.61) by the product of the matrices due to the substrate slab and the substrate–air interface, i.e. by

$$\exp(i\delta_3)/t_3 \begin{vmatrix} 1 & 0 \\ 0 & \exp(-2i\delta_3) \end{vmatrix} \begin{vmatrix} 1 & r_3 \\ r_3 & 1 \end{vmatrix} \tag{5.62}$$

$$= \exp(i\delta_3)/t_3 \begin{vmatrix} 1 & r_3 \\ r_3 \exp(-2i\delta_3) & \exp(-2i\delta_3) \end{vmatrix} . \tag{5.63}$$

The following matrix is then obtained

$$\frac{\exp[i(\delta_2 + \delta_3)]}{t_1 t_2 t_3} \times \begin{vmatrix} 1 + r_1 r_2 \exp(-2i d_2) + r_3 \exp(-2i d_3) [r_2 + r_1 \exp(-2i d_2)] & r_3 + \exp(-2i\delta_3) [r_2 + r_1 \exp(2i\delta_2)] \\ r_1 + r_2 \exp(-2i d_2) + [r_1 r_2 + \exp(-2i d_2) r_3 \exp(-2i d_3)] & r_3 [r_1 r_2 r_3 \exp(-2i\delta_2)] + [r_1 r_2 \exp(-2i\delta_2) \exp(-2i\delta_3)] \end{vmatrix} \tag{5.64}$$

This leads to the following expressions for r and t

$$r = \frac{r_1 + r_2 \exp(-2i\delta_2) + [r_1 r_2 + \exp(-2i\delta_2)] r_3 \exp(-2i\delta_3)}{1 + r_1 r_2 \exp(-2i\delta_2) + r_3 \exp(-2i\delta_3) [r_2 + r_1 \exp(-2i\delta_2)]} \tag{5.65}$$

and

$$t = \frac{t_1 t_2 t_3}{\exp[i(\delta_2 + \delta_3)] [1 + r_1 r_2 \exp(-2i\delta_2)] + r_3 \exp(-2i\delta_3) [r_2 + r_1 \exp(-2i\delta_2)]} . \tag{5.66}$$

Once again these results are identical to those in equations (5.27) and (5.28).

The effects of further layers on the optical functions may be found in the same manner by multiplying by the product of the matrices of the layer and its second interface. The matrix technique is thus a convenient process to calculate the optical properties of a multilayer film.

References

Abelés F 1953 *Rev*. Opt. **32** 257

—— 1957 *J. Opt. Soc. Am.* **47** 473

—— 1963 *Prog. Opt.* **II** 249

Allen T H 1985 *J. Opt. Soc. Am.* A **2** P92

Anderson G S, Mayer W N and Wehner G K 1962 *J. Appl. Phys.* **32** 2991

Arada M 1986 *J. Quantum Electron.* **22** 29

Azzam R M A 1985 *J. Opt. Soc. Am.* A **2** 480

Behrndt K H 1966 *Phys. Thin Films* **3** 1

Bennett H E and Bennett J M 1967 *Phys. Thin Films* **4** 1

Bennett J M and Dancy J H 1981 *Appl. Opt.* **20** 1785

Boehme H 1982 *Optik* **62** 27

Bomchil G, Halimaoui A and Hernio R 1989 *Appl. Surf. Sci.* **41/42** 604

Born M and Wolf E 1975 *Principles of Optics* (Oxford: Pergamon) 5th edn

Campbell D S 1976 *Thin Solid Films* **32** 3

Carey R, Thomas B W J and Newman D M 1985 *J. Phys. D: Appl. Phys.* **18** 269

Caswell H L 1963 *Solid State Tech.* (December)

Cox J T, Hass H and Ramsay J B 1964 *J. Phys.*, Paris **25** 250

Dingle R, Wiegmann W and Henry C H 1974 *Phys. Rev. Lett.* **33** 827

Dirks A G and Leamy H J 1977 *Thin Solid Films* **47** 219

Dupuis R D 1984 *Science* **226** 623

Eschbach H L and Verhoyen F 1974 *Thin Solid Films* **21** 237

Feist W M, Steele S R and Reading D W 1969 *Phys. Thin Films* **5** 237

Franks J 1981 *Thin Solid Films* **86** 219

Gates A D and Robins J L 1987 *Thin Solid Films* **149** 113

Girard A and Delisle C 1982 *Opt. Acta* **29** 611

Hass G and Ramsey J B 1969 *Appl. Opt.* **8** 1115

Hayfield P C S and White G W T 1964 *Ellipsometry in the Measurement of Surfaces and Thin Films* ed E Passaglia, R R Stromberg and J Kruger NBS Miscellaneous Publication 256 (Washington. DC: US Government Printing Office)

Heavens O S 1959 *J. Sci. Instrum.* **26** 951

—— 1965 *Optical Properties of Films* (New York: Dover)

Holland L 1961 *Vacuum Deposition of Thin Films* (London: Chapman and Hall)

Hunt R P 1967 *J. Appl. Phys.* **38** 1652

Hurley R E and Williams E W 1982 *Thin Solid Films* **92** 99

Jacobson R 1975 *Phys. Thin Films* **8** 51

Joyce B A 1985 *Rep. Prog. Phys.* **48** 1637

Kojami K, Mitsunaja K and Kyuma E 1980 *Japan. J. Appl. Phys.* **19** L735

Lewis B and Anderson J S 1978 *Nucleation and Growth of Thin Films* (New York: Academic)

Macleod H A 1982 *Proc. SPIE Int. Soc. Opt. Eng.* **325** 21

Maissel L 1 1966 *Phys. Thin Films* **3** 61

Martin P J, Netterfield R P and Sainty W G 1984 *J. Appl. Phys.*s **55** 235

Martin P J and Netterfield R P 1986 *Prog. Opt.* **XXIII** 113

Matthews J W 1967 *Phys. Thin Films* **4** 137

Mattox E M 1964 *Electrochem. Tech.* **2** 295

McNeil J R, Barron A C, Wilson S R and Herrmann W C 1984 *App. Opt.* **23** 252
McNeil J R, Al-Jumaily G A, Barron A C and McNally J J 1985 *Proc 5th Int. conf. on Ion and Plasma Assisted Techniques, Munich 1985* ed H Oechsner (Edinburgh, CEP Consultants Ltd; p 63
Meyer H 1950 *Physik Dunner Schichten*. Wissenschaftliche Verlagsgesellschaft mbH Stuttgart, Teil 1 pp 207, 218
Nag A and Ward L 1971 *J. Phys. D: Appl. Phys.* **4** 829
Olsen L O, Smith C S and Critenden E C 1945 *J. Appl. Phys.* **16** 425
Panish M B 1980 *Science* **208** 916
Parker E C A 1988 *The Physics and Technology of MBE*, (New York: Plenum)
Patten J W, Moss R W, McClanahan E D and Pawlewiez W T 1981 *Thin Solid Films* **83** 3
Perenboom J A A J, Wyder P and Maier F 1981 *Phys. Rep.* **78** 173
Piegari A and Masetti E 1985 *Thin Solid Films* **124** 249
Sakaki H 1992 *Surf. Sci.* **267** 623
Samson J A R, Padur J P and Sharma A 1967 *J. Opt. Soc. Am.* **57** 966
Schmeisser H 1974 *Thin Solid Films* **22** 83
Schopper H 1952 *Z. Phys.* **131** 215
Seraphin B O 1962 *J. Electrochem. Soc.* **109** 877
Smith H M and Turner A F 1965 *Appl. Opt.* **4** 147
Takagi T 1982 *Thin Solid Films* **92** 1
Teer D G 1988 *The Physics of Ionised Gases SPIG'88 ed L Tanovic, N Konjevic and N Tanovic (New York: Nova Scince Inc.) p 349*
Thun R E 1963 *Phys. Thin Films* **1** 187
Turner J A. Birtwistle J K and Hoffmann G R 1963 *J. Sci. Instrum.* **40** 557
Velfe H D, Stenzil H and Krohn M 1982 *Thin Solid Films* **98** 115
Venables J A, Spiller G D T and Hanbrucken M 1984 *Rep. Prog. Phys.* **47** 399
Weisbuch C 1993 *J. Crystal Growth* **127** 742
Wood C E C 1980 *Phys. Thin Films* **11** 35
Wolter H 1937 *Z. Phys.* **105** 260
Younger P R 1985 *J. Vac. Sci. Tech.* **A 3** 588

6 Determination of the Optical Constants of Thin Films

6.1 Introduction

The main groups of experimental techniques used for measuring the optical constants of bulk materials, namely photometry, polarimetry and ellipsometry, may also be employed, in the same or modified forms, when using thin-film specimens. In addition, the optical functions associated with the transmitted radiation are no longer negligibly small for metal films and so a wider range of techniques becomes available. Thus, in photometry, the transmittance, T, in either the s- or p-state of polarisation may be used. In polarimetry, the phase changes on transmission, χ_s and χ_p, may be determined, whilst transmission ellipsometry is also a successful technique.

A number of review articles have been published concerning the measurement of the optical properties of thin films and these may be referred to for greater experimental detail. Thus, Heavens (1964, 1965) gave critical analyses of several techniques for measurements in this field. Bousquet and Rouard considered the measurement techniques under three headings: spectrophotometry, interferometry and polarimetry (Bousquet and Rouard 1960, Rouard and Bousquet 1964). They believed that normal-incidence techniques were preferable because of the generally non-homogeneous nature of the films and the presence of a transition layer at the film–air interface. This layer becomes particularly troublesome for measurements taken near the Brewster's angle, rendering of dubious quality those results in which this angle has been involved. Anisotropy is also a problem with films where the crystallites might be uniaxial with their optic axes at right angles to the substrate; again normal-incidence methods would overcome this difficulty. These authors were also of the opinion that polarimetry methods were the most difficult to use with non-homogeneous and non-isotropic films. Abelès (1971), however, considered ellipsometry to be a very sensitive method for the study of very thin films on glass substrates. On the other hand, he noted that the photometric functions are not well discriminated, particularly at normal incidence, and from this point of view a preference for oblique incidence would be indicated. Abelès also considered the granular nature of the films and its effect on the optical constants; this matter is discussed in more detail in Chapter 8.

6.2 Photometric techniques

If it is required to determine the film thickness, d, in addition to n and k, three independent optical functions must be measured. Many investigations have utilised (i) the reflectance, R, at the air or vacuum side of the film, (ii) the reflectance, R', at the substrate side and (iii) the transmittance, T, all measured at the same angle of incidence in either the s- or p-state of polarisation. Experimental techniques similar to those described in Chapter 2 may be used and figure 6.1 shows a suitable arrangement, due to Kuhn and Wilson (1950), by which such measurements may be made; the angle of incidence can be made variable as an additional experimental parameter. Another absolute method for measuring R and T for thin films has been described by Fray *et al* (1963) and is shown in figure 6.2. Radiation from the exit slit, E, of the spectrometer falls on the plane mirror, which is used to switch the light path between the concave mirrors M_1 and M_2. In the M_1 position, the detector, D_1, measures the reflectance, R, of the specimen, S,

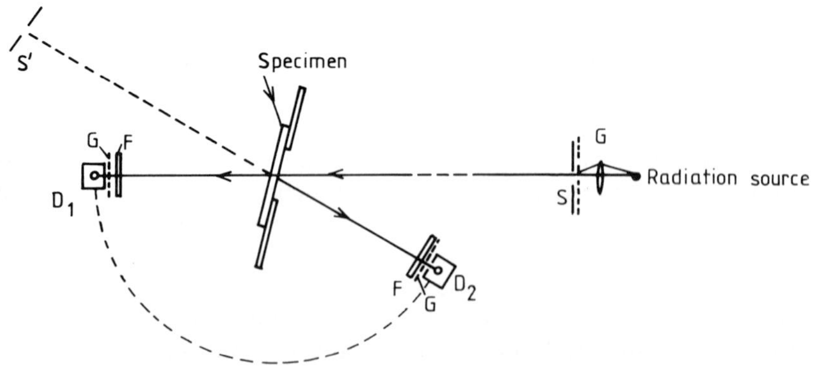

Figure 6.1 Kuhn and Wilson's apparatus for R, R' and T. S, slit; S′, virtual image of slit formed by specimen; D_1, D_2, positions of detector for measurements of T and R; F, filter: G, ground glass screen. (From Kuhn H and Wilson B A 1950 *Proc. Phys. Soc.* B **63** 746.)

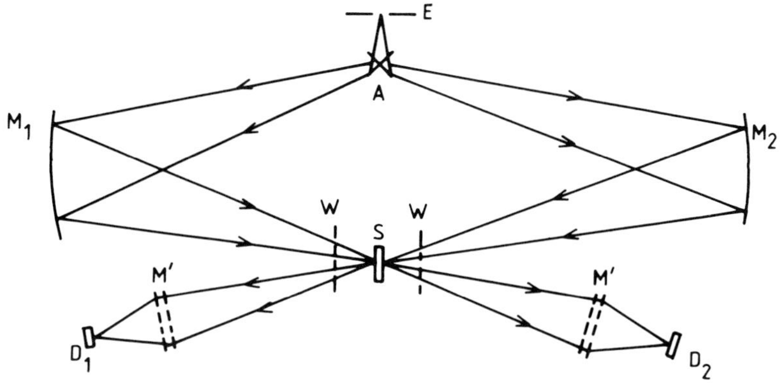

Figure 6.2 Apparatus of Fray *et al* for R, R' and T. E, exit slit of spectrometer; A, pair of plane mirrors; M_1, M_2, concave mirrors; M', representations of the focusing mirrors; W, windows, D_1, D_2, detectors; S, specimen. (From Fray S J *et al* 1963 *J. Sci. Instrum.* **40** 387.)

whilst D_2 measures the transmittance. When switched to M_2, D_1 measures the transmittance and D_2 measures the reflectance R' from the substrate side.

As with bulk specimens, having measured the optical functions, expressions for which were given in explicit form in §5.5, it is in making the reverse calculation to find n and k where difficulties arise as, in general, the equations cannot be solved analytically. Considerable simplification is possible, however, if measurements are made at normal incidence, which is another reason for preferring this configuration. However, there may be several pairs of values of n and k which satisfy the measured functions, these being usually found by some kind of iteration procedure; conversely, the iteration processes may fail to find the correct values of n and k. Multiple and missing solutions are amongst the major problems in this field.

Malé (1964) devised a geometrical method for calculating n and k from measurements of R, R' and T. A large number of graphs were constructed where the calculated values of R, R' and T were plotted against the film thickness for different pairs of n and k. Then, starting with the experimentally determined value of T, the corresponding values of R, R', n and k were read off each graph. A second set of graphs was then prepared plotting these values of R and R' against k for the different values of n. This enabled the various combinations of n and k corresponding to the measured values of R and of R' to be read off and then plotted on a third graph. The intersection of the two curves due to R and R' on this graph then gave the values of n and k which satisfied all three parameters. This process could also be used to determine the thickness of the film by referring back to the first set of graphs. The whole process was clearly exceedingly tedious, and in any case might not be very accurate owing to the lack of discrimination between the three optical functions leading to the final intersection being at a very acute angle. Later in this chapter (§6.5) the inherent accuracies of the techniques for measuring n and k of thin films will be considered in detail, and it will be seen that in combinations involving R and R' long, thin error parallelograms are generated and there are few regions of n, k space where high accuracy can be expected; in addition, the combination of R and T, although giving more extensive regions of high accuracy, is still subject to some limitations. Malé's technique may be used to solve for only two optical functions if the thickness can be determined separately, and this will bring about a considerable simplification in the processing of the results.

A similar photometric method was used by Murmann (1933), who measured R and R' and also, independently, the thickness of the film. A set of graphs was prepared of calculated values of R and R' against n for different values of k and, marking in the measured values of these two functions, the interpolated sets of pairs of n and k could be read off. As with Malé's method, a final graph of R and R' in the n, k plane was needed to determine n and k. In general it was found that these final curves intersected twice and some additional information was needed to find the unique solution. This might take the form of previous knowledge of the approximate solution or else the transmittance could be measured and a similar search carried out to find where all three lines crossed.

Computational methods of seeking solutions to this kind of problem by iterative techniques have now superseded geometrical solutions of the type described above. In one of these processes (Ward *et al* 1969), n, k and d are varied to seek a minimum in the function

$$(X - X_c)^2 + (Y - Y_c)^2 + (Z - Z_c)^2$$

where X, Y, Z are the measured values of the optical functions and X_c, Y_c, Z_c are the values calculated from the appropriate form of Fresnel's equations. Iterative procedures have also been described by Harris and Loeb (1955), Bennett and Booty (1966) Hunter (1965) and Casset (1977).

Paulick (1986) has described in algorithm for use in solving for n and k for thin films from experimental values of R and T; several steps were involved. Firstly, on a chosen n, k grid, values of T were computed. Then, using inverse interpolation, values of k were found which generated T_{exp} at each value of n on the mesh; this produced a set of $k(nT_{exp})$. At each value of n, $R(n, k(n))$ was computed and a cubic spline interpolation method was used with the equation: $R(n, k(n)) = R(n)$. Finally, the solution of the polynomial $R_n = R_{exp}$ gave the values of n and k. Buffeteau and Desbat (1989) have also reported a computer program to calculate the optical constants of thin films from measurements of R and T. Three stages are involved: (*a*) a matrix formulation to calculate values of R and T of the multilayer system (*b*) an iterative Newton–Raphson technique to estimate values for n and k

and (*c*) a fast Kramers–Kronig transform which improved the overall accuracy.

Bueche (1948) measured the reflectance in the s-state of polarisation at five different angles of incidence and calculated σ and ε (hence n and k) from a set of previously compiled tables. Avery (1952), using non-normal incidence, measured the ratio of the reflectance in the s- and p-states and also solved for n and k by a graphical method. Abelès (1952), using the same experimental technique for very thin films, expanded the equations for the optical functions as power series in d/λ. He was able to obtain and solve a cubic equation for n in terms of the above ratio.

The work of Schultz is interesting because he used two different specimens in the determination of n and k. n was measured by a reflection method at 45° (Schultz and Tangherlini 1954) whilst k was found from measurements of the transmittance (Schultz 1951). The results of this work must be open to some suspicion as the two specimens were unlikely to be identical either in texture or in surface finish.

Another photometric technique which is applicable to films on a transparent substrate is attenuated total reflection (ATR). In this technique an electromagnetic wave is made to propagate along the interface of two semi-infinite media, i.e. a guided surface wave. Maxwell's equations only have a solution for this case when one of the media, called the surface active medium, has a negative dielectric constant, ε, and when p-polarised radiation is used. The second medium, known as the surface inactive medium, usually has a positive dielectric constant whose value, ε_2, does not change significantly over the wavelength range used. It may be shown (Abelès 1976) that the dispersive relationship between the wave number, K, and the angular frequency, ω, is of the form

$$K^2 = (\omega/c)^2 \varepsilon\varepsilon_i/(\varepsilon + \varepsilon_i). \qquad (6.1)$$

In order to excite these surface waves, a reflective prism is employed whose dielectric constant, ε_p, is greater than or equal to ε_i; in addition, the angle of incidence, θ_p, must be larger than the critical angle for total reflection. Equation (6.1) may be rewritten in the form

$$K = (\omega/c)\varepsilon_p^{1/2} \sin \theta_p. \qquad (6.2)$$

This equality may be realised either by varying ω at constant θ_p or θ_p at constant ω. The resonance is detected as a maximum in R_p. Under the conditions of the experiment where large angles of incidence are used $R_s \simeq 1$ so that $R_p = \tan^2 \theta_p$. Two separate configurations are possible for ATR. Figure 6.3(*a*) shows Kretschmann's (1971) method where the specimen is in contact with the base of the prism which is the surface inactive medium. Otto's (1968) technique is illustrated in figure 6.3(*b*), showing a narrow gap between the specimen and the prism; the surface inactive medium is now the air in the gap.

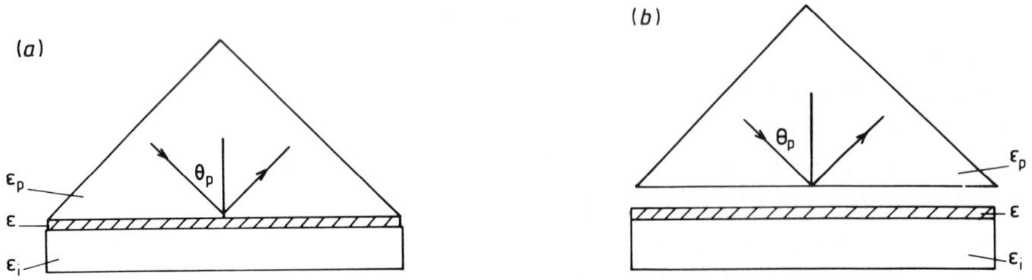

Figure 6.3 Attenuated total reflection. (*a*) Kretchmann's method, (*b*) Otto's method.

Harrick (1967) has described an ATR technique in which the radiation is incident from the substrate side, the value of R being measured at two different angles of incidence, $\theta_1 < \theta_c$ and $\theta_2 > \theta_c$.

Denton *et al* (1972) used the functions $(1 \pm R)/T$ in calculating the optical constants of thin films, claiming that these led to simplified calculations; however, the problem of multiple solutions still remained.

A three-stage process for the unambiguous determination of the optical constants of thin films from photometric measurements has been proposed by McPhedran *et al* (1984). The first stage would use a bivariate optimisation process based on the functions $(1 \pm R)/T$; in the second stage a plane-variate approach would be used based on an algorithm for locating the complex zeros of analytic functions. The third stage would employ an evolution curve giving the single-wavelength reflectance as a function of the thickness.

Dobrowolski *et al* (1983) calculated the optical constants of a thin film by an inverse synthesis technique, the quantities in the dispersion equation being adjusted until a good fit was obtained between the measured and calculated values of reflectance and transmittance. The sensitivity of the method could be increased by feeding in more data than the minimum needed to solve the equations; this not only reduced susceptibility to errors of measurement but also made the possibility of multiple solutions less likely.

An algebraic (rather than numeric) method for extracting n, k and d from measurements of reflectance and transmittance was presented by Case (1983). He claimed that no initial guess of the values of n, k and d was required nor were any multiple solutions produced. The technique was primarily designed for use with a stack of thin-film layers; there was no restriction on the number of films in the stack, but the unknown film had to be the last layer adjacent to the output medium.

Beauchamp and Rancourt (1980) measured the refractive index of germanium by a wedged film technique. A wedge-shaped film was deposited on a strongly absorbing substrate; the substrate was rotated and a 90° shutter placed in the beam was spun at half the angular velocity. Scanning along the wedge, the reflectance at a given wavelength showed a series of interference maxima and minima from whose spacing the refractive index could be calculated.

An optical wave guide was used by Sasaki *et al* (1980) in measuring the absorption coefficients of thin films. They applied the technique to both low-index polyurethane and to high-index zinc selenide.

Singleton and Shirkey (1983) measured the optical constants of mica up to 25 μm using the interference effect produced by reflections at the top and bottom surfaces of the film. The thickness was determined by optical interference and then, by plotting the reflectance versus thickness, a unique curve was obtained from which the optical constants could be found by curve fitting.

Kretschmann (1969) measured the transmittance at two different angles of incidence and was then able to calculate the optical constants of the film material. He produced sets of curves from which it was possible to select the best angle of incidence to use for particular values of n and k. The technique was applied to thin silver films about 60 nm in thickness.

Manuficier *et al* (1976) have described a method for obtaining n, k and the thickness of a thin transparent film surrounded by a non-absorbing medium. An interference method was used to produce fringes in transmission and then the envelopes of T_{max} and T_{min} were drawn; the values of T_{max} and T_{min} can be expressed as functions of the wavelength and the refractive index, absorption coefficient and thickness of the film and the equations may be solved for n; hence k and thickness may be computed.

There have been several interesting modifications to the straightforward R, T method for finding the optical constants of thin films deposited on a substrate. Nagendra and Thutupalli (1983) deposited a thin transparent film of known optical properties on a metal substrate; measurements of the reflectance of the combination enabled the optical constants of the metal to be determined. Hjortsberg (1981) covered half the substrate with a highly reflecting metal layer before depositing his unknown film; measurements were then made of the reflectance over the metallic half and the transmittance over the clear half, in both cases at near normal incidence. Providing the optical properties of the metal layer were known, the optical constants of the film could then be calculated. Gupta (1988) measured the reflectance of his films with and without a non-absorbing overcoating by a material of known n; this is really a modification of the method of immersion photometry. A further refinement was suggested by Siqueiros *et al* (1988) where reflectance measurements were made with overcoats of two different thicknesses ($\lambda/4$ and $\lambda/2$); this method only works at one wavelength but it was claimed that 'missing' solutions were eliminated. Paneyovov and Konstantinov (1991) measured R and T from both sides of the film; an algebraic technique involving numerical interpolation was used to extract n and k.

6.3 Polarimetry

Methods for the measurement of the change of phase on reflection have been described by Bennett (1964) and by Barokat and Mokhtar (1963). They both used semitransparent films deposited on optical flats, the fringe shift between the coated and uncoated regions of the flats being measured by an interferometer. Bennett noted that this phase change was very sensitive to the optical constants of the film and the technique could therefore be used in the study of ageing or of oxide formation of films.

Shaalan (1983) also claimed that the phase change on reflection was a sensitive parameter for finding either the film thickness or the optical constants; high sensitivities would be obtained when the quantity nd/λ had a magnitude of about 0.5. For example, with selenium, a change of 1° in the phase would take place from a change in n of 0.004 or from a change in thickness of 1.5 Å.

Ghezzo (1967) described a graphical method for the determination of the optical properties of a transparent film on a substrate by plotting Δ and Ψ in terms of n and also the phase change on transmission, χ.

Seraphin (1962) employed a modified form of Malé's graphical method in using the phase change on reflection at a double layer as a basis of calculating the optical constants.

6.4 Mixed photometry and polarimetry

The optical constants of a film may be determined from measurements of a combination of transmittance and phase change on transmission. Casset (1979) describes one such method in which the transmittance was measured by photometry whilst the phase angle was determined by means of a Michelson interferometer. A similar technique was also described by Joensen *et al* (1973). It was claimed in both these studies that transmittance measurements would be more representative of the body of the film than reflectances which give information mainly about the surface.

6.5 Ellipsometry

Johnson and Bashara (1971) used the method of multiple-angle ellipsometry to measure the optical constants and thicknesses of very thin films. In theory, in order to determine all three parameters, measurements at only three different angles are necessary, but these authors found that better results were obtained by using five angles of incidence, namely 45, 55, 65, 75 and 85°.

The effect of a thin surface film of contaminant on the determination of optical constants by ellipsometry was investigated by Burge and Bennett (1964). The presence of such a film may be difficult to detect but it can cause large errors if left uncompensated. The variation of Δ and Ψ with angle of incidence cannot be used to detect films of this kind as these functions are almost the same for the film-free and film-covered surfaces. The relationship, $R_s^2 = R_p$ at 45° is also not of any significant usefulness for this purpose. By considering the exact equations in each case, they showed that Archer's (1962) formula was likely to be more accurate than that of Drude in dealing with such films.

Theeten (1981) discussed the use of ellipsometry in thin-film analysis. In real-time analysis, measurements are made at a fixed wavelength and angle of incidence, the emphasis being on fast automatic data acquisition and *in situ* examination. On the other hand, spectroscopic ellipsometry requires a longer time to obtain full spectra and consequently it is important that the optical properties of the samples have stabilised before commencing this technique.

Further examples of the application of ellipsometric techniques to thin films may be found in the work of Fragstein and Kampermann (1963) Hodgson (1955) and Beattie and Conn (1955).

6.6 Accuracy studies on techniques for *n* and *k* for thin films

Estimates of the inherent accuracies obtainable in the measurement of n and k for thin films when using the three basic techniques of photometry, polarimetry and ellipsometry may be made by employing the same method as in §2.9. In the case of thin films, since the relative thickness, $x = 2\pi d/\lambda$, is also a variable, it is necessary to include in the error equations terms involving the partial derivatives of the optical functions X and Y with respect to x as below:

$$\pm \Delta X = \frac{\partial X}{\partial n}\Delta n + \frac{\partial X}{\partial k}\Delta k + \frac{\partial X}{\partial x}\Delta x + \frac{\partial X}{\partial \theta}\Delta\theta \tag{6.3}$$

$$\pm \Delta Y = \frac{\partial Y}{\partial n}\Delta n + \frac{\partial Y}{\partial k}\Delta k + \frac{\partial Y}{\partial x}\Delta x + \frac{\partial Y}{\partial \theta}\Delta\theta. \tag{6.4}$$

In the first instance, however, it will be assumed that both Δx and $\Delta\theta$ are zero and that only the variations of X and Y with n and k need be considered. A method of dealing with the errors in x is described in §6.7.

As shown in Chapter 2, solution of equations (6.3) and (6.4) yields the four corners of the error parallelogram centred on the original coordinates of n and k. The partial derivatives used in these equations may be obtained by appropriate differentiation of the expressions in §5.5 for the various optical functions of a thin film on a substrate. The values of ΔX and ΔY

have again been set at ± 0.01 for the photometric functions and ± 0.01 rad for the polarimetric functions; the sides of the parallelograms may be scaled proportionately either up or down if different accuracy levels are used. Similarly, an 'accurate' parallelogram has again been defined as one whose longer diagonal was less than 0.6 umts of n or k.

The survey covers values of d/λ between 0.001 and 0.20. The very small values of d/λ may be regarded as representing long wavelengths (i.e. the infrared or microwave regions) rather than ultra-thin films which are difficult to produce in uniform thickness; however, the large values of d/λ may be taken to indicate either thick films or short wavelengths, i.e. in the ultraviolet region.

6.6.1 Photometric functions

(a) (R, T)

The pattern of error parallelograms most characteristic of combinations of the reflectance with the transmittance is shown in figure 6.4 which is for

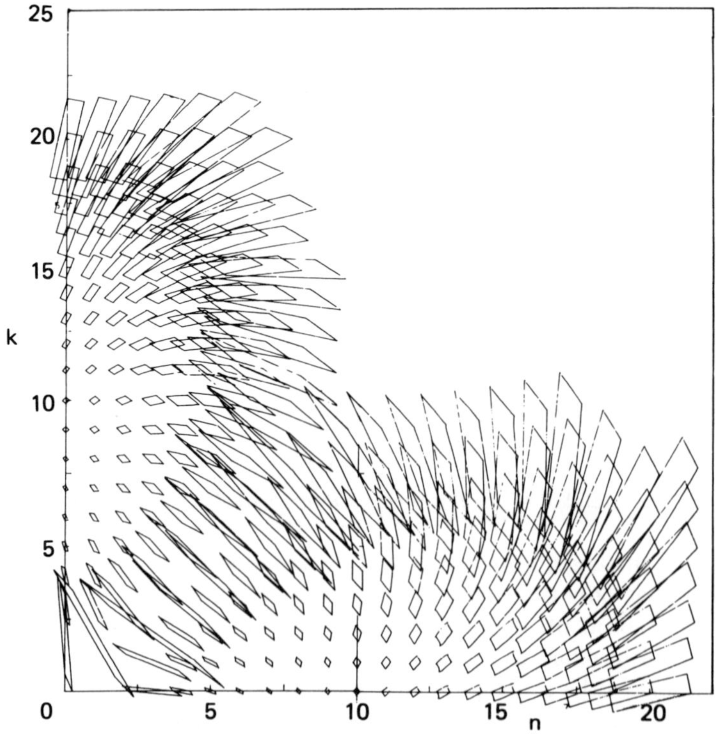

Figure 6.4 Accuracy pattern due to (R_s, T_s) at 15°, $d/\lambda = 0.005$. (From Ward L 1985 *Optica Acta* **32** 156.)

(R_s, T_s) at an angle of incidence of 15°. The individual parallelograms tend to be diamond shaped, formed by the small-angle intersection between the contours due to R which are roughly elliptical in form, and those due to T which are lenticular (see figure 6.5(a)). These two sets of contours are almost parallel at $n = k$ thus producing very long, narrow parallelograms in this region. Close to the n and k axes, however, the intersections take place at angles which are less acute and a characteristic double-lobe pattern of high accuracy is produced.

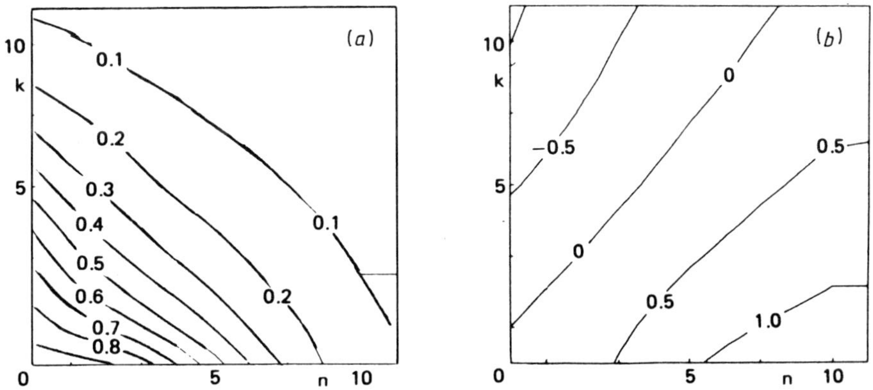

Figure 6.5 Contours of optical functions. (a) T_s and (b) χ_s. (From Ward L 1985 *Optica Acta* **32** 159.)

This basic pattern may expand, contract or rotate as the optical parameters are varied. Thus, as the thickness is increased, R also increases causing the contours of constant R to contract towards the origin. At the same time T decreases, which again leads to a contraction of the contours of constant T as their magnitudes decrease with n and k. The net result is that the accurate lobes move inwards towards the origin as illustrated in figure 6.6 which shows a sequence of patterns due to increasing film thicknesses for the combination (R_s, T_s), similar results being obtained with (R_p, T_p); for large values of d/λ the two lobes coalesce to give the same pattern as that obtained with solid specimens (figure 2.13). The shift with d/λ of the lobe on the k-axis is interesting; the lobe moves towards larger values of k as the value of d/λ decreases, i.e. as either d decreases or λ increases. This lobe covers the optical properties of metals such as copper, silver and gold (see figure 4.2) which have small values for n and whose values of k also increase rapidly with wavelength, thus indicating that photometric methods for determining n and k should be practicable for these materials.

Increasing the angle of incidence at constant d/λ produces a small but steady shift towards the origin for (R_s, T_s) and a similar shift away from the origin for (R_p, T_p). As in Chapter 2, these movements may again be

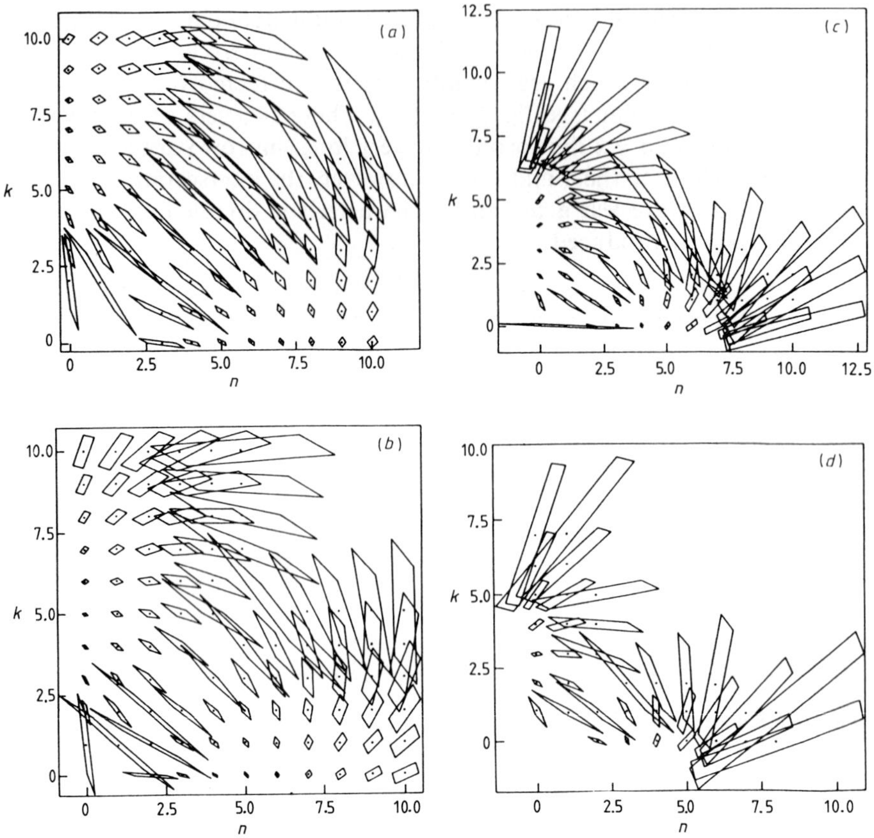

Figure 6.6 Sequence of accuracy patterns for (R_s, T_s) at $\theta = 60^\circ$ at different values of d/λ: (*a*) 0.005, (*b*) 0.01, (*c*) 0.02, (*d*) 0.04. (From Ward L 1985 *Opt. Laser Technol.* 265 (October).)

associated with the changes with angle of the photometric functions; R_s increase with θ whilst T_s decreases and thus the pattern will contract whilst the decrease in R_p and the increase in T_p with θ will cause the pattern to expand.

(b) (R_s, R_p)

R_s varies only slowly with θ, and consequently the contours due to $R_{s\theta_1}$ and $R_{s\theta_2}$ intersect at very acute angles and no significant areas of high accuracy are generated.

However, (R_s, R_p) is a viable combination, particularly at angles of incidence greater than 45°. The most extensive areas of high accuracy are obtained at $\theta = 75^\circ$ when a pattern similar to figure 6.6 is obtained. The error parallelogram pattern remains virtually unchanged when $d/\lambda > 0.08$, and thus these combinations may be used for solid materials as well as thin films in the region $0 < n < 3, 0 < k < 3$ as discussed in §2.9.1.

The combination $(R_{p\theta_1}, R_{p\theta_2})$ is also viable for $\theta_1 < 30^\circ$ and $\theta_2 > 60^\circ$, producing the characteristic double-lobe pattern.

(c) *Use of* R', *the reflectance from the substrate side*

Instead of R_s and R_p it is possible with films on a transparent substrate to use the reflectance measured from the substrate side, R_s' and R_p'. The behaviour of these functions is very similar to that for R_s and R_p respectively, and no significant differences are observed in the error diagrams. Combinations such as (R_s, R_s') and (R_p, R_p') only show regions of accuracy, in the form of double-lobe patterns, when the angles of incidence are greater than 60° and when the values of d/λ are 0.04 or more.

(d) (T_s, T_p)

No viable high-accuracy areas are generated for combinations of these functions as the two sets of contours run almost parallel generating very long and narrow error parallelograms.

(e) R_p/R_s *and* T_p/T_s *at two angles of incidence*

Another possible method using photometric techniques is that of the ratio R_p/R_s measured at two different angles of incidence; this has been used in several investigations, e.g. Avery (1952). The accuracy studies for these functions show that the error patterns are also in the form of double lobes and their accurate regions are similar to those of the combination (R_s, R_p). Although it is theoretically possible to employ the ratio T_p/T_s, the accurate regions are very small as would be expected from the studies on (T_s, T_p).

In all the examples of the double-lobe accuracy pattern obtained when using the photometric functions, plots of the logarithm of the limits of the accurate region on the n or k axis against $\log(\lambda/d)$ are linear. The upper and lower limits of the accurate regions yield separate lines, but both of them have the same slope of 1/3. Figure 6.7 illustrates this relationship and also acts as a summary of the variation of the limits of the accurate regions with angle of incidence for the p- and s-states of polarisation.

Boehme (1981) differentiated the expressions for reflectance and transmittance with respect to n, k and d/λ, and hence derived expressions for the relative uncertainties $\Delta n/n$, $\Delta k/k$ and $\Delta d/d$. His results indicated that it would be possible to have only small uncertainties over a large range of film parameters if the reflectance and transmittance were measured by p-polarised light at large angles of incidence, in broad agreement with the conclusions reached in this section.

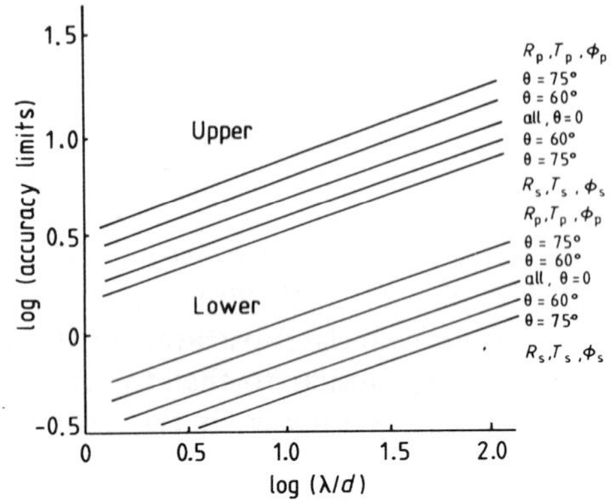

Figure 6.7 Log (accuracy limits) versus log(λ/d).

6.6.2 Polarimetry

(a) (ϕ, χ)

The contours of constant phase change on transmission, χ, are almost straight lines running radially from near the origin (figure 6.5(b)) and those of constant phase change on reflection, ϕ, are also roughly radial but curvilinear (fiure 1.2(c, d)). It would therefore be expected that the combinations (ϕ, χ) would show few regions of accuracy as the two sets of contours are again almost parallel, especially at $n = k$. This is in fact the case and, as with the photometric functions, two small lobes of high accuracy are produced on the n and k axes. In general, however, there is little to recommend these combinations for the determination of the optical constants.

(b) (ϕ_1, ϕ_2) *and* (χ_1, χ_2)

A similar argument applies to these two combinations which are also unsuitable for high-accuracy measurements.

6.6.3 Mixed photometry and polarimetry measurements

The shapes of the contours due to (i) the photometric functions (elliptical) and (ii) the polarimetric functions (radial) imply that combinations of one photometric and one polarimetric function might produce a wide region of high accuracy where the two sets of contours cross at large angles. Figure 6.8 shows the error parallelogram array for (R_s, ϕ_s) at $d/\lambda = 0.02$ and $\theta = 15^\circ$. The high-accuracy region is now in the form of an annular quadrant whose boundaries are determined by the R_s contours which become widely separated at both large and small values of n and k.

As d/λ increases, the contours of constant R_s contract inwards and thus the quadrant also contracts towards the origin, the inner quadrant of inaccuracy disappearing for $d/\lambda > 0.04$.

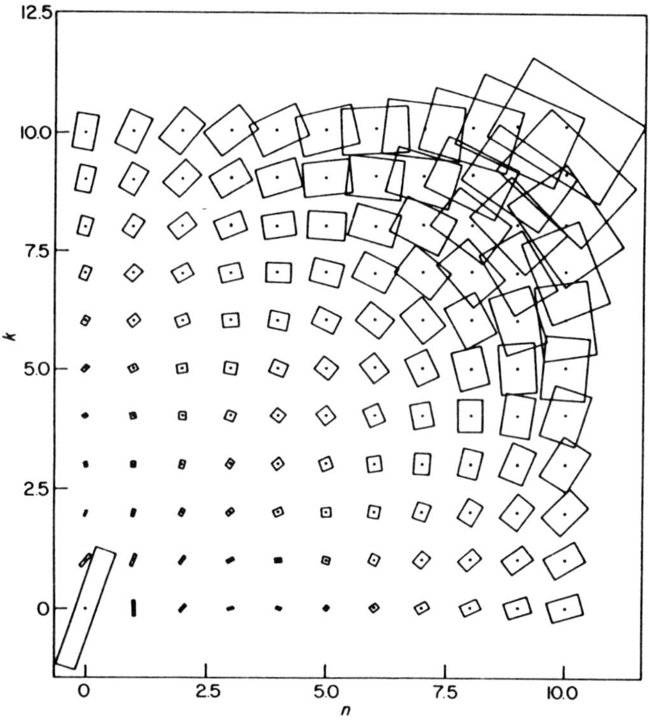

Figure 6.8 Error parallelogram array for (R_s, ϕ_s) at 15° and $d/\lambda = 0.02$ showing annular quadrant pattern. (From Ward L 1985 *Opt. Laser Technol.* 266 (October).)

The variation of the pattern with θ is again determined by the change in R_s with θ; as this function increases with θ the pattern also contracts.

(a) (R_p, ϕ_p)

This combination also produces an annular pattern which, at small values of θ and d/λ, is almost identical with that for (R_s, ϕ_s); the variation of the pattern with d/λ is also similar to that for the combination (R_s, ϕ_s). However, two important differences arise when the changes with angle of incidence are considered. Firstly, as R_p decreases with increase in θ, the quadrant pattern tends to expand as θ increases. Secondly, for $\theta > 60°$ the singularity discussed in Chapter 2 makes its appearance in the contours of ϕ_p where the lines of equal ϕ_p converge very strongly in a small region. This generates a set of long narrow parallelograms radiating from a particular point in *n*, *k* space. Figure 6.9 shows how the centre of the singularity varies with d/λ and θ for values of d/λ between 0.001 and 0.04. If the singularity falls inside the accurate area it seriously affects the accuracy attainable.

Because of the presence of the singularity, the use of this combination may be restricted to angles of less than 60°.

As with the photometric functions, it was found that the boundaries of the accurate quadrant were proportional to $(\lambda/d)^{1/3}$ and could also be accommodated on figure 6.7.

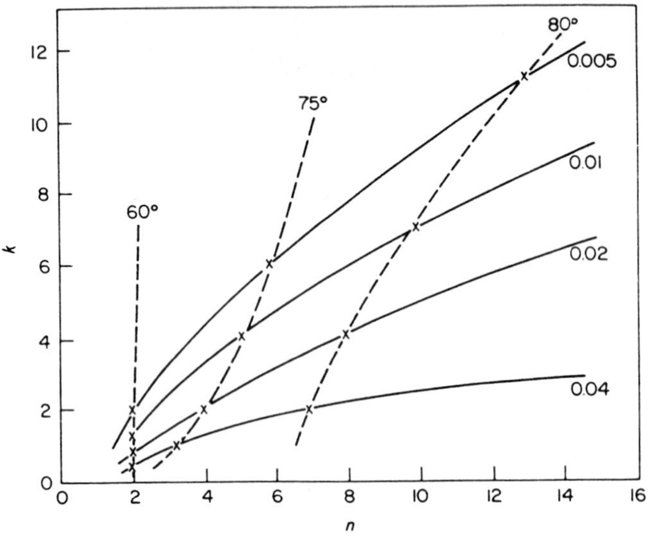

Figure 6.9 Variation with d/λ and θ of singularity in ϕ_p. (From Ward L 1985 *Opt. Laser Technol.* 267 (October).)

(b) (T, χ)

This combination also produces extensive accurate regions as the contours of constant χ_s and χ_p are also roughly radial whilst the transmittance, T, generates a pattern whose contours are lenticular. The individual error figures are in the form of long rectangles, the long sides being due to T and the short ones to χ. The accurate regions, which are also annular quadrants, are thus largely determined by the transmittance terms. Figure 6.10 shows the patterns for (T_s, χ_s) and (T_p, χ_p) at different angles of incidence for d/λ equal to 0.02. Both combinations produce similar patterns for $\theta < 30°$ but differ considerably at larger angles of incidence due to the different behaviour with θ of the dominant transmittance terms. The pattern of (T_s, χ_s) becomes smaller as T_s decreases with θ; on the other hand that of (T_p, χ_p) increases in size up to about 75° (the pseudo Brewster's angle) because T_p increases with θ, but then rapidly decreases as T_p falls towards zero. It may be noted that χ_p does not exhibit a singularity such as that found with ϕ_p and hence there is no restriction on its use. This was the combination used successfully by Casset (1979).

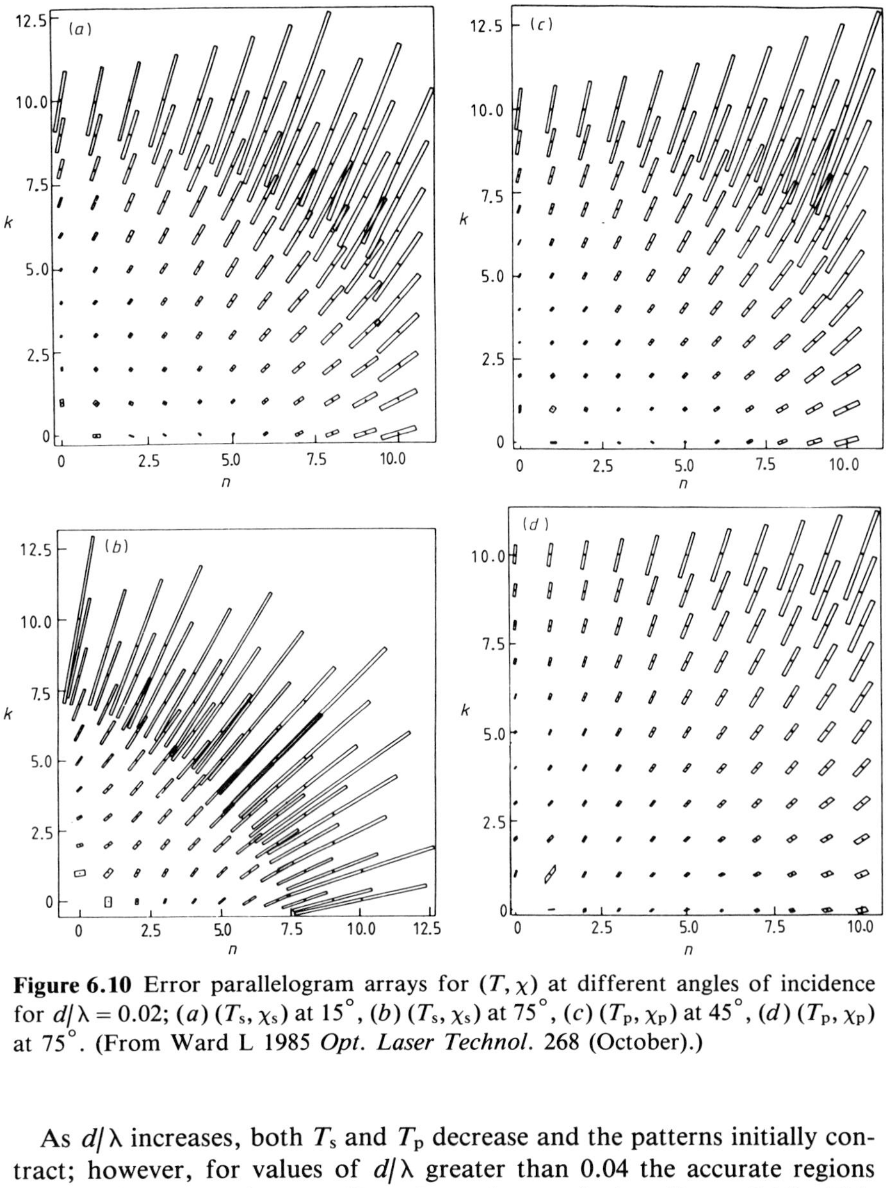

Figure 6.10 Error parallelogram arrays for (T, χ) at different angles of incidence for $d/\lambda = 0.02$; (*a*) (T_s, χ_s) at 15°, (*b*) (T_s, χ_s) at 75°, (*c*) (T_p, χ_p) at 45°, (*d*) (T_p, χ_p) at 75°. (From Ward L 1985 *Opt. Laser Technol.* 268 (October).)

As d/λ increases, both T_s and T_p decrease and the patterns initially contract; however, for values of d/λ greater than 0.04 the accurate regions develop into trapezia extending along the n-axis (see figure 6.11). This happens because the value of χ is almost completely determined by n, so the contours of constant χ become parallel to the k-axis.

The annular quadrants of high accuracy associated with the mixed optical functions makes them very suitable for measurements on materials which have roughly equal values of n and k, for example iron, cobalt, nickel and zinc (see figure 4.2) and which fall in regions of low accuracy for the

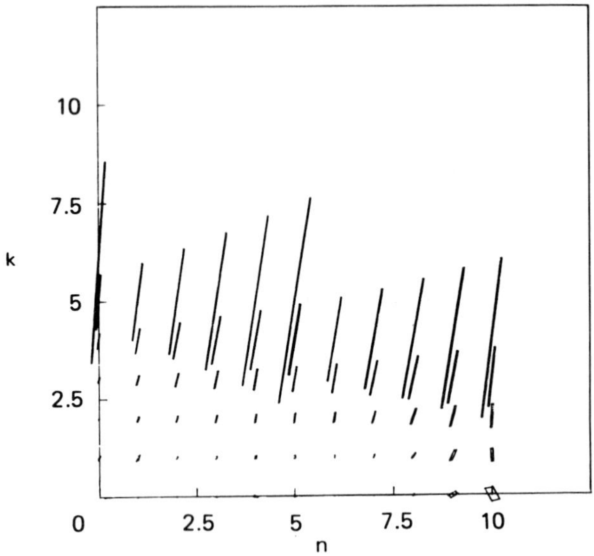

Figure 6.11 Trapezoidal error pattern due to $(T, \chi)(d/\lambda = 0.08, \theta = 60^\circ)$. (From Ward L 1985 *Optica Acta* 32 164.)

photometric functions. Detailed results for mixed functions have been given by Ward (1984, 1985).

6.6.4 *Ellipsometry*

In ellipsometry the optical functions measured are Ψ, the phase difference between the p- and s-components and $\Delta = \arctan(R_p/R_s)^{1/2}$ for reflection or $\arctan(T_p/T_s)^{1/2}$ for transmission. The accuracy patterns produced are also of the annular quadrant form, the size of the accurate quadrant being smallest at small angles of incidence and increasing with angle. The coverage of these functions is thus similar to that for the mixed functions.

(a) *Ellipsometry by reflection*

At small angles of incidence $R_p \simeq R_s$ and $\phi_p \simeq \phi_s$; although intersecting at large angles, the two sets of contours are each very widely spaced and consequently no accurate regions are observed. As the angle of incidence increases the s- and p-components of the photometric and polarimetric functions diverge and hence the contour lines become more closely spaced so that accurate regions are formed. However, when $\theta > 60^\circ$ the singularity associated with ϕ_p becomes apparent, introducing restrictions in the use of these functions. Figure 6.12 shows a sequence of error patterns for these functions at different angles of incidence.

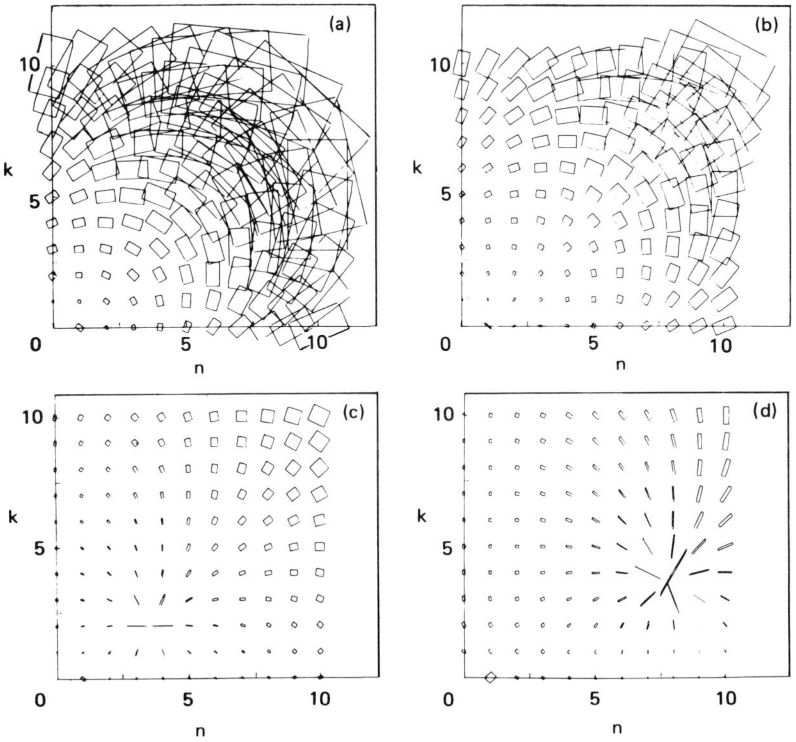

Figure 6.12 Sequence of error patterns for reflection ellipsometry at $d/\lambda = 0.02$ for four values of θ: (*a*) 30°, (*b*) 45°, (*c*) 75°, (*d*) 85°. (From Ward L 1985 *Optica Acta* **32** 163.)

As d/λ increases the ratio R_p/R_s becomes smaller for a given angle of incidence and the contour pattern contracts to give smaller regions of accuracy.

(b) *Ellipsometry by transmission*

The results at small angles of incidence are very similar to those for reflection ellipsometry, and for similar reasons namely the lack of discrimination between T_s and T_p and between χ_s and χ_p. The best results are obtained at the highest possible angle of incidence; in this case there is no singularity to inhibit the use of these functions.

The variation with d/λ is also similar to that for reflection ellipsometry. Limits to the use of this technique are set, however, by the thickness of the film material as the intensity of the transmitted radiation falls off rapidly with thickness, causing problems in making experimental measurements.

Table 6.1 gives a summary of the results for the ellipsometric combinations at different angles of incidence and for a range of values of d/λ.

Table 6.1 Radii of quadrants of accuracy in n and k for ellipsometry of thin films: (a) reflection ellipsometry; (b) transmission ellipsometry.

	$\theta = 15°$	30°	45°	60°	75°	85°
$f/\lambda = 0.002(a)$	—	3–6	5–10	5	5	6–12
$0.01(a)$	—	0–4	0–6	0–9	1–11	2–12
$0.10(a)$	0–1	0–3	0–4	0–6	0–9	0–9
$0.001(b)$	—	—	0–2	0–3	0–3	0–4
$0.01(b)$	—	—	0–2	0–5	0–7	1–8
$0.10(b)$	—	0–1	0–3	0–7	0–8	0–9

Vuye and Lopez-Rios (1982) have also discussed the precision obtainable in ellipsometric determinations of the optical constants of thin films, noting that the accuracy will be extremely small when the dielectric constant of the film is equal to that of the substrate.

6.7 Accuracy studies for n, k and d/λ

The accuracy studies for n and k described in section 2.6.6 may be extended to cases where it is required to calculate n, k and also the relative thickness x ($x = 2\pi d/\lambda$) of a film from the measurement of three optical functions; let them be called P, Q and S.

It is first necessary to deduce expressions for the errors produced in n, k and x by experimental errors ΔP, ΔQ and ΔS in the three optical functions P, Q and S. The following relations apply:

$$\begin{aligned}\Delta P &= \Delta n \cdot \partial P/\partial n + \Delta k \cdot \partial P/\partial k + \Delta x \cdot \partial P/\partial x \\ &= \Delta n \cdot P_n + \Delta k \cdot P_k + \Delta x \cdot P_x\end{aligned} \tag{6.5a}$$

$$\begin{aligned}\Delta Q &= \Delta n \cdot \partial Q/\partial n + \Delta k \cdot \partial Q/\partial k + \Delta x \cdot \partial Q/\partial x \\ &= \Delta n \cdot Q_n + \Delta k \cdot Q_k + \Delta x \cdot Q_x\end{aligned} \tag{6.5b}$$

$$\begin{aligned}\Delta S &= \Delta n \cdot \partial S/\partial n + \Delta k \cdot \partial S/\partial k + \Delta x \cdot \partial S/\partial x \\ &= \Delta n \cdot S_n + \Delta k \cdot S_k + \Delta x \cdot S_x\end{aligned} \tag{6.5c}$$

where $P_n = \partial P/\partial n$, $P_k = \partial P/\partial k$, $P_x = \partial P/\partial x$ etc; in this analysis, it is assumed that these partial differentials are constant over small changes.

These equations may also be expressed in the form of an error matrix as below;

$$\begin{vmatrix} \Delta P \\ \Delta Q \\ \Delta S \end{vmatrix} = \begin{vmatrix} P_n & P_k & P_x \\ Q_n & Q_k & Q_x \\ S_n & S_k & S_x \end{vmatrix} \cdot \begin{vmatrix} \Delta n \\ \Delta k \\ \Delta x \end{vmatrix}. \tag{6.6}$$

The equations may be solved analytically to give the following expressions for Δn, Δk and Δx:

$$\Delta n = (\pm \Delta P(S_x \cdot Q_k - Q_x \cdot S_k) \pm \Delta Q(P_x \cdot S_k - S_x \cdot P_k) \pm \Delta S(Q_x \cdot P_k - P_x \cdot Q_k))/Y \quad (6.7a)$$

$$\Delta k = (\pm \Delta P(Q_x \cdot S_n - S_x \cdot Q_n) \pm \Delta Q(S_x \cdot P_n - P_x \cdot S_n) \pm \Delta S(P_x \cdot Q_n - Q_x \cdot P_n))/Y \quad (6.7b)$$

$$\Delta x = (\pm \Delta P(Q_n \cdot S_k - Q_k \cdot S_n) \pm \Delta Q(P_k \cdot S_n - P_n \cdot S_k) \pm \Delta S(P_n \cdot Q_k - P_k \cdot Q_n))/Y \quad (6.7c)$$

where

$$Y = S_k(P_x \cdot Q_n - Q_x \cdot P_n) + S_n(P_x \cdot Q_k - Q_x \cdot P_k) + S_x(P_k \cdot Q_n - P_n \cdot Q_k). \quad (6.7d)$$

The errors in n, k and d/λ produced by given uncertainties in any combination of three optical functions chosen from R_s, R_p, T_s, T_p, ϕ_{rs}, ϕ_{rp}, ϕ_{ts} and ϕ_{tp} may readily be computed. For given starting values of n, k and d/λ, if all the $\pm$ values of Δn, Δk and Δx are plotted graphically, an eight cornered figure, the error parallelepiped, is produced whose geometrical centre is located at n, k and d/λ. The smaller the size of the error parallelepiped, the more sensitive will be the combination at that point; in practice, most of the individual parallelepipeds are needle or prism shaped i.e. long and thin with the long axis lying almost in the n–k plane. Arrays of error parallelepipeds may be then plotted for ranges of n and k and for discrete values of d/λ and the angle of incidence.

There are 56 ways of choosing three optical functions from the eight available (not counting different angles) and all of them have been processed (Ward 1992a, b) in order to discover which combinations exhibit the best regions of good sensitivity. The ranges of n and k were each 0 to 10, of d/λ 0.005 to 0.16, and of θ, 5 to 85°. The values of ΔP, ΔQ and ΔS were set initially at 0.001 and an arbitrary standard of sensitivity was adopted in which parallelepipeds having a maximum length less than 1.0 unit were defined as sensitive. In order to study a particular region in more detail the values of ΔP, ΔQ and ΔS could be increased appropriately.

It was found that the arrays could be broadly classified into two main patterns; (1) where there were two lobes consisting of small parallelepipeds representing high sensitivity, these being located symmetrically on the n and k axes respectively, with the individual parallelepipeds lying parallel to the outline of the lobes and (2) a single quadrant of high sensitivity centred about a point at or close to the origin of the n, k coordinates and from which the parallelepipeds were oriented radially. Figure 6.13(a) shows an example of the type 1 pattern and is for $R_pT_p\phi_{rs}$ at $d/\lambda = 0.001$ and an angle of incidence of 60°, whilst an example of the type 2 pattern is shown in figure 6.13(b) and this is for $T_s\phi_{rs}\phi_{ts}$ at $d/\lambda = 0.001$ and 75°.

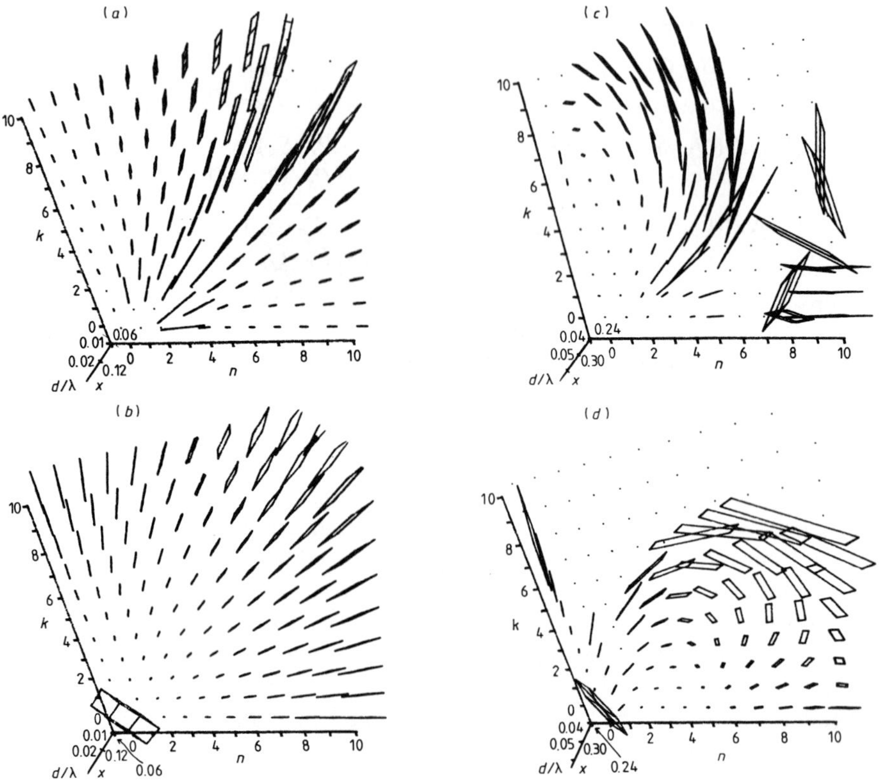

Figure 6.13 (*a*) Two-lobe sensitivity pattern. (*b*) Quadrant sensitivity pattern. (*c*) *k*-lobe sensitivity pattern. (*d*) *n*-lobe sensitivity pattern.

6.7.1 *Two-lobe pattern*

The two-lobe pattern was found to be exhibited by 35 of the combinations examined as below, these being further divided into sub-groups (*a*), (*b*) and (*c*): (*a*) $R_sR_p\phi_{rs}$, $R_sR_p\phi_{rp}$, $T_s\phi_{rs}\phi_{rp}$, $T_p\phi_{rs}\phi_{rp}$, $\phi_{ts}\phi_{tp}\phi_{rs}$, $\phi_{ts}\phi_{tp}\phi_{rp}$; (*b*) $R_sT_p\phi_{tp}$, $R_sT_p\phi_{ts}$, $R_sT_s\phi_{ts}$, $R_sT_s\phi_{tp}$, $R_pT_p\phi_{tp}$, $R_pT_p\phi_{ts}$, $R_pT_p\phi_{rs}$, $R_pT_s\phi_{tp}$, $R_pT_s\phi_{ts}$, $R_pT_s\phi_{rp}$, $R_sT_s\phi_{rp}$; (*c*) $R_sR_p\phi_{ts}$, $R_sR_p\phi_{tp}$, $T_sT_p\phi_{rs}$, $T_sT_p\phi_{rp}$, $\phi_{ts}\phi_{rs}\phi_{rp}$, $\phi_{tp}\phi_{rs}\phi_{rp}$, $R_sT_s\phi_{rs}$, $R_pT_s\phi_{ts}$, $R_sT_p\phi_{rp}$, $R_sT_p\phi_{rs}$, $R_s\phi_{rs}\phi_{rp}$, $R_p\phi_{rs}\phi_{rp}$, $R_sT_sT_p$, $R_pT_sT_p$, $T_sT_p\phi_{ts}$, $T_sT_p\phi_{tp}$, $R_sR_pT_s$, $R_sR_pT_p$.

In general, for $d/\lambda = 0.01$ and for angles of incidence less than 15°, the areas of the regions of high sensitivity for all these combinations were small and none of them could be deemed useful for optical measurements. However, the sizes of these regions increased with angle of incidence up to about 75° after which there was a sharp decrease. Many of the combinations would thus be usable over a wide range of values of *n* and *k* at angles of incidence of 60 or 75°.

As d/λ increased to 0.04 and 0.08 the regions of good sensitivity in groups (*a*) and (*b*) increased in size slightly but also became increasingly asymmetrical, ending up as a single lobe on either the *n*- or *k*-axis. The combinations in group (*a*) above tended to a *k*-axis lobe (figure 6.13(*c*)) whilst those in group (*b*) produced *n*-axis lobes (figure 6.13(*d*)). Those in group (*c*) maintained the two-lobe pattern which decreased in size as d/λ increased. Some of these combinations would be optically viable at values of d/λ up to 0.08, the choice depending on the type of substance being investigated; thus, a metal ($k > n$) would be best studied by a combination giving a *k*-lobe, group (*a*), whilst an *n*-lobe combination, group (*b*), would be more suitable for an insulator or semiconductor ($n > k$).

There were two classes of combinations which produced the two lobe pattern. Class (i) contained those combinations where both the s- and p-components of the same optical function were present; at small angles of incidence such functions are not well discriminated. As the angle of incidence increases, these optical functions become better discriminated and the size of the lobes then expands to larger values of *n* and *k*. Class (ii) contained combinations in which both the optical functions *R* and *T* in either state of polarisation were present; the surfaces of constant *R* and constant *T* run almost parallel at small angles of incidence resulting in very small lobes of high sensitivity. Once again, increase in angle of incidence leads to better discrimination between *R* and *T*.

Among the best combinations from this group, unable at angles of incidence of 60° and for $0.01 < d/\lambda < 0.08$ were; $R_sT_p\phi_{tp}$, $R_sT_p\phi_{ts}$, $R_sT_s\phi_{rp}$, $R_sT_s\phi_{rs}$, $R_pT_p\phi_{tp}$, $R_pT_p\phi_{ts}$, $\phi_{ts}\phi_{tp}\phi_{rs}$ and $\phi_{ts}\phi_{tp}\phi_{rp}$. At $d/\lambda = 0.01$ their lobes of sensitivity were as follows: $2 < n < 9$, $0 < k < 4$ and symmetrically on the *k*-axis.

The worst of the two lobe combinations from the point of view of sensitivity were: $T_s\phi_{rs}\phi_{rp}$, $T_p\phi_{rs}\phi_{rp}$, $T_sT_p\phi_{rs}$, $T_sT_p\phi_{rp}$, $R_sT_sT_p$ and $R_pT_sT_p$.

6.7.2 *Single quadrant pattern*

The remaining 21 combinations were found to exhibit the single quadrant sensitivity pattern and this was considered the most suitable for general use. Those in which the single quadrant pattern was shown were thus: $R_p\phi_{rs}\phi_{ts}$, $R_p\phi_{rp}\phi_{ts}$, $T_p\phi_{rp}\phi_{ts}$, $T_p\phi_{rs}\phi_{ts}$, $R_pT_p\phi_{rp}$, $R_s\phi_{rs}\phi_{ts}$, $R_p\phi_{rp}\phi_{tp}$, $T_p\phi_{tp}\phi_{rp}$, $T_p\phi_{tp}\phi_{rs}$, $T_s\phi_{rs}\phi_{ts}$, $R_s\phi_{rp}\phi_{ts}$, $R_s\phi_{ts}\phi_{tp}$, $R_p\phi_{ts}\phi_{tp}$, $R_p\phi_{tp}\phi_{rs}$, $T_p\phi_{ts}\phi_{tp}$, $T_s\phi_{ts}\phi_{tp}$, $T_s\phi_{tp}\phi_{rp}$, $T_s\phi_{tp}\phi_{rs}$, $T_s\phi_{ts}\phi_{rp}$, $R_s\phi_{tp}\phi_{rs}$ and $R_s\phi_{tp}\phi_{rp}$. The single quadrant pattern was generated by combinations of one photometric (*R* or *T*) and two polarimetric (ϕ_r or ϕ_t) optical functions; the only exception being $R_pT_p\phi_{rp}$.

At small values of d/λ, e.g. 0.01, the sensitive regions for this group have a radius measured from the origin of coordinates of about 10 units of *n* or *k* and remain about the same size as the angle of incidence increases from 5 to 60°; then there is a rapid contraction of the pattern up to 85°. With increase in thickness, the sensitive regions tend to decrease in size until at $d/\lambda = 0.08$ they are only small and lie close to the origin of coordinates.

However, for some combinations, the pattern was found to expand with thickness e.g. $R_s\phi_{rs}\phi_{ts}$. The best combination of this group was $R_p\phi_{rp}\phi_{tp}$ which would be viable at all angles and relative thicknesses up to $d/\lambda = 0.08$; other good combinations were $T_s\phi_{ts}\phi_{tp}$ and $T_p\phi_{ts}\phi_{tp}$.

6.7.3 *Other accuracy studies*

The accuracies obtainable when using various combinations of three photometric functions, including the internal reflectances R_s' and R_p' and also the same functions at different angles of incidence, have been considered by Hansen (1973) who set up error matrices for the variation of the functions with n, k and d/λ. He evaluated all of the matrix elements for a given combination and then examined each matrix to see which had the lowest set of values of the elements. Amongst his conclusions were that two combinations frequently used to determine n, k and d/λ, TR_sR_s' and TR_pR_p' at 45°, were not very sensitive; however good sensitivity was obtainable with $TR_{s45}R_{s75}'$ and $TR_{p45}'R_{s75}'$. These four combinations were also tested by Ward who found rather different results; $TR_{s45}R_{s45}'$ and $TR_{s45}R_{s75}'$ both produced very poor two-lobe patterns, whilst $TR_{p45}R_{p45}'$ and $TR_{p45}'R_{s75}'$ gave good two-lobe patterns which developed with thickness into the n-lobe variety.

Elizelda *et al* (1986) have also discussed the suitability of photometric (R and T) and polarimetric (PHI and DEL) optical functions by using them in the actual determination of n, k and d/λ of real films. They concluded that solutions could be obtained using either (*a*) R, T and DEL (two lobe pattern) or (*b*) T, DEL and PHI (single quadrant pattern) but that the two sets of results were inconsistent. Consistent values of n, k and d/λ could only be obtained by using all four optical functions in conjunction with a three-stage iteration technique.

6.8 Summary

When using two optical functions, the largest accurate regions, in the form of annular quadrants, are obtained with the mixed photometric and polarimetric functions especially (R_s, ϕ_s), (T_s, χ_s) and (T_p, χ_p). The accurate quadrants stretched all the way from the k- to the n-axes and covered the optical constants of many materials in the ultraviolet, visible and near-infrared spectra. However, the polarimetric functions are not easily measured and these combinations are not commonly used in practice.

Ellipsometry gives equally good results at high angles of incidence, particularly in the transmission form which does not suffer from the singularity present in the reflection version.

Photometry is a relatively straightforward technique and gives accurate regions along the k-axis which are useful for long-wavelength studies;

however, it gives very poor accuracy when $n \simeq k$, a region which includes many interesting materials.

If it is required to measure all three optical parameters (n, k and d), two groups each of three optical functions are capable of giving accurate results. Group one, which produces a two lobe pattern of accurate error parallelepipeds, consists of combinations of R, T and either ϕ_r or ϕ_t. Group two, producing a single quadrant error pattern, involves combinations of one photometric and two polarimetric optical functions.

References

Abelès F 1952 *Proc. Phys. Soc.* B **65** 996
—— 1963 *Prog. Opt.* **II** 265
—— 1971 *Phys. Thin Films* **6** 151
—— 1976 *Surf. Sci.* **56** 237
Archer R J 1962 *J. Opt. Soc. Am.* **52** 971
Avery D C 1952 *Proc. Phys. Soc.* B **65** 425
Azzam R M A and Bashara N M 1973 *Appl. Opt.* **1** 203
—— 1977 *Ellipsometry and Polarised Light* (Amsterdam: North-Holland) pp 153–268
Barokat N and Mokhtar S 1963 *J. Opt. Soc. Am.* **53** 1963
Beattie J R and Conn G K T 1955 *Phil. Mag.* **46** 989
Beauchamp W T and Rancourt J D 1980 *Appl. Opt.* **19** 3239
Bennett H E and Bennett J M 1967 *Phys. Thin Films* **4** 1
Bennett J M 1964 *J. Opt. Soc. Am.* **54** 612
Bennett J M and Booty M J 1966 *Appl. Opt.* **5** 41
Boehme H 1981 *Optik* **62** 27
Bousquet A and Rouard P 1960 *J. Phys. Radium* **21** 873
Bueche F 1948 *J. Opt. Soc. Am.* **38** 806
Buffeteau T and Desbat B 1989 *Appl. Spectrosc.* **43** 1027
Burge D K and Bennett H E 1964 *J. Opt. Soc. Am.* **54** 1428
Case W E 1983 *Appl. Opt.* **22** 1832
Casset J 1977 *Thin Solid Films* **41** 315
—— 1979 *J. Opt. Soc. Am.* **69** 725
Denton R E, Campbell R D and Tomlin S G 1972 *J. Phys. D. Appl. Phys.* **5** 852
Dobrowolski J A. Ho F C and Waldorf A 1983 *Appl. Opt.* **22** 3191
Elizelda E, Frigeris J M and Rivory J 1986 *Appl. Opt.* **25** 4557
Fragstein C V and Kampermann H 1963 *Z. Phys.* **173** 39
Fray S J, Goodwin A R and Quarrington J E 1963 *J. Sci. Instrum.* **40** 387
Ghezzo M 1967 *J. Opt. Soc. Am.* **57** 1427
Gupta M F 1988 *Appl. Opt.* **27** 954
Halford J H, Chin F K and Norman J E 1973 *J. Opt. Soc. Am.* **63** 786
Hanson W N 1973 *J. Opt. Soc. Am* **63** 793
Harrick N J 1967 *Internal Reflection Spectrometry* (New York: Interscience)
Harris L and Loeb A L 1955 *J. Opt. Soc. Am.* **45** 179

Heavens O S 1964 *Phys. Thin Films* **2** 193
—— 1965 in *Optical Properties of Thin Solid Films* (New York: Dover) p 46
Hemming D 1969 *Appl. Opt.* **8** 207
Hjortsberg A 1981 *Appl. Opt.* **20** 1254
Hodgson J H 1955 *Proc. Phys. Soc. B* **68** 593
Hunter W R 1965 *J. Opt. Soc. Am.* **55** 1197
Hunter W R and Hass G 1974 *J. Opt. Soc. Am.* **64** 429
Joensen P, Irwin J C, Cochran J F and Curzon A E 1973 *J. Opt. Soc. Am.* **63** 1556
Johnson J A and Bashara N M 1971 *J. Opt. Soc. Am.* **61** 457
Kretschmann E 1969 *Z. Phys.* **221** 346
Kretschmann K 1971 *Z. Phys* **241** 313
Kuhn H and Wilson B A 1950 *Proc. Phys. Soc.* B **63** 745
Lindquist R E and Ewald A W 1983 *J. Opt. Soc. Am.* **53** 247
McPhedran R C, Botten L C, McKenzie D R and Netterfield R P 1984 *Appl. Opt.* **23** 1197
Manuficier J C, Gasiot J and Fillari J p 1976 *J. Phys. E: Sci-Instrum.* **9** 1002
Malé D 1964 *J. Phys., Paris* **25** 74
Meetens F P, Theroux P and Plumb R C 1963 *J. Opt. Soc. Am.* **53** 788
Mishima T and Kao K C 1981 *Appl. Opt.* **20** 3719
Murmann H 1933 *Z. Phys.* **80** 161
Nagranda C L and Thutupalli G K M 1983 *Appl. Opt.* **22** 587
Otto A 1968 *Z. Phys* **216** 398
Panayotov V and Konstantinov I 1991 *Appl. Opt.* **30** 2795
Paulick T C 1986 *Appl. Opt.* **25** 562
Rouard P and Bousquet A 1964 *Prog. Opt.* **IV** 145
Sasaki K. Takahashi H, Kudo Y and Suzuki N 1980 *Appl. Opt.* **19** 3018
Schultz L G 1951 *J. Opt. Soc. Am.* **41** 1047
Schultz L G and Tangherlini F R 1954 *J. Opt. Soc. Am.* **44** 372
Sequiros J M, Rigalado L E and Machorro R 1988 *Appl. Opt.* **27** 4260
Seraphin B O 1962 *J. Opt. Soc. Am.* **52** 912
Shaalan M S 1983 *J. Phys. D. Appl. Phys.* **16** 419
Singleton E B and Shirkey C T 1983 *Appl. Opt.* **22** 185
Theeten J B and Aspnes D E 1981 *Ann. Rev. Mater. Sci.* **11** 97
Vuye G and Lopez-Rios T 1982 *Appl. Opt.* **21** 2968
Ward L 1982 *J. Phys. D: Appl. Phys.* **15** 1361
—— 1984 *J. Phys. D: Appl. Phys.* **17** 1781
—— 1985 *J. Phys. D: Appl. Phys.* **18** 291
—— 1992a *Opt. Laser Tech.* **24** 97
—— 1992b *Opt. Laser Tech.* **24** 285
Ward L, Nag A and Dixon L C W 1969 *J. Phys. D: Appl. Phys.* **2** 301

7 Results for Thin Films

7.1 Introduction

The measured optical constants of thin films tend to be different from those of the same materials in bulk form and this has been demonstrated to be largely due to the dislocations, grain boundaries, lattice defects and voids which are more prevalent in the films. These effects are particularly troublesome at the vacuum and substrate interfaces and some workers have regarded such regions as 'transition zones'. Considerable efforts have been made over the years to grow continuous films which have the same optical properties as the parent bulk materials: deposition on heated substrates, using rapid deposition rates, and annealing the films after deposition have all had beneficial effects towards this end and, in the last few years, the objective has been attained for many materials.

The survey of results in this chapter covers thin films in all states of aggregation and is arranged in approximate historical order. Once again, it is convenient to discuss the experimentally determined values of the optical constants under similar groupings of materials to those used for bulk materials in Chapter 4. e.g. alkalis, noble metals, ferromagnetic materials, dielectrics and semiconductors.

7.2 Alkali metals

Rouard and Meesen (1977) carried out a comprehensive review of the optical properties of thin metal films, concentrating particularly on the alkali metals. They found that it was important to separate the surface and volume effects caused by various types of imperfections. Volume effects are caused by vacancies, lattice imperfections and atomic impurities whilst surface effects are due to the anomalous skin effect and to surface irregularities, the latter allowing the coupling of surface plasmons with the incident electromagnetic radiation. It was suggested by these authors that the best techniques for the measurement of the optical constants of films would be those using the metal-substrate interface rather than that between vacuum and metal which was subject to progressive contamination even in the best vacuum conditions. A high rate of deposition (at least 10 nm s^{-1}) was also necessary in order to minimise the oxidation of the film particles as they passed from source to substrate.

Sodium films were investigated in the ultraviolet up to 200 nm by Sutherland *et al* (1967), the values of n being obtained from (i) interference studies and (ii) determination of the critical angle. The real part of the dielectric constant was found to be consistent with the nearly free electron theory. The growth of an oxide layer on the surface of the film was shown not to change the positions of the interference maxima and minima from which n was determined. In addition, an effective volume plasmon was discovered at an energy of 5.69 eV (218 nm) and this was in good agreement with the value obtained from experiments in which $\mathrm{Im}(1/\varepsilon)$ was measured and also with Wood's result for solid sodium.

Sutherland and Arakawa (1968) carried out a similar study on potassium between 3.96 and 9.69 eV (128 to 331 nm). Using a nearly free electron model, ε_1 can be expressed in the form:

$$\varepsilon_1 = (1 + 4\pi K a_1) - (4\pi K e^2/m_0\omega^2). \tag{7.1}$$

Thus, setting $\omega = u/\hbar$,

$$\varepsilon_1 U^2 = (1 + 4\pi K a_1)U^2 - (4\pi K e^2\hbar^2/m_0). \tag{7.2}$$

A graph of $\varepsilon_1 U^2$ against U^2 should therefore be linear with a slope of $1 + 4\pi K a_1$ and an intercept of $4\pi K e^2 h^2/m_0$; figure 7.1(a) shows the plot of these functions. The optical mass, determined from the intercept, was found to be 1.06 ± 0.03 electron masses; it was also found necessary to include a volume plasmon at 3.72 ± 0.05 eV (333 nm) which is at a slightly lower energy than Wood's value for bulk potassium.

Rasigni and Rasigni (1973), using lithium films deposited at 10^{-10} Torr, found the surface was covered by a two-dimensional array of granular elevations which gave rise to strong absorption between 1.5 and 2.1 eV. In addition, there was an absorption peak in the ultraviolet due to interband transitions. Fleischmann (1963), working with several of the alkali metals found peaks in the transmittance measured with p-polarised radiation incident at 55°. These were as follows: potassium 400 nm, rubidium 500 nm and caesium 550 nm, all these values being at wavelengths about 100 nm longer than Wood's results for hulk materials again most probably due to the granular structure of the surfaces. Similar absorption was found by Payan and Roux (1969) with sodium potassium and rubidium films between 3.0 and 4.3 eV.

Whang *et al* (1971, 1972) measured the optical properties of the alkali metals between 4 and 10.7 eV when deposited on calcium fluoride and

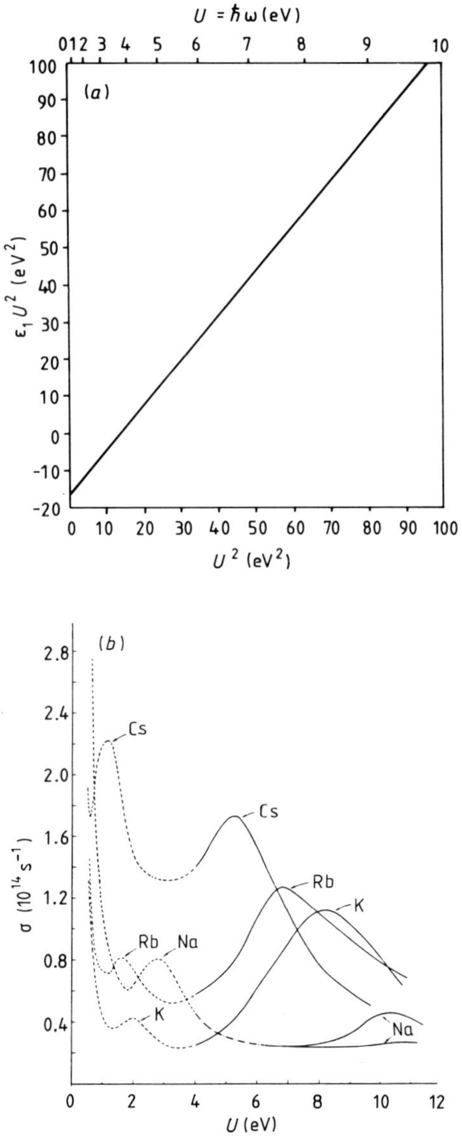

Figure 7.1 (*a*) $\varepsilon_1 U^2$ versus U^2 for potassium. (From Sutherland J C and Arakawa E T 1968 *J. Opt. Soc. Am.* **58** 1081.) (*b*) σ versus U for films of various alkali metals. (From Whang U S *et al* 1972 *Phys. Rev.* B **6** 2116.

magnesium fluoride substrates. n was determined by the critical angle method for total internal reflection at the metal-substrate interface, whilst k was obtained from the slope of the curve of R against angle of incidence measured at the critical angle. For caesium, the real part of the dielectric constant differed from that predicted by the nearly free electron theory at the lower energies because of the proximity of interband transitions and plasmon-assisted transitions and, at the higher photon energies, due to transitions from inner electron shells. A value for m_0 of 1.05 ± 0.05 was obtained and the plasma energy was found to be 2.87 ± 0.07 eV (435 nm), which is close to Wood's value for the bulk material. Plots of σ versus U (figure 7.1(*b*)) showed two strong absorption peaks, that at the lower energy moving to lower energies and becoming stronger as the atomic number increased. The higher-energy peaks occurred about 1.5 eV above the interband transitions.

Calcott and Arakawa (1974) measured the reflectance as a function of angle of incidence in their work on lithium films. At energies above 6 eV reflectance from the vacuum and substrate sides produced the same values of n and k, but this was not the case below 6 eV; this effect was attributed to surface roughness. The plasma frequency of lithium calculated from the condition: $\varepsilon_1 = 0$ was 6.7 eV, whilst that from $\mathrm{Im}(1/\varepsilon)$ gave 7.1 eV indicating the proximity of an interband transition.

Other results for the alkali metals have been reported by Schopper (1963), Wagnerberger (1963), Mayer and El Naby (1963), Deitschel (1963), Muller (1963) and El Naby (1963).

7.3 Alkaline earths group

A series of articles dealing with the optical properties of thin films of the alkaline earths was published by Rivoira and Blanc (1967a, b, c). Barium (1967a) exhibited absorption bands at 0.75 and 4.5 eV (0.27 to 1.65 μm) whilst strontium (1967b) had an abnormal band in the infrared but otherwise their optical properties agreed with results for the bulk materials. Calcium (1967c) displayed a long-wavelength absorption band at 690 nm and there was also evidence for another, interband, absorption band below 230 nm.

Fisher *et al* (1966) studied evaporated films of barium by measuring the reflectance as a function of angle of incidence, a graphical method being used to obtain values of ε_1 and ε_2. A plot of $\mathrm{Im}(1/\varepsilon)$ showed a peak at 7.0 eV caused by a volume plasmon, whilst that of $\mathrm{Im}[1/(\varepsilon + 1)]$ exhibited a peak due to a surface plasmon at 5.0 eV.

Beryllium films have been investigated by Serymac and Robin (1976) between 2 and 25 eV using a Kramers–Kronig technique operating on reflectance measurements. Maxima were found in n at 2.0 eV and in k at

5.6 eV; there were also plasmons at 13.6 eV (surface) and 19.2 eV (volume). The same technique was used by Endriz and Spicer (1970) for strontium films between 1 and 11.6 eV.

7.4. Noble metals

The noble metals have been investigated by many authors partly because they can be obtained in a very pure state and partly because they form a set of materials which all have one s-electron outside fully filled d-subshells and so may be expected to behave similarly to classical free-electron metals.

Gold has received special attention as it does not readily combine with atmospheric gases and can therefore be used as the model for a clean surface. However, Pletzoder and Steinmann (1968) claimed that even gold surfaces were subject to contamination by small quantities of condensed water vapour and that this might account for the observed fluctuations in the values of the optical constants between different reported measurements.

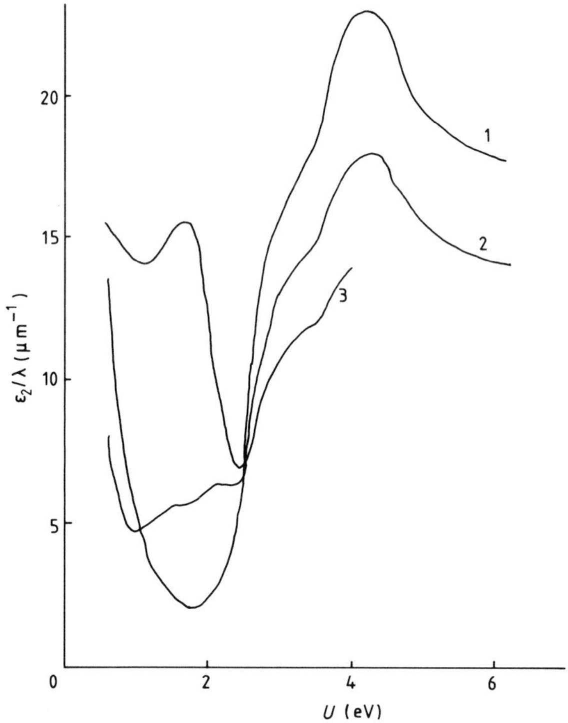

Figure 7.2 ε_2/λ versus U for gold films: 1. well crystallised film; 2, polycrystalline film; 3, island film. (From Thèye M-L 1970 *Phys Rev.* B **2** 3064.)

Thèye (1970) made a thorough investigation of semitransparent gold films (thicknesses 10–25 nm) by means of reflectance and transmittance measurements over the range 0.5 to 6.0 eV (0.2 to 2.5 μm). The state of the film could be varied from minute polycrystals to well organised crystals by adjusting the deposition conditions. In figure 7.2, ε_2/λ is plotted against the photon energy for different film structures showing that the polycrystalline films give considerably lower values than those exhibiting some organised crystal structure. The results for the well crystallised films which had also been annealed were in good agreement with data for bulk specimens. In the range 0.5 to 1.5 eV a Drude-type equation gave a good approximation to the experimental results, leading to an average value of m_0 of 1.05 m. The electron relaxation time τ_{eff} could best be represented as a variable of the form $1/\tau_{\text{eff}} = 1/\tau_0 + b\omega^2$, the final term arising from electron–electron collisions. Figure 7.3. in which $1/\tau_{\text{eff}}$ is plotted against U^2, illustrates this relationship for two of the films. The inter band contribution, obtained by subtracting the Drude values from the experimental results, showed good agreement with Hodgson's results for bulk gold (see figure 7.4 which shows ε_2 versus U). The absorption edge in gold occurred round 2.35 eV and the optical properties in this region are dominated by the transitions from the flat, narrow d-band which lies a few electron volts below the Fermi energy. Impurities in the gold caused some supplementary absorption below the absorption edge.

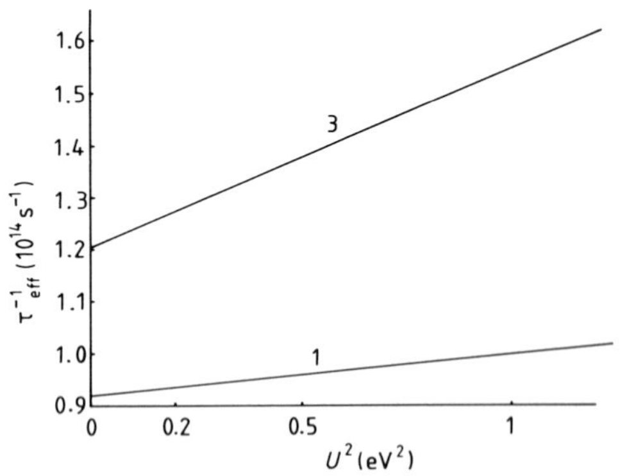

Figure 7.3 $1/\tau_{\text{eff}}$ versus U^2 for gold films; 1, well crystallised film: 3 Island film. (From Thèye M-L 1970 ***Phys. Rev.*** B **2** 3068.)

Aspnes *et al* (1980) investigated the effects on the optical properties of different samples of gold. The major discrepancies between the results above an energy of 2.5 eV (500 nm) could be accounted for by a single-parameter model representing voids in an otherwise homogeneous medium. Ellipsometric measurements made on thin-film specimens which had been

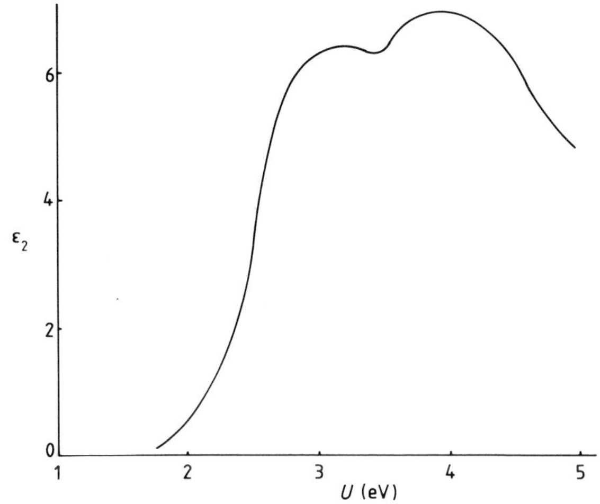

Figure 7.4 ε_2 interband versus U for gold films. (From Thèye M-L 1970 *Phys. Rev.* B **2** 3071.)

characterised by transmission electron microscopy supported this void model but also showed the importance of measuring both ε_1 and ε_2 in order to separate the volume from the surface effects. Differences in the data at energies below the band gap arose from at least two mechanisms; (i) grain size effects in samples with a large volume fraction of imperfections and (ii) increased surface scattering in annealed samples probably due to thermal grooving. Their best data for energies below the band gap, measured on specimens evaporated by electron beam heating on to cleaved rock-salt substrates, showed, unobstructed by the Drude tail, the linearly increasing d-band to Fermi level ($X_5 - U_F$) transition threshold near the X point in the Brillouin zone at 1.8 eV. They obtained a value of ε_2 at 1.5 eV (840 nm) which was identical to that calculated from the electron scattering lifetime derived from the known conductance of bulk metal, indicating that the grain size and the surface quality of their specimens were comparable with bulk materials.

Other contributions on gold films have been made by Richard (1962a, b), Spencer and Givens (1964), Barker (1973), Davey and Pankey (1968) and Casset (1973).

Silver films have been investigated by Huebner *et al* (1964) who measured the transmittances at various angles of incidence in unpolarised light between 250 and 370 nm. Discrepancies had been observed between the optical emission of electron-bombarded silver films and the absorption resonance calculated using the optical constants of bulk silver; better agreement was obtained by employing the optical constants of the silver films which had been calculated from these reflectances. Their results are shown in figure 7.5; a maximum occurs in the transmittance at 320 nm corresponding to a minimum in k at the same wavelength and a maximum in n at 300 nm. The damping term, ε_2, for films was smaller than for bulk silver. The condition for collective electron oscillations occurred at 329 nm (3.77 eV). MacAlister and Stern (1963) also discovered a surface plasmon for silver films at 3.8 eV.

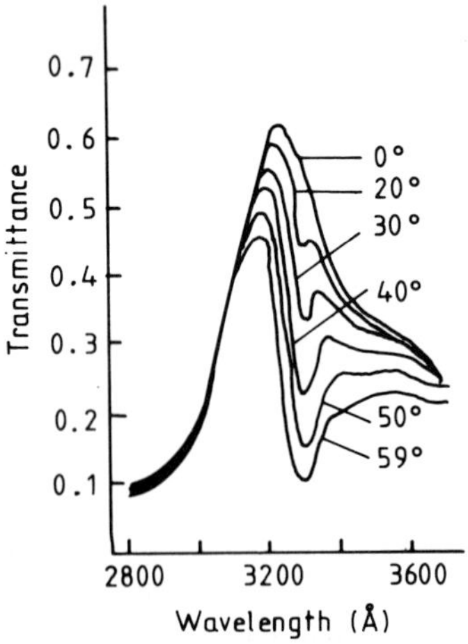

Figure 7.5 T versus λ for thin silver films. (From Huebner R H *et al* 1964 *J. Opt. Soc. Am.* **54** 1436.)

The effects of annealing on the reflectance of silver films were investigated by Vijayakumar and Purnushotaman (1981); the subsequent reduction in reflectance could be explained in terms of the growth of small hillocks on the films.

The optical constants of silver chemically deposited on borosilicate glass have been derived by Abass *et al* (1987) over the energy range 1.6–6.0 eV (0.20–0.78 μm) from measurements of reflectance and transmittance. It was claimed that the values of n and k obtained were the same as those for vacuum deposited films over the same wavelength range; free electron absorption was observed above 0.4 μm whilst interband absorption began to appear below 0.33 μm.

The optical effects of grain boundaries were studied by Liebert and Thomas (1984) using films of copper, silver and gold in the energy range 0.55 to 4.40 eV (0.28 to 2.55 μm). By means of selective reflection, it was possible to investigate only those surface areas which contained the grooves due to grain boundaries. At an energy of 0.70 eV (1.77 μm) absorption was only found in areas containing grain boundaries and this could be described in terms of a Lorentz oscillator of this energy with a half-width of 0.2 eV.

The effect was probably caused by lattice defects and dislocations which had concentrated close to the grain boundaries.

Additional results for thin silver films have he en reported by Barakat and Mokhtar (1963) and by Shklyarevskii and Korneeva (1971).

The reflectances of front surface mirrors of gold and silver with different overcoatings such as SiO, SiO_2 and Al_2O_3 were considered by Hass *et al* (1982). It was shown that the coatings had little effect on the reflectance near normal incidence, but caused a large decrease in R_p for angles of incidence above 45°. In view of this effect, the possibility was considered of using such arrangements as polarising devices.

7.5 Aluminium

Aluminium has also been the subject of considerable investigation regarding its optical properties, largely because of its use for the coating of mirror surfaces by deposition in vacuum. One of the objectives of these studies has been to be able to predict the reflectance of the finished mirrors, but this is made difficult by the readiness with which aluminium oxide is formed in the presence of the slightest amount of air. The highest values of reflectance have been obtained by rapid deposition in ultrahigh vacua of at least 10^{-10} Torr.

Bates and Bradley (1967) measured the reflectances and transmittances of evaporated aluminium films in the ultraviolet and investigated the effects of the evaporation conditions, including the rate of evaporation and the content and pressure of the residual atmosphere. Their results are illustrated in figure 7.6 which shows the relation between functions of reflectance and

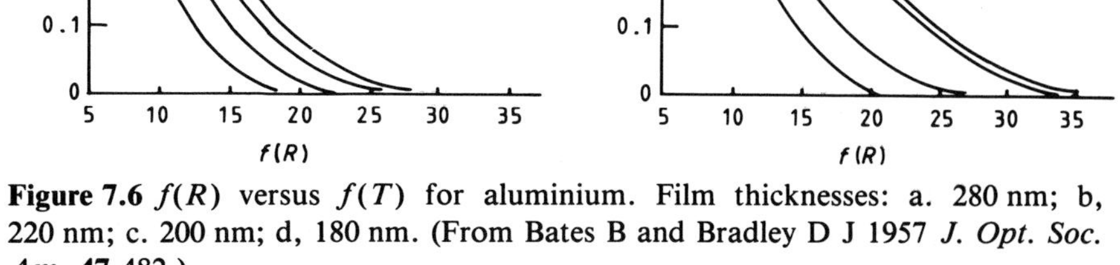

Figure 7.6 $f(R)$ versus $f(T)$ for aluminium. Film thicknesses: a. 280 nm; b, 220 nm; c. 200 nm; d, 180 nm. (From Bates B and Bradley D J 1957 *J. Opt. Soc. Am.* **47** 482.)

transmittance for films produced under different vacuum conditions. The film deposited at the lower pressure clearly gives the higher values of both reflectance and transmittance indicating a decrease in absorption.

Hass and Waylonis (1961) also used extremely fast evaporation rates (50 nm s^{-1}) in their work on thick aluminium films in the ultraviolet and visible spectra. For the opaque films the optical constants were determined by Drude's method whilst the semitransparent films were investigated by measurements of R, R' and T.

Vehse *et al* (1967) measuring the reflectance at normal incidence, extended work by Madden *et al* (1963) from 200 nm down to 79 nm where the reflectance of aluminium is changing rapidly with wavelength. Applying free-electron theory, they were able to explain the values of the normal-incidence reflectance and transmittance in terms of two parameters, ω_p (equal to $5.55 \times 10^{14}\,s^{-1}$) and τ ($1.1 \times 10^{-15}\,s^{-1}$). Their results are shown in figure 7.7 in which the reflectance is plotted against the wavelength for different values of the relaxation time. The relaxation time for the experimental films fell to 0.7×10^{-15} s at 135 nm and then remained constant

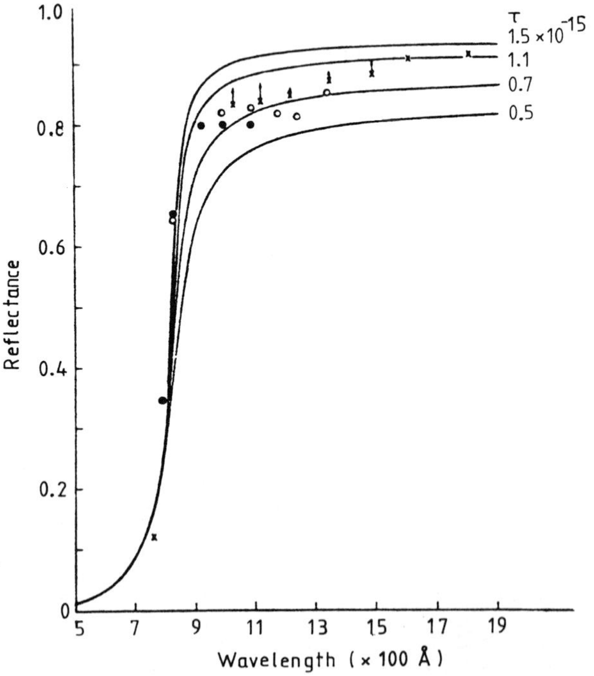

Figure 7.7 Reflectance of aluminium versus wavelength: o. Vehse *et al*; x, Madden *et al*; full curves calculated using free-electron model for $\omega_p = 855$ Å and various values of τ. (From Vehse R C *et al* 1967 *J. Opt. Soc. Am.* **57** 551.)

down to 90 nm; this value was in good agreement with that of Hunter (1962, 1964a, b) who was working at about the same wavelength. It was also noticed that, due to progressive oxidation, the reflectance over this spectral range decreased linearly with time after deposition in a vacuum of 3×10^{-8} Torr.

Another study of the effects of the vacuum deposition conditions on the optical properties of aluminium films was made by Halford *et al* (1973). By controlling the conditions carefully, it was claimed that values of Δ, n and R could be predicted from measurements of the terminal pressure and the terminal deposition rate as illustrated in figure 7.8. A review of the literature on aluminium films carried out by these authors supported the proposition that the optical properties were most seriously affected by oxidation in the volume of the film formed during the deposition process rather than by later surface oxidisation. The surface oxide grew rapidly in air to reach a stable thickness of about 4–5 nm; as the oxide layer increased in thickness the value of Δ decreased. High evaporation rates gave higher values of both Δ and R.

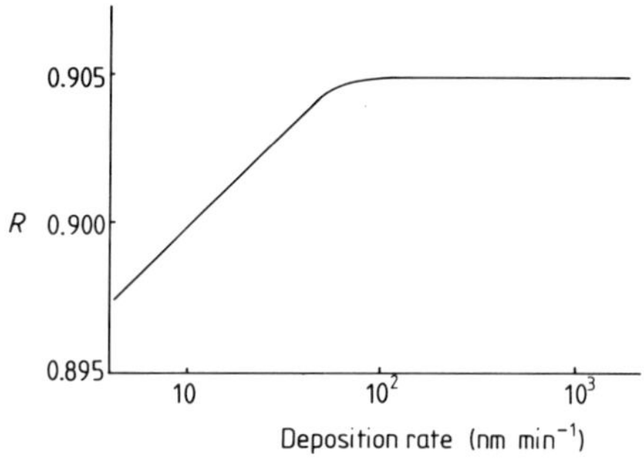

Figure 7.8 R versus deposition rate for aluminium films. (From Halford J H *et al* 1973 *J. Opt. Soc. Am.* **63** 788.)

Fane and Neal (1970) made ellipsometric measurements on aluminium films deposited on glass substrates which were kept at temperatures of −196 or 25 °C. They obtained a wide range of values for n and k which could only be explained if oxidation were taken into account. Their values of n and k for unoxidised films were larger than any previously published at that time; values of n of less than 1.5 and k of less than 6.0 at a wavelength of 549 nm were said to indicate the presence of oxidation. The values of n, which is not sensitive to the film structure, were in good agreement with values calculated from the free-electron theory.

Inagaki *et al* (1985) were able to demonstrate the existence of surface plasmons in a free-standing aluminium film on which regular shallow corrugations had been ruled.

Wesolowska and Dobierzewska-Mozrzymas (1963) have reported the results of an investigation on films of indium. Another study of this metal was made by Theye and Devant (1973) between 0.5 and 6.2 eV when interband transitions were found at 0.5 and 1.5 eV.

7.6 Transition elements

Lenham (1967) and also Lenham and Treherne (1967) measured the absorption of low-energy photons in eleven transition metals, the interband and intraband contributions being empirically separated. For the intraband transitions, a plot of log absorption versus log λ would be linear for any relationship of the form of Drude's equation; this linearity was only observed at the longer wavelengths as the interband components of the absorption extended to very low photon energies (0.1 to 0.4 eV, i.e. 3 to 10 μm) in the transition metals.

Nestell and Christy (1980) surveyed the BCC elements vanadium, niobium, tantalum, chromium, molybdenum and tungsten from measurements of reflectance and transmittance at normal incidence on rapidly evaporated thick films deposited on heated substrates. They claimed that the optical spectra could be accounted for in terms of direct transitions between the d-bands as was shown in Chapter 4 for the same materials in bulk form.

A wide ranging survey of optical properties covering many elements in thin film form has been carried out by Windt *et al* working between 2.4 and 121.6 nm. The elements investigated were Ti, Zr, Nb, Mo, Ru, Rh, Pd, Ag, Hf, Ta, W, Re, Ir, Os, Pb and Au (1988a); C, diamond, Al, and Si (1988b). Thick films (100 nm) of each material were deposited by electron beam evaporation on to the (111) plane of silicon substrates and the optical technique used was reflectance versus angle of incidence. Independent values of the thickness and roughness of the film were used in a least-squares fit process to find n and k. Tables of results are given for each of these elements and also the uncertainties in terms of maximum and minimum values for n and k. The results up to 10 nm were compared with those calculated by Henke's method (1981, 1982) and agreement was generally very good.

7.6.1 Ferromagnetic metals

The elements of iron, nickel and cobalt have been extensively studied in thin-film form because of their commercial use in tape recording and as potential computer storage elements. An early study was that made by Ornstein and van der Veen (1936, 1938, 1939) on iron films. Bueche (1948) carried out measurements on iron, nickel, cobalt and manganese in the visible and near-infrared regions using the reflectances at several angles of incidence. He was able to calculate σ and ε and claimed that his results were in good agreement with other work on these metals. Clemens and Jaumann

(1963) studied films of iron, nickel and cobalt over the region 0.43 to 2.50 μm by measuring the rotation of the plane of polarisation and the ellipticity of reflected light from both sides of the films and also the transmittances and the reflectance from both sides. These six measurements enabled them to calculate both the optical and magneto-optical constants of the materials. Their results for iron are shown in figure 7.9, indicating different values of n and k from the measurements taken on the two sides; a peak in n is observed at about 1.4 μm. The curve for k in cobalt showed a change in slope at 1.2 μm and there was a possibility that a similar change took place for nickel at 0.7 μm, although this was near the extreme wavelength of their measurements. Nag and Ward (1971) calculated n and k for nickel for wavelengths between 1.0 and 2.0 μm and film thicknesses from 5 to 100 nm. Over this region, both n and k were independent of d and approximated to the bulk values. Sonntag *et al* (1969) used synchrotron radiation to measure the absorption coefficients of iron and cobalt between 4 and 60 nm whilst, over a similar wavelength region, Wehenkel and Gauthe (1974) calculated e_1 and e_2 from electron energy loss measurements. Both sets of results are in fair agreement and also agree well with those calculated by Henke's technique. Yu *et al* (1968) who worked between 0.1 and 5 μm measured the normal incidence reflectance of polycrystalline cobalt and then calculated n and k by means of Kramers–Kronig analysis; in the region of wavelength overlap, their results join up well with those of Wehenkel and Gauthe. A well known and often quoted paper covering both iron and cobalt at wavelengths between 0.2 and 2.0 μm was presented by Johnson and Christy (1974) who measured the reflectance and transmittance for film thicknesses of 20 to 50 nm. Their values of n for cobalt were smaller than those of Yu *et al* probably because of more rapid evaporation; the ones for

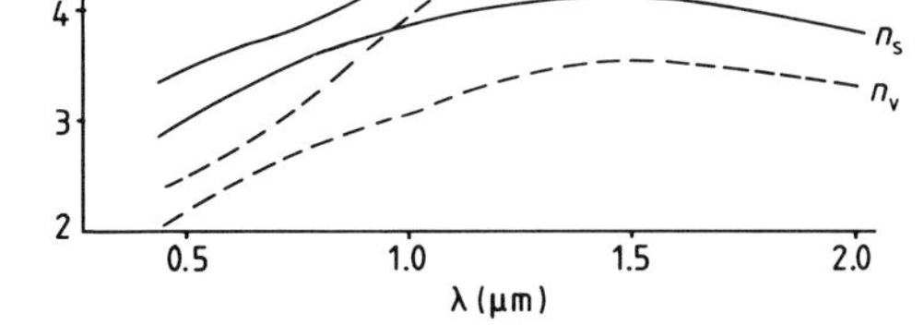

Figure 7.9 n and k versus λ for iron films (v indicates vacuum side and s substrate side). (From Clemens K H and Jaumann J 1963 *Z. Phys.* **173** 137.)

iron were considerably smaller than those reported by Yolken and Kruger (1965) and this may be put down to better prepared surfaces. Even smaller values of *n* were obtained for cobalt by Viguier *et al* (1986) working between 0.25 and 1.25 μm. Kudryavtsev and Lezhnenko (1981) have reported results for both HCP and FCC cobalt films, changing from one form to the other by annealing at 400° C.

Additional authors for the ferromagnetic metals are Reid (1962), Corciovei (1960), Alessandrini (1964) and Germer and Hartmann (1960).

7.6.2 Titanium sub-group

Molybdenum films were investigated by Raymo and Juenker (1963) in the visible and infrared spectral regions. The values of *n* and *k* obtained were consistently smaller than those for the bulk material and it was suggested that this might be due to the polycrystalline structure of the films.

Idczak (1964) has reported a set of values for *n* and *k* for chromium films between 400 and 1000 nm. R, R' and T were measured and Malé's method was employed to calculate the optical constants. She also (Idczak 1967) used double metallic films of chromium and silver, the upper film being partially transparent whilst the lower one was opaque. Chromium has also been studied by Bruce and Clothier (1974) who measured the phase changes in R and T by means of a scanning interferometer, the thicknesses (0–36 nm) being determined by Talystep. Henderson and Weaver (1965, 1966) used both ellipsometry and Male's method in their study of thin films of chromium and copper. The two sets of results could only be made to agree by considering the surface to be composed of ellipsoidal particles, a topic which is discussed further in Chapter 8.

The transmittance of titanium in the vacuum ultraviolet region was measured by Rustgi (1964) who observed absorption due to transitions between the 3p and conduction bands between 32 and 38 eV.

Thin films of vanadium, niobium and molybdenum were studied by Chassaing *et al* (1976) between 0.32 and 5.50 eV (0.2 to 3.8 μm) using two sets of measurements: (i) R, R' and T at normal incidence and (ii) T_s and T_p at angles of incidence of 20, 45 and 70°. For energies above 1.0 eV, all the features in the plots of σ versus U could be explained in terms of interband transitions, a table of which was provided.

Weaver and Olson (1976) studied the 4d transition metals in the form of self-supporting films between 20 and 250 eV. They found evidence for transitions of the type 4p–4d which were analogous to those for 4d–4f in the rare earths. At the highest energies there was optical structure related to the 3d transitions.

Gravier *et al* (1979) have given information on the BCC transition metals vanadium, niobium, tantalum, chromium, molybdenum and tungsten in the energy range 0.32 to 5.6 eV. The results, some of which are shown in figure 7.10, are presented in terms of the conductivity plotted against the photon energy. These curves were compared with histograms of the conductivity calculated from the densities of states. Good agreement was also claimed between the results for thin films and those for bulk materials, both single-crystal and polycrystalline.

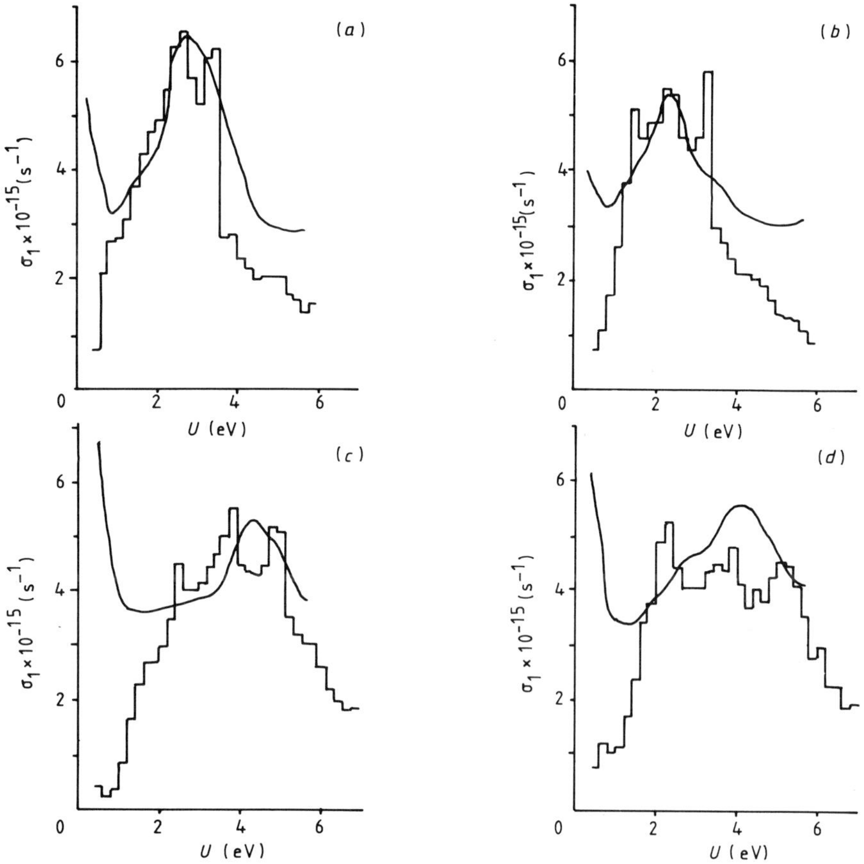

Figure 7.10 σ versus U for films of (a) vanadium, (b) chromium, (c) niobium and (d) molybdenum. (From Gravier P *et al* 1979 *Thin Solid Films* **57** 95.)

7.6.3 Platinum sub-group

Jacobus *et al* (1963) have published a set of results for platinum films which had been prepared by evaporation by induction heating, measurements of the reflectance being made at 58, 74 and 122 nm. No change in reflectance was observed after the specimens were stored for a prolonged period in air, indicating the inert nature of this metal. Their results, which are illustrated in figure 7.11, show the beneficial effects of higher substrate temperatures

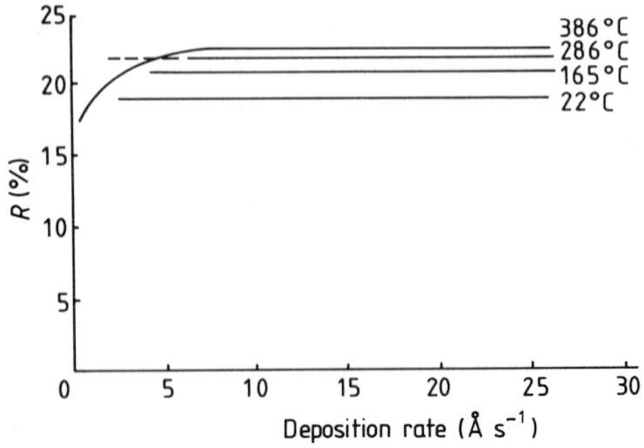

Figure 7.11 *R* versus deposition rate for platinum films deposited at different substrate temperatures. (From Jacobus G F *et al* 1963 *J. Opt. Soc. Am.* **53** 1086.)

on the reflectance. Another investigation on platinum was made by Hunter *et al* (1979) when a strong peak in $\mathrm{Im}(1/\varepsilon)$ was observed at 40 nm (31 eV). In this case, the optical properties were not altered significantly by changes in the temperature of the substrate during deposition.

Palladium has been studied by Malé and Trompette (1957) from measurements of R, R' and T on films up to 300 nm thick, and by Abelès *et al* (1963) who used photometry to measure the optical properties of palladium films between 190 and 2500 nm.

Rhodium has received attention from Coutler *et al* (1973) over the wavelength region between 400 and 700 nm. Another investigation on this metal was carried out by Hoffman and Coutts (1976) who studied the effects of oxygen on the optical properties of thin films of rhodium. It was found that in oxygen at a pressure between 10^{-6} and 10^{-4} Torr a more uniform crystallite size was deposited due to control by the adsorbed gas.

Ahrndt *et al* (1984) carried out an interesting experiment in which samples of rhodium films, of two different thicknesses and all prepared in a single run, were sent to seven measuring laboratories. A number of different optical techniques were employed including reflection ellipsometry, wideband spectrophotometry and Strong's method for reflectance. Two methods of calculating the optical constants were used: Bennett and Booty's iteration technique (1966) and an envelope method (Manuficier *et al* 1976). The results from all the laboratories were in reasonably good agreement but different by a few parts in a thousand.

Aspnes and Craighead (1986) have continued the analysis of the results of these rhodium films. They concluded that the scatter in the values of the optical constants could be explained by errors in the thickness values which must be known to ± 1 Å; to obtain the best results, values by both photometric and ellipsometric techniques were needed. It was possible to fit the results to an effective medium theory (see chapter 8); the modified theory due to Sen *et al* (1981) was a better fit than Bruggemann's original theory (1935).

Ruthenium was investigated by Cox *et al* (1974) in the ultraviolet between 30 and 200 nm. The films were evaporated by electron bombardment and deposited on both glass and quartz substrates at temperatures of 40 and 300 K. Variation of the rate of deposition between 1 and 80 $\text{Å}\,s^{-1}$ had little effect on the reflectance values, presumably because ruthenium is less prone to oxidisation than metals such as aluminium where the reflectance is strongly affected by the deposition rate. This was further borne out as there was little evidence of the effects of oxygen films on the optical properties.

Rhenium, tungsten and osmium were also studied by Cox *et al* (1972, 1973) over the same range of wavelengths, the reflectance being measured at number of angles of incidence. The results for osmium are shown in figure 7.12. It is interesting to note the reduction in reflectance which occurred on increasing the thickness of osmium. This effect, which was caused by the interference between waves reflected at the top and bottom surfaces of the osmium layer, was greatest when the film was approximately half a wavelength in thickness (580 Å).

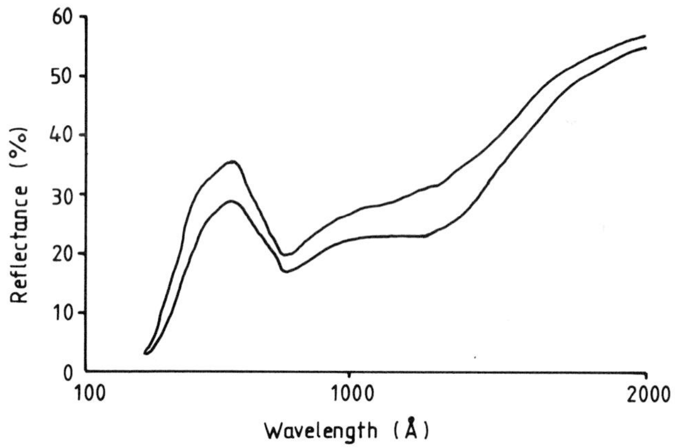

Figure 7.12 Reflectance of evaporated osmium films. Upper curve, $d = 170$ Å; lower curve, $d = 580$ Å. (From Cox J T *et al* 1973 *J. Opt. Soc. Am.* **63** 437.)

A study of iridium was made by Samson *et al* (1967), stimulated by the use of this metal for high-reflectance mirrors in the ultraviolet spectrum where the reflectance of aluminium falls off; the reflectance and transmittance were determined between 25 and 150 nm. On the basis of the 6s electrons, the transmittance would be expected to increase rapidly at the plasma resonance at 89 nm, but the measured value was 65 nm. This difference was due to the involvement of the 5d electrons which have a plasma resonance at 42 nm.

Iridium, deposited on glass substrates, has been studied by Osantowski *et al* (1974), a Kramers-Kronig type analysis being employed between 0.22 and 40.0 μm. Hass *et al* (1967) used the reflectance versus angle of incidence technique in their measurements on iridium at various temperatures. Little change in reflectance was observed on exposure to air and the highest value of reflectance was obtained when the substrate temperature was above 300 °C, a possible explanation being that such films contained the smallest number of grain boundaries.

Sigrist *et al* (1987) measured the near normal incidence reflectance of polycrystalline films of scandium between 0.22 and 10 eV. (0.13 to 5.7 μm); they then used Kramers–Kronig analysis to extract values of e_1 and e_2. Good agreement was obtained between their interband component of e and the electron energy bands of scandium proposed by Das (1976).

7.7 Other metals

A set of optical constants for selenium was obtained by Reale (1976) from measurements of reflectance, transmittance and, independently, the thickness. He found that selenium behaved in the typically metallic fashion and that its reflectance increased as the lattice constant decreased.

Tin was investigated by Codling *et al* (1966) between 80 and 100 nm (12 to 150 eV). X-ray edges were discovered at 23.8 and 24.9 eV and were identified with the 4s and 4p electron energy bands. There was also a broad absorption feature near 19 nm caused by electron transitions from the 4d to the conduction band. Another study on tin, using near-normal reflectance, has been reported by McRae *et al* (1967). They analysed the results over the range 0.1 to 27.5 eV, finding interband transitions at 1.2, 3.0 and 24.5 eV, a volume plasmon at 13.4 eV and a surface plasmon at 9.2 eV.

Harris and Piper (1963) measured the reflectance and transmittance in the far infrared for epitaxial bismuth films on Al_2O_3 substrates. The films ranged in thickness from 22 to 900 nm and all of them showed a maximum in the reflectance curve at about 30 μm with a corresponding minimum in transmittance at the same wavelength. n and k are equal at this point, indicating a volume plasmon. Their results are shown in figure 7.13.

Barakat and El-Kadry (1970) investigated antimony films at a wavelength of 588 nm by measuring the phase changes on reflection at the air–antimony and antimony–glass interfaces. For thicknesses greater than 30 nm the films behaved like bulk materials, but at smaller thicknesses they behaved like semiconductors.

A study was made of thick antimony films on Al_2O_3 substrates by Harris and Corrigan (1964). Using measurements of reflectance and transmittance between 4 and 140 μm, the optical and electrical constants were calculated and hence the complex dielectric constant and the relaxation times. In antimony, conduction takes place by both electrons and holes so the simple Drude theory would not be expected to hold. k showed a minimum centred around about 7 μm whilst n had a flat minimum centred round about 30 μm.

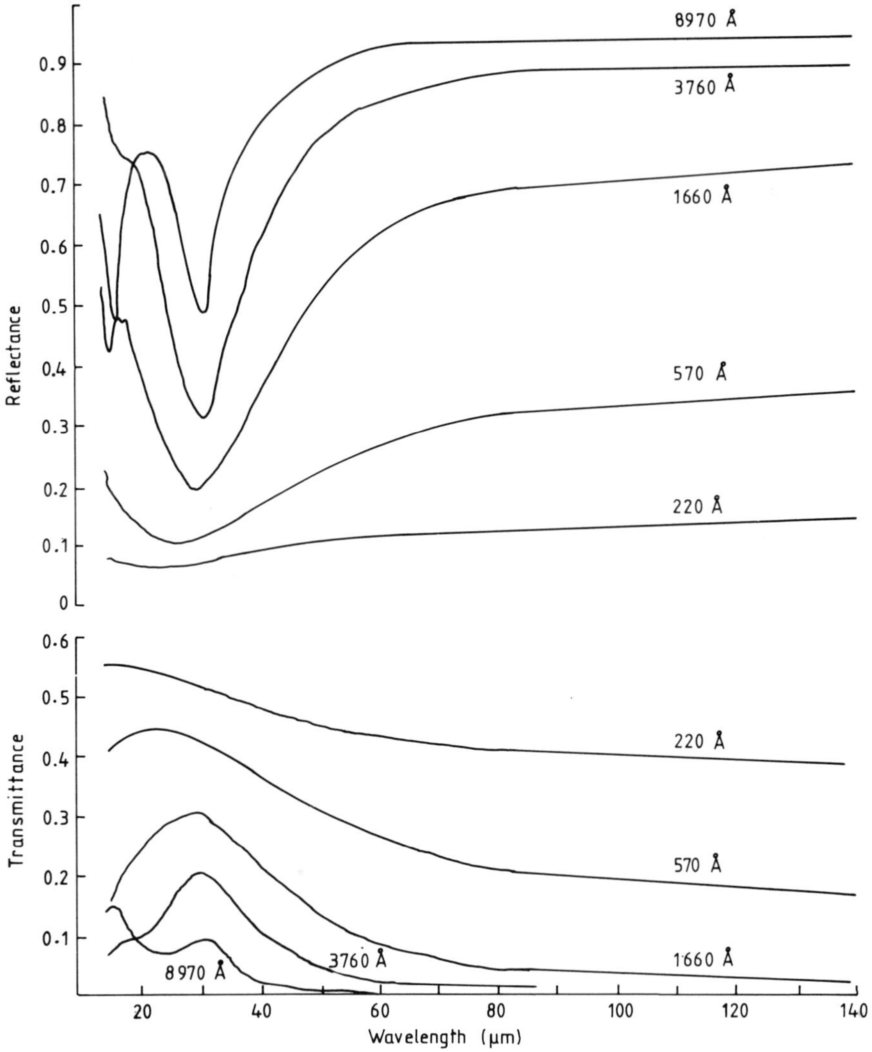

Figure 7.13 R and T versus λ for bismuth films of various thicknesses. (From Harris L and Piper J 1963 *J. Opt. Soc. Am.* **53** 1274.)

Toots and Marton (1969) carried out measurements on films of antimony and bismuth between 10 and 26 eV (48–124 nm). The results for antimony could be fitted to an inverted Drude–Sellmeier type equation, but this was not possible with bismuth.

Petrakian (1972) has reported results for the rare earth groups. All the lanthanides have an absorption band at around 5.7 eV (218 nm), probably due to transitions between the d-bands. On the other hand, there was no evidence of transitions from the 4f bands. The heavy lanthanides also produced absorption in the range 2.5 to 3.5 eV, again due to a direct transition. In figure 7.14 the reflectance of a set of films of the heavy lanthanide thulium are plotted against the wavelength; the two peaks are clearly visible for all the films.

Muller (1963) has reported work on lead films; Martinuzzi (1961) has given results for gallium and arsenic; gallium has also been studied by Lenham (1967).

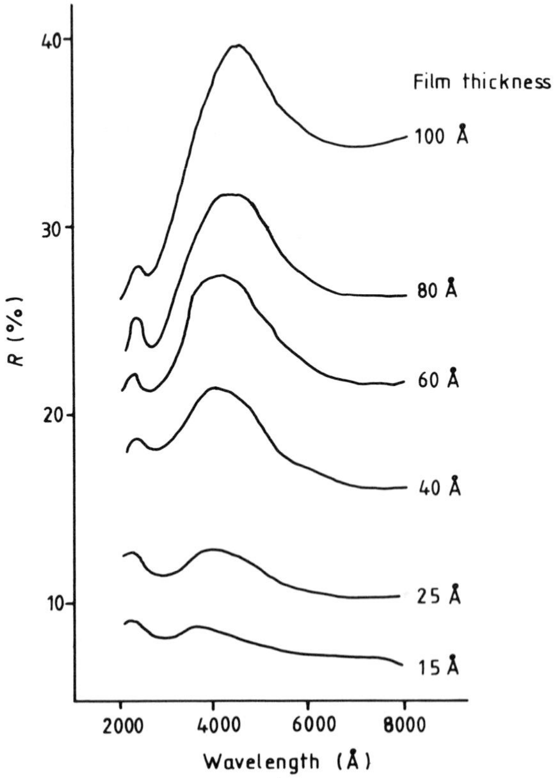

Figure 7.14 Reflectance versus wavelength of six thin thulium films. (From Petrakian J P 1972 *J. Opt. Sol. Am.* **62** 401.)

7.8 Alloys

Harrison and Baumeister (1973) made composite films by the simultaneous evaporation of silver and chromium, and the optical constants were determined by the pseudo Brewster's angle method. It was found that the values of n and k were roughly proportional to the percentage composition of the mixtures.

Silver–gold films were studied by Irani *et al* (1971) using a Kramers-Kronig analysis of reflectance data over the energy range 2.0 to 11.5 eV (100 to 600 nm). The extrapolations used in the analysis were as follows: as $U \to 0$, the reflectance was made asymptotic to 1 the relation being assumed to be linear; between 11.5 and 40 eV the results of other workers were employed, whilst beyond 40.0 eV an empirical power law was assumed.

Harris and Washwell (1972) investigated the optical properties of films of mixtures of bismuth and tellurium, reflectance and transmittance measurements being made between 15 and 70 μm. A minimum, which was identified with a plasma resonance, was observed in the reflectance and this moved towards shorter wavelengths as the proportion of tellurium increased.

Aagard (1974) studied the optical constants of co-deposited bismuth and manganese films by means of measurements of the reflectance and transmittance. These films, overcoated with silicon dioxide, were considered as possible memory devices.

Amorphous films of gold–germanium alloys were investigated by Nguyen Van Dong *et al* (1982) at energies from 0.6 to 4.5 eV. Drude's model was used and good fits were obtained for ε_1 at the higher values of energy and for ε_2 at small energies.

7.9 Dielectrics

The refractive indices of films of dielectric materials are especially important because of their uses in a variety of commercial applications such as blooming, interference filters and microcircuit capacitor elements. Weaver (1962) has given a comprehensive review of the properties of films of dielectric materials covering such topics as preparation, measurement of the dielectric constant, the optical properties, structure and water absorption. More recent reviews covering similar areas have been given by Ritter (1975) and Pulker (1979).

Powell and Morad (1978) measured the transmittance of unbacked films of silicon dioxide in the vacuum ultraviolet spectrum, a band gap of 8.0 ± 0.2 eV being observed. Weinberg *et al* (1979) measured the transmittance of amorphous silicon dioxide (glass) films between 7 and 14 eV (88 to 176 nm). Their results, together with those of Powell and Morad, are illustrated in figure 7.15; the band gap of silicon dioxide was calculated to be 9.3 eV.

Heavens and Smith (1957) have given detailed results for the optical properties of films of MgF_2, CaF_2, ZnS and LiF determined by a Brewster's angle technique and by reflectances.

Zukic *et al* (1990) have published graphical data of n and k versus λ for thin films of BaF_2, CaF_2, LaF_3, HfO_2, AlO_3 and SiO_2 between 120 and 230 nm using photometry.

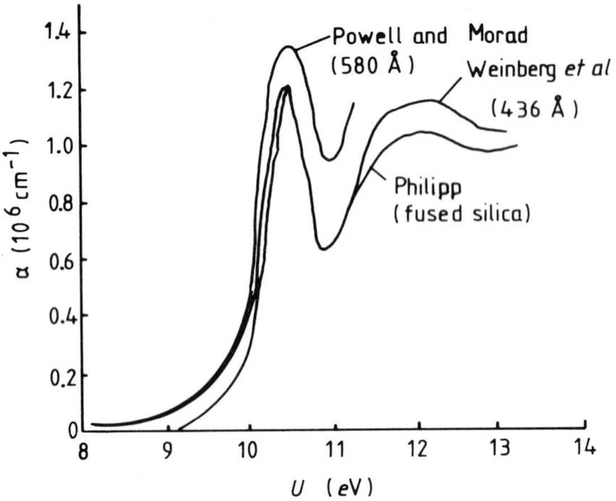

Figure 7.15 Results of Weinberg *et al* and Powel and Morad for SiO_2 films. (From Weinberg Z A *et al* 1979 *Phys. Rev.* B **19** 341.)

Hacskaylo (1964) determined the refractive indices of dielectric films from measurements of the Brewster's angle and claimed that they were accurate to between 0.0002 and 0.0006. The method was independent of both the thickness of the film and the refractive index of the substrate.

An investigation of dielectric films on dielectric substrates has been reported by Ruiz-Urbita *et al* (1971). The reflectance was measured as a function of angle of incidence and the maxima and minima in reflectance which were exhibited were used to calculate the refractive indices.

7.10 Semiconductors

Archer (1962), working with silicon films, made ellipsometric measurements from which he could calculate n, k and d. He claimed that the values of these quantities bore out the predictions of the exact theory of the optics of thin films.

The optical properties of germanium in the far ultraviolet were measured by Marton and Toots (1967) using the reflectance versus angle of incidence technique. The thicknesses of the films were between 49 and 160 nm and optical measurements were made with the specimens kept in an ambient pressure of 5×10^{-7} Torr. Their values of ε_1, ε_2, n and k as functions of photon energy are shown in figure 7.16. The function $-\mathrm{Im}(1/\varepsilon)$ showed a peak at 16.1 eV (78 nm) with a half width of 3.9 eV (320 nm); using a sum rule for the electron oscillator strengths between 8 and 25 eV gave a value for the plasmon energy of 13.7 eV (92 nm) and it was claimed that this was in better agreement with plasma theory than any previous measurements on germanium.

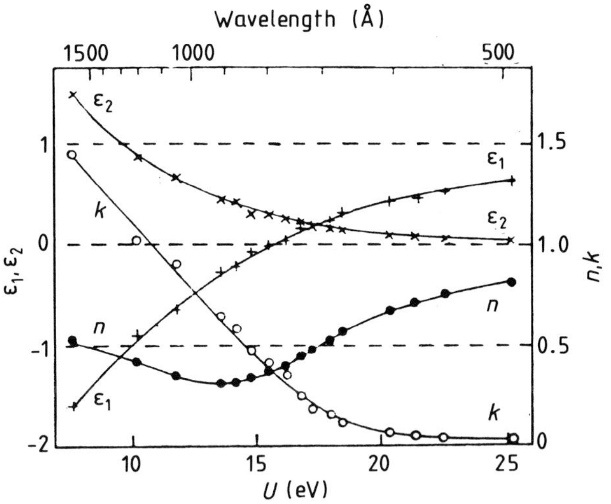

Figure 7.16 n, k, ε_1 and ε_2 versus U for germanium films. (From Marton L and Toots T 1967 *Phys. Rev.* **160** 603.)

Another set of optical constants for germanium films was obtained by Wales *et al* (1967). The films were deposited on sapphire, glass and quartz substrates and n was determined from the wavelengths at which the maxima and minima in transmittance occurred due to interference effects in the film. It was found that n was about 5% larger than for bulk germanium and was relatively independent of the rate of deposition and the ambient pressure.

A set of values of n and k for germanium films in the near infrared was reported by Nag and Ward (1971). n was observed to decrease with film thickness and, as the thickness increased, it was asymptotic to the values for bulk germanium.

Goldschmidt (1984) measured the optical properties of single-crystal germanium up to 500 °C by means of the ratio R_p/R_s at the pseudo Brewster's angle. He claimed that the variation in thickness of the surface oxide was small enough to be ignored when using this technique, although this would not be the case with ellipsometry.

Optical constants for cadmium sulphide films have been reported by Gottesman and Ferguson (1954) who plotted n against λ for different evaporation rates, noting that, even at the highest rates (90 nm min^{-1}), the values of n were considerably less than those for bulk material, as illustrated in figure 7.17.

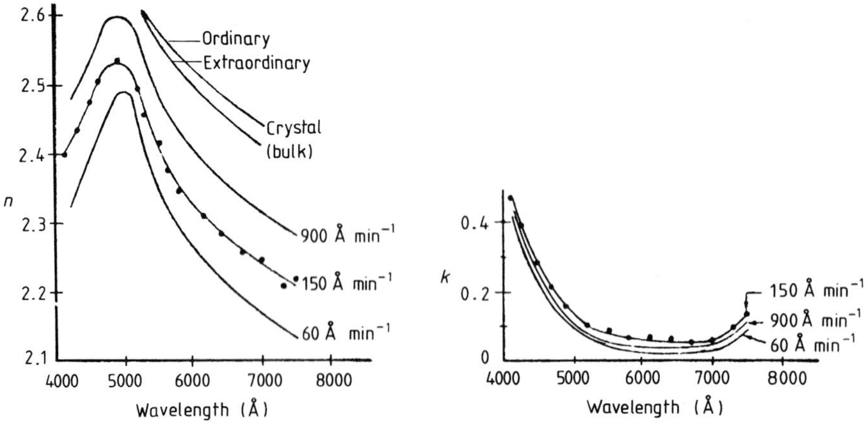

Figure 7.17 The optical constants of CdS films. Measured values are shown for one rate of deposition to illustrate precision. (From Gottesman J and Ferguson W F C 1954 *J. Opt. Soc. Am.* **44** 369.)

Wendlund (1962) also studied the changes in the optical properties of evaporated films of cadmium sulphide due to (i) the substrate temperature, (ii) the source temperature and (iii) the heating of the film. As the temperature of the substrate was increased from 50 to 100 °C the transmittance increased whilst the proportion of amorphous material decreased. The transmittance also increased as the source temperature decreased again combined with a decrease in the amount of amorphous film material. The effects of heating on the film were complicated, depending on the evaporation conditions. When heated above 350 °C, there was a sharp increase in reflectance for films which had been deposited on substrates heated above 60 °C, but this effect was not observed for substrate temperatures below 60 °C.

Khawaja and Tomlin (1975) investigated cadmium and zinc sulphides in the range 250 to 2500 nm (0.5 to 5 eV) through measurements of the reflectance and transmittance at normal incidence; they also took into account the surface roughness. The results for cadmium sulphide could be explained as follows: direct transitions only in the region 2.43 to 2.83 eV, followed at higher energies by a mixture of direct and indirect transitions between electron energy bands which were assumed to be parabolic in shape. A similar explanation applied to zinc sulphide.

The structure and optical properties of RF sputtered films of cadmium sulphide were reported by Martil de la Plaza *et al* (1984). The grain size of the films could be controlled by either the target potential or the substrate temperature. The band gap was determined to be between 2.23 and 2.38 eV, which was close to the value for a single crystal.

Soera and Seryduk (1960) found a complicated structure in the curve of reflectance versus wavelength near the main absorption edge of CdS and this was attributed to the presence of excitons. Hall and Ferguson (1955) measured R and T between 0.6 and 14 μm; n decreased steadily with wavelength in this region over which CdS behaves as a normal dispersive medium.

Results of n and k for films of CdS are included in figure 4.19 in chapter 4.

The optical properties of thin films of ZnS, ZnSe and ZnTe have been reviewed in the paper of Li (1984).

The absorption coefficient of evaporated films of zinc selenide between 36 and 150 eV was measured by Cardona and Haensel (1970) using synchrotron radiation; good agreement was obtained between their derived values of k and those calculated by Henke's method (1982). In the spectral region 0.3–2.0 μm Thutupalli and Tomlin (1976) measured R and T at temperatures of 300 and 700 K. At 300 K only cubic crystals were found but raising the temperature led to the appearance of the hexagonal form.

For zinc telluride, in the X-ray region, Cardona and Haensel found an absorption doublet at 13.3 and 13.9 nm corresponding to the M-transition in Zn. Gauthe (1959) and also Tomoda and Mannami (1969) used the electron energy loss technique over the photon energy range 2–15 eV (83–625 nm) finding a series of absorption peaks due to excitation of (*a*) the valence electron plasmons and (*b*) cation d-electrons and anion s-electrons. Using transmittance measurements, Miloslavski *et al* (1975) worked between 0.44 and 1.0 μm finding the main energy gap at 0.526 μm; this gap size was confirmed by Chu *et al* (1986) working with ZnTe films on glass and ceramic.

Values of n and k for thin films of ZnS, ZnSe and ZnTe are included in figure 4.20.

Zemal *et al* (1965) have given sets of results for epitaxial films of PbS, PbSe, PbTe and SnTe; another series of values for PbS are those by Schoolar and Dixon (1965).

References

Aagard R L 1974 *J. Opt. Soc. Am.* **64** 1456
Abass A K, Mukhtar D A and Adnan N R 1987 *Acta Phys. Hung.* **61** 281
Abelès F, Lostis P and Wilmanns L 1963 *J. Opt. Soc. Am.* **53** 1332
Ahrndt D P *et al* 1984 *Appl. Opt.* **23** 3571
Alessandrini E I 1964 *J. Appl. Phys.* **31** 2085
Archer R J 1962 *J. Opt. Soc. Am.* **52** 970
Aspnes D E, Kinsbron E and Bacon D D 1980 *Phys. Rev.* B **21** 3290
Aspnes D E and Craighead H G 1986 *Appl. Opt.* **25** 1299
Barakat N and El-Kadry N 1970 *J. Opt. Soc. Am.* **60** 570L
Barakat N and Mokhtar S 1963 *J. Opt. Soc. Am.* **53** 1153
Barker A S 1973 *Phys. Rev.* B **8** 5418
Bates B and Bradley D J 1967 *J. Opt. Soc. Am.* **57** 481
Bennett J M and Booty M J 1966 *Appl. Opt.* **5** 41
Brown H M and Brodie D E 1972 *Can. J. Phys.* **50** 2502
Bruce C F and Clothier W K 1974 *J. Opt. Soc. Am.* **64** 823
Bruggemann D A G 1935 *Ann. Phys., Lpz* **24** 636
Bueche F 1948 *J. Opt. Soc. Am.* **38** 806
Calcott T A and Arakawa E T 1974 *J. Opt. Soc. Am.* **64** 839
Cardona M and Haensel R 1970 *Phys. Rev.* B **1** 2605
Casset J 1973 *Thin Solid Films* **18** 99
Chassaing G, Gravier P and Sigrist M 1976 *Thin Solid Films* **35** L25
Chu T L, Chu S S, Firszt F and Herrington C 1986 *J. Appl. Phys.* **59** 1259
Clemens K H and Jaumann J 1963 *Z. Phys.* **173** 135
Codling K, Madden R P, Hunter W R and Angel D W 1966 *J. Opt. Soc. Am.* **56** 189
Corciovei A 1960 *Czech. J. Phys.* **10** 917
Coutler J K, Hass G and Ramsey J B 1973 *J. Opt. Soc. Am.* **63** 1149
Cox J T, Hass G and Jacobus G F 1961 *J. Opt. Soc. Am.* **51** 714
Cox J T, Hass G, Ramsey J B and Hunter W R 1972 *J. Opt. Soc. Am.* **62** 781
—— 1973 *J. Opt. Soc. Am.* **63** 435
—— 1974 *J. Opt. Soc. Am.* **64** 423
Davey J E and Pankey T 1968 *J. Appl. Phys.* **36** 2571
Das S G 1976 *Phys. Rev.* B **13** 3978
Deitschel H 1963 *Z. Phys.* **174** 136
El Naby M H 1963 *Z. Phys.* **174** 269
Endriz T G and Spicer W E 1970 *Phys. Rev.* B **2** 1466
Fane R W and Neal W E J 1970 *J. Opt. Soc. Am.* **60** 790
Fisher E I, Fujita I and Weissler G L 1966 *J. Opt. Soc. Am.* **56** 1560
Fleischmann R 1963 *Z. Phys.* **174** 102
Francombe W H and Johnson J E 1969 *Phys. Thin Films* **5** 143
Gauthe B 1959 *Phys. Rev.* **114** 1265
Germer L H and Hartmann C D 1960 *J. Appl. Phys.* **31** 2085
Goldschmidt D 1984 *Proc. Int. Soc. Opt. Eng.* **452** 110
Gottesman J and Ferguson W F C 1954 *J. Opt. Soc. Am.* **44** 368
Gravier P, Chassaing C and Sigrist M 1979 *Thin Solid Films* **57** 93

Hacskaylo M 1964 *J. Opt. Soc. Am.* **54** 198
Halford J H, Chin F K and Norman J E 1973 *J. Opt. Soc. Am.* **63** 786
Hall J F and Ferguson W F C 1955 *J. Opt. Soc. Am.* **45** 714
Harris L and Corrigan F R 1964 *J. Opt. Soc. Am.* **54** 1437
Harris L and Piper J 1963 *J. Opt. Soc. Am.* **53** 1271
Harris L and Washwell E R 1972 *J. Opt. Soc. Am.* **62** 1162
Harrison D H and Baumeister P W 1973 *J. Opt. Soc. Am.* **63** 1281
Hass G, Heaney J B and Hunter W R 1982 Phys. *Thin Films* **12** 2
Hass G, Jacobus G F and Hunter W R 1967 *J. Opt. Soc. Am.* **57** 758
Hass G and Waylonis J E 1961 *J. Opt. Soc. Am.* **51** 719
Heavens O S and Smith S D 1957 *J. Opt. Soc. Am.* **47** 469
Henderson G and Weaver C 1965 *J. Opt. Soc. Am.* **55** 596
—— 1966 *J. Opt. Soc. Am.* **56** 1551
Henke B L 1981 *AIP Conf. Proc.* **75** 146
Henke B L, Lee P, Tanoka R, Shimabukura R and Fujikawa B K 1982 *At. Data Nucl. Data Tables* **27** 1
Hoffman D M and Coutts M D 1976 *J. Opt. Soc. Am.* **66** 163
Huebner R H, Arakawa E T, MacRae R A and Hamm R N 1964 *J. Opt. Soc. Am.* **54** 1434
Hunter W R 1962 *J. Opt. Soc. Am.* **52** 1310
—— 1964a *J. Phys., Paris* **25** 154
—— 1964b *J. Opt. Soc. Am.* **54** 15
Hunter W R, Angel D W and Hass G 1979 *J. Opt. Soc. Am.* **69** 1695
Hutcheson E T, Hass G and Cox J T 1972 *Appl. Opt.* **11** 2245
Idczak E F 1964 *Opt. Spectrosc.* **17** 501
—— 1967 *Opt. Spectrosc.* **22** 507
Inagaki T, Matosuga M. Arakawa E T and Goudonnet J P 1985 *Phys. Rev.* B **31** 2548
Irani G B, Hurn T and Wooten F 1971 *J. Opt. Soc. Am.* **61** 128
Jacobus G F, Madden R P and Canfield L R 1963 *J. Opt. Soc. Am.* **53** 1084
Johnson P B and Christy R W 1974 *Phys. Rev.* B **9** 5056
Khawaja E E and Tomlin S G 1975 *J. Phys. D: Appl. Phys.* D **8** 581
Kudryavtsev Yu Y and Lezhnenko I V 1981 *Sov. Phys. Solid State* **23** 249
Lenham A P 1967 *J. Opt. Soc. Am.* **57** 1967
Lenham A P and Treherne D M 1967 *J. Opt. Soc. Am.* **57** 476
Li H H 1984 *J. Phys. Chem. Ref. Data* **13** 103
Liebert M and Thomas L K 1984 *Z. Metallkd.* **75** 273
MacAlister A J and Stern E A 1963 *Phys. Rev.* **132** 1599
McRae R A, Arakawa E T and Williams M W 1967 *Phys. Rev.* **162** 615
Madden E P, Canfield L R and Hass G 1963 *J. Opt. Soc. Am.* **53** 620
Male D and Trompette J 1957 *J Phys. Radium* **18** 128
Manuficier J C, Gasiot J and Fullard J P 1976 *J. Phys. E: Sci. Instrum.* **9** 1002
Martil de la Plaza I, Gonzales-Diaz G, Sanchez-Quesada F and Rodriguez-Vidal M 1984 *Thin Solid Films* **120** 31
Martinuzzi S 1961 C. R. *Acad. Sci., Paris* **252** 1314
Marton L and Toots T 1967 *Phys. Rev.* **160** 602

Mayer H and El Naby M H 1963 *Z. Phys.* **174** 289
Miloslavskii V K and Nabockina E N 1975 *Opt. Spectosc.* **38** 565
Muller H 1963 *Z. Phys.* **174** 149
Nag A and Ward L 1971 *J Phys. D: Appl. Phys.* **4** 829
Nestell J E and Christy R W 1980 *Phys. Rev.* B **21** 3173
Nguyen Van Dong V, Fisson S and Thèye M-L 1982 *Thin Solid Films* **89** 315
Ornstein L S and van der Veen J H 1936 *Physica* **3** 289
—— 1938 *Physica* **5** 175
—— 1939 *Physica* **6** 439
Osantowski J F, Hunter W R and Hass G 1974 *J. Opt. Soc. Am.* **64** 1383
Payan R and Roux D 1969 *Opt. Commun.* **1** 37
Petrakian J P 1972 *J. Opt. Soc. Am.* **62** 401
Pletzoder K and Steinmann W 1968 *J. Opt. Soc. Am.* **58** 588
Powell R J and Morad M 1978 *J. Appl. Phys.* **49** 2499
Pulker H K 1979 *Appl. Opt.* **18** 1969
Rasigni G and Rasigni M 1973 *J. Opt. Soc. Am.* **63** 775
Raymo C T and Juenker D W 1963 *J. Opt. Soc. Am.* **53** 1332
Reale C 1976 *J. Opt. Soc. Am.* **66** 1416
Reid F J 1962 *Rev. d'Opt.* **41** 566
Richard J 1962a *C. R. Acad. Sci., Paris* **254** 2762
—— 1962b *C. R. Acad. Sci., Paris* **255** 291
Ritter E 1975 *Phys. Thin Films* **8** 1
Rivoira R and Blanc R 1967a *C. R. Acad. Sci,. Paris* **265** 940
—— 1967b *C. R. Acad. Sci., Paris* **265** 1482
—— 1967c *C. R. Acad. Sci., Paris* **264** 634
Rouard P and Meesen A 1977 *Prog. Opt.* **XV** 77
Ruiz-Urbita M, Sparrow E M and Eckert E R G 1971 *J. Opt. Soc. Am.* **61** 1392
Rustgi O P 1964 *J. Opt. Soc. Am.* **54** 1387
Samson J A R, Padur J D and Sharma A 1967 *J. Opt. Soc. Am.* **57** 966
Schoolar R B and Dixon J R 1965 *Phys. Rev.* **137** A667
Schopper H 1963 *Z. Phys.* **174** 125
Sen P N, Scala C and Cohen M H 1981 *Geophys.* **46** 781
Serymac A and Robin S 1976 *Solid State Commun.* **19** 343
Shklyarevskii I N and Korneeva T 1971 *Opt. Spectrosc.* **31** 44
Sigrist M, Chassaing O, Francois J C, Antonangeli F, Zema N and Piacentini M 1987 *Phys. Rev.* B **35** 3760
Soera T Y and Seryduk V V 1960 *Opt. Spectosc.* **9** 210
Sonntag B, Haensel R and Kunz C 1969 *Solid State Commun.* **7** 597
Spencer W T and Givens M P 1964 *J. Opt. Soc. Am.* **54** 1337
Sutherland J C and Arakawa E T 1968 *J. Opt. Soc. Am.* **58** 1080
Sutherland J C, Arakawa E T and Hamm R N 1967 *J Opt. Soc. Am.* **57** 645
Thèye M-L 1970 *Phys. Rev.* B **2** 3060
Thèye M-L and Devant G 1969 *Thin Solid Films* **41** 205
Thutupalli G K M and Tomlin S G 1976 *J. Phys. D: Appl. Phys.* **9** 1639
Toots J and Marton L 1969 *J. Opt. Soc. Am.* **59** 1305
Tomoda T and Mannami M 1969 *J. Phys. Soc. Japan* **27** 1204
Vehse R C. Arakawa E T and Stanford J L 1967 *J. Opt. Soc. Am.* **57** 551

Viguier C, Cros A, Humbert H, Ferrin C, Thomas O, Madar R and Senateur J P 1986 *Solid State Commun.* **60** 923
Vijayakumar K P and Purnshotaman C 1981 *Thin Solid Films* **82** 225
Volken H T and Kruger J 1965 *J. Opt. Soc. Am.* **55** 842
Wagnerberger W 1963 *Z. Phys.* **174** 161
Waldron J P and Juenker D W 964 *J. Opt. Soc. Am.* **54** 204
Wales J, Lovitt G J and Hill R A 1967 *Thin Solid Films* **1** 137
Weaver C 1962 *Adv. Phys.* **11** 83
Weaver J H and Olson C G 1976 *Phys. Rev.* B **14** 3251
Weinberg Z A, Rubloff G W and Bassous E 1979 *Phys. Rev.* B **19** 3107
Wehenkel C and Gauthe B 1974 *Phys. Status Solidi* B **64** 515
Wendlund P H 1962 *J. Opt. Soc. Am.* **52** 581
Wesolowska C and Dobierzewska-Mozrzymas E 1963 *Acta. Phys. Pol.* **24** 65
Whang U S, Arakawa E T and Callcott T A 1971 *J. Opt. Soc. Am.* **61** 740
—— 1972 *Phys. Rev.* B **6** 2109
Windt D I, Cook W C, Scott M, Arendt P, Newman B, Fisher R F and Swartzlander A B 1988 *Appl. Opt.* **27** 246, 279
Wood R W 1933 *Phys. Rev.* **44** 153
Yu A J-C, Donovan T M and Spicer W E 1968 *Phys. Rev.* **167** 670
Zemal J N, Jensen J D and Schooler R E 1965 *Phys. Rev.* **140** 4330
Zukic M, Torr D G, Spann J F and Torr M R 1990 *Appl. Opt.* **29** 4284

8 The Optics of Discontinuous Films

8.1 Introduction

The study of discontinuous films is particularly interesting for two main reasons: (i) their optical properties are considerably different from those of both bulk materials and continuous films, principally because of the presence of additional anomalous absorption bands, and (ii) the shapes of the islands of which the films consist provide clues to the processes by which films are built up.

Figure 8.1 shows a series of electron micrographs of discontinuous films of increasing density. The thinnest films consist of a large number of small, roughly spherical islands; as the amount of deposited material increases, the number of islands decreases by coalescence, their shape becomes flatter and their outline irregular. Eventually they join together to form a complicated network.

One of the first observations on discontinuous films was made by Faraday when he noticed that brilliant colours were obtained when very thin films of gold and silver were viewed in white light. He suggested, correctly, that this effect was caused by the granular nature of the films; it has been shown subsequently that the complementary colours reflected or transmitted by the films are produced by selective absorption due to an anomalous band in the visible spectrum which is observed neither in bulk materials nor in continuous films of the same material.

Since then, a large amount of experimental evidence has been accumulated on the exact forms of the absorption bands for many metals. Typical results are those of McKenna and Ward (1981) who investigated discontinuous films of silver; figure 8.2 shows $(I - R - T)\lambda/T$ plotted against the wavelength for a series of films of different thicknesses (corresponding to the films illustrated in figure 8.1). The anomalous absorption peak becomes

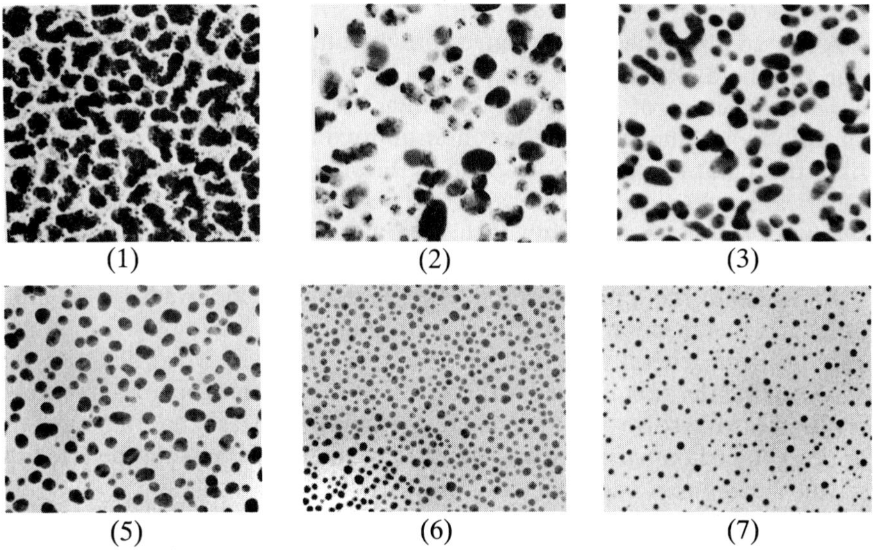

Figure 8.1 Electron micrographs of a series of discontinuous films. (Numbers refer to curves in figure 8.2.)

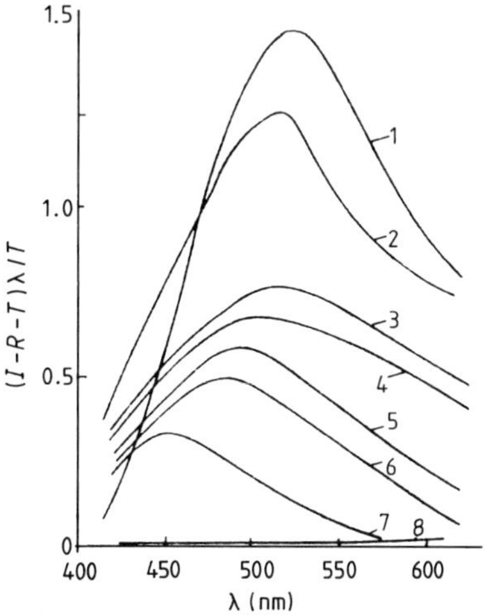

Figure 8.2 $(I - R - T)\lambda/T$ versus λ for silver films. (Numbers refer to films in figure 8.1.) (From McKenna W T and Ward L 1981 *Phys. Status Solidi* a **68** K 11.)

larger and also moves to longer wavelengths as the thickness of the film increases (this is also associated with larger sized film particles).

A second phenomenon associated with discontinuous films is the percolation threshold. It has been found that very thin metal films are extremely poor conductors of electricity, and that the conduction process starts abruptly when the substrate becomes approximately half covered by the film; this point is called the percolation threshold. The connection between electrical conduction and optical properties has been firmly established in this book and some discussion of this phenomenon is therefore of relevance to this chapter.

In general, the propagation of an electromagnetic wave in a nonhomogeneous medium involves the scattering of the wave by the inhomogeneities, the general treatment of which was given by Mie (1908). If the wavelength is large compared with the size of the inhomogeneities, the wave is unable to resolve the individual scattering centres and the medium appears to be homogeneous with an effective dielectric constant denoted by ε' and having real and imaginary parts, ε_1' and ε_2'.

There are two main classes of theoretical approach to the optical behaviour of discontinuous films: (i) the discrete island model where the response of each island to the applied field is considered and (ii) the effective medium theory in which one island is considered, the rest of the film being averaged into a homogeneous medium.

8.2 Discrete island theories

The first theoretical model to attempt to explain the anomalous absorption bands was propounded by Maxwell-Garnett (1904, 1906) who treated a discontinuous film as a dispersion of small metal spheres embedded uniformly in the ambient material which had a dielectric constant of ε_a. The application of an external electric field causes the spheres to become electrically polarised and so generate their own local fields. Assuming these local fields to be of the Lorentz type, the effective dielectric constant of the matrix can be calculated by means of the Clausius–Mosotti equation which was discussed in §1.5.1. This relates the effective dielectric constant of the film, ε', to the dielectric constant of the bulk material, ε, and also to the volume fraction of the metal, q, defined as the proportion of the volume of the film actually occupied by metal. Thus

$$\frac{\varepsilon' - \varepsilon_a}{\varepsilon' + 2\varepsilon_a} = \frac{q(\varepsilon - \varepsilon_a)}{\varepsilon + 2\varepsilon_a}. \tag{8.1}$$

Substituting $\varepsilon = \varepsilon_1 - i\varepsilon_2$ and $\varepsilon' = \varepsilon_1' - i\varepsilon_2'$ leads to:

$$\varepsilon_1' = \frac{\varepsilon_a[\varepsilon_1(1+2q)/(1-q)+2\varepsilon_a][\varepsilon_1+\varepsilon_a(2+q)/(1-q)] + [\varepsilon_2^2(1+2q)/(1-q)]}{[\varepsilon_1+(2+q)/(1-q)\varepsilon_a]^2+\varepsilon_2^2} \tag{8.2}$$

and

$$\varepsilon_2' = \frac{9q\varepsilon_a^2\varepsilon_2}{(1-q)^2\{[\varepsilon_1+\varepsilon_a(2+q)/(1-q)]^2+\varepsilon_2^2\}}. \tag{8.3}$$

From equation (8.3) it is seen that ε_2' will have its maximum value when $\varepsilon_1 \simeq -\varepsilon_a(2+q)/(1-q)$; as q increases, the corresponding negative value of ε_1 must also increase. Thus, the maximum in ε_2' will move towards longer wavelengths where the negative values of ε_1 are larger.

Doremus (1966) developed the Maxwell-Garnett theory by introducing expressions for ε_1 and ε_2 based on Drude's free-electron theory. This led to the following expression for the wavelength, λ_m, at which the absorption maximum occurs:

$$\lambda_m = \lambda_c\left(\varepsilon_0 + \frac{\varepsilon_a(2+q)}{1-q}\right)^{1/2} \tag{8.4}$$

where $\lambda_c = 2\pi c/\omega_p$.

The bandwidth, given by $[\varepsilon_0 + \varepsilon_a(2+q)/(1-q)]\,c/2\sigma$, also increases with q.

When the absorption peak falls within the visible spectrum, the film will exhibit the complementary colours observed by Faraday.

Doremus claimed that his treatment gave good results when applied to copper, and his model is still used by many workers in this field. However, although qualitatively satisfactory, this formulation does not agree well with all sets of experimental data and several important modifications have subsequently been made to the basic theory.

For instance, when using Drude's equation, the electron relaxation time between collisions, τ_c, is involved. When the film islands are small, their dimensions may be less than the normal mean free path of the electrons in that particular material at the temperature involved and the movements of the electrons will then be constrained by reflections at the surfaces of the islands. Consequently a smaller value of τ_c must be used depending on the size of the islands. This modification is known as the optical size effect and will have its maximum impact when the islands are very small; modifications to include this effect have been introduced by Karlsson and Beckmann (1967) and also by Kreibig and Fragstein (1969) who, dealing with spherical particles, showed that the effective mean free path, l_e, was related to that in bulk materials, l_0, and the radius of the particle, r, by the equation

$$1/l_e = 1/l_0 + 1/r. \qquad (8.5)$$

8.2.1 Shapes of islands and interactions between them

The other important modifications to the Maxwell-Garnett theory have taken place in two separate directions: (i) the shapes of the island particles and (ii) the nature of the electrical interactions between the particles. David (1939) suggested that the matrix particles should be regarded as ellipsoids of revolution, so allowing a wide range of possible particle shapes to be introduced ranging from oblate (flat discs) to prolate (tall columns). David used the following expression for the polarisability of an ellipsoid:

$$\alpha = \frac{V\varepsilon_0(\varepsilon - \varepsilon_a)}{1 + (\varepsilon - \varepsilon_a)f} \qquad (8.6)$$

Here, V is the volume of the particle and f is a shape factor which determines the self-polarising effect of the ellipsoid; its magnitude depends on the ratio of the half axes of the ellipsoid. For an ellipsoid of revolution two values of f are required, these being generally referred to the orientation of the particle on the substrate. Thus, parallel to the plane of the substrate the shape factor is $f_\parallel$ whilst at right angles to the substrate, i.e. along the axis of revolution of the ellipsoid, the shape factor is $f_\perp$. David's expressions for $f_\perp$ and $f_\parallel$ are as follows:

$$f_\parallel = \tfrac{1}{2}w(w^2 + 1)\left[\tfrac{1}{2}\pi - \tan^{-1} w - \left(\frac{w}{w^2 + 1}\right)\right] \qquad (8.7)$$

where $w = 1/[(b^2/a^2) - 1]$, b and a being the semi-axes of the ellipsoid of revolution (see figure 8.3), and

$$f_\perp = 1 - 2f_\parallel. \qquad (8.8)$$

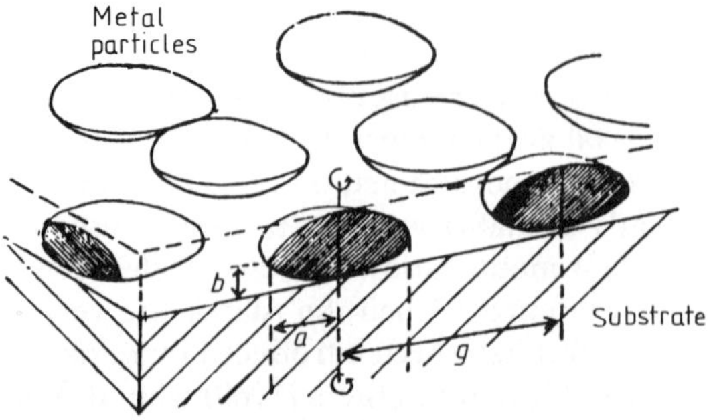

Figure 8.3 Model of a discontinuous thin film. (From Jarrett D N and Ward L 1976. *J. Phys. D: Appl. Phys.* **9** 1521).

Figure 8.4 shows the variation of $f_\perp$ and $f_\parallel$ with b/a; for a sphere, $b/a = 1$ and $f_\perp = f_\parallel = 1/3$, the value used by Maxwell-Garnett.

In deriving his expression for the effective dielectric constant, David did not introduce the electrical interactions between the particles, claiming that the behaviour of the film could be accounted for entirely in terms of q and f. He gave:

$$\varepsilon_2' = \frac{\varepsilon_2 \varepsilon_a^2 q}{[\varepsilon_a + (\varepsilon_1 - \varepsilon_a) f]^2 + (f \varepsilon_2)^2}. \tag{8.9}$$

This equation predicts a maximum in ε_2' when $\varepsilon_1 \simeq -\varepsilon_a(1/f - 1)$; the position of the peak is thus controlled entirely by the value of the shape factor. As the particles grow larger, $f_\parallel$ will decrease and, again, the negative value of ε_1 will have to increase as was the case in Doremus' treatment. It may be noted that David's theory predicts two peaks in ε_2' due to the two values of f; these would be observed in p-polarised radiation but not that polarised in the s-plane. Again, although partially successful in explaining the behaviour of the anomalous absorption band, David's theory does not provide the complete solution in all cases.

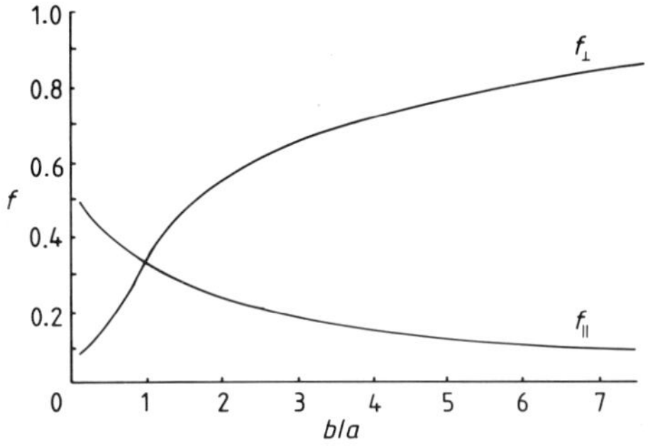

Figure 8.4 $f_\parallel$ and $f_\perp$ versus b/a.

Schopper (1952) developed this theory a stage further by introducing a Gaussian distribution of the ellipsoidal shapes into David's model. This did not affect the wavelength at which maximum absorption took place, but it enabled the width of the absorption band to be explained by a suitable choice of the Gaussian parameters. Other forms of the distribution function of particle shapes have been used in later studies, for example by Dobierzewska-Mozrzymas *et al* (1985) who employed a log-normal type of distribution. The effect of a similar distribution of the inter-island spacings has also been considered by the same authors.

Maxwell-Garnett had considered the interactions between the islands to be Lorentzian, but Yamaguchi *et al* (1972) assumed that they were of the same form as those between static dipoles. They calculated the interaction term to be equal to $0.50 d_m/g$ where d_m is the mass thickness, i.e. the thickness the film would have if the material were uniformly spread, and g is the average separation of the islands.

The concept of retarded dipole–dipole interactions (i.e. alternating dipoles) between the island particles was introduced by Yamaguchi *et al* (1974) in order to explain the shapes of the absorption curves, although the size effect was also needed to account for the broadening of the plasma resonance peaks.

Jarrett and Ward (1976) derived a general equation for the behaviour of discontinuous films incorporating some of the ideas discussed above; details of this are now given.

Consider an assembly of ellipsoidal particles, N per unit volume, and with a polarisability term defined by $\alpha = P/NE_t$ where E_t is the total electric field acting on the particle and P is the dipole moment per unit volume. Thus α is also equal to p/E_t, p being the dipole moment of each particle. Now,

$$E_t = E_a + fP/\varepsilon_0 \tag{8.10}$$

and hence the applied field,

$$E_a = E_t(1 - fN\alpha/\varepsilon_0). \tag{8.11}$$

In addition, from electrostatic considerations, the electrical displacement, D, for the particles in a vacuum environment is given by:

$$D = E_a\varepsilon_0\varepsilon = E_a\varepsilon_0 + P \tag{8.12}$$

where ε is the dielectric constant of the bulk material from which the particle is made (equation (1.1)). If the particles are embedded in a medium of relative permittivity ε_a, ε in equation (8.12) must be replaced by $\varepsilon/\varepsilon_a$. Hence

$$E_a\varepsilon_0\varepsilon/\varepsilon_a = E_a\varepsilon_0 + P \tag{8.13}$$

giving

$$E_a = \frac{P\varepsilon_a}{\varepsilon_0(\varepsilon - \varepsilon_a)} \tag{8.14}$$

so from equation (8.10),

$$\begin{aligned} E_t &= \frac{P\varepsilon_a}{\varepsilon_0(\varepsilon - \varepsilon_a)} + \frac{fP}{\varepsilon_0} \\ &= \frac{P[\varepsilon_a + f(\varepsilon - \varepsilon_a)]}{\varepsilon_0(\varepsilon - \varepsilon_a)}. \end{aligned} \tag{8.15}$$

Hence

$$\alpha = P/\mathrm{NE}_t = \frac{\varepsilon_0(\varepsilon - \varepsilon_a)}{N[\varepsilon_a + f(\varepsilon - \varepsilon_a)]}. \tag{8.16}$$

The effective permittivity of the film is then given by

$$\varepsilon' = \varepsilon_a\left(1 + \frac{N\alpha E_t}{\varepsilon_0 E_a}\right) \tag{8.17}$$

$$= \frac{\varepsilon_a}{\varepsilon_0}\left(\varepsilon_0 + \frac{N\alpha}{1-(fN\alpha/\varepsilon_0)}\right) \tag{8.18}$$

$$= \frac{\varepsilon_a}{\varepsilon_0}\left(\varepsilon_0 + \frac{N\alpha}{1-A\alpha}\right) \tag{8.19}$$

where A, which is equal to fN/ε_0, is a term representing the interaction between the particles.

When the particles are made from an absorbing material, both ε and α must be represented by complex quantities, namely $\varepsilon = \varepsilon_1 - \mathrm{i}\varepsilon_2$ and $\alpha = \alpha_1 - \mathrm{i}\alpha_2$.

Substituting the expression for α in equation (8.16) into equation (8.19) produces:

$$\varepsilon_1' - \mathrm{i}\varepsilon_2' = \frac{\varepsilon_a}{\varepsilon_0}\left(\frac{\varepsilon + NV\varepsilon_0(\varepsilon_1 - \varepsilon_a - \mathrm{i}\varepsilon_2)}{\varepsilon_a + f(\varepsilon_1 - \varepsilon_a) - \mathrm{i}f\varepsilon_2}\right) \tag{8.20}$$

$$= \frac{\varepsilon_a}{\varepsilon_0}\left(\frac{\varepsilon_0 + NV\varepsilon_0(\varepsilon_1 - \varepsilon_a - \mathrm{i}\varepsilon_2)}{\varepsilon_a + (\varepsilon_1 - \varepsilon_a)(f - AV\varepsilon_0) - \mathrm{i}\varepsilon_2(f - AV\varepsilon_0)}\right). \tag{8.21}$$

A new shape factor F may now be defined given by $F = f - AV\varepsilon_0$ which leads to:

$$\varepsilon_1' - \mathrm{i}\varepsilon_2' = \frac{\varepsilon_2}{\varepsilon_0}\left(\frac{\varepsilon_0 - NV\varepsilon_0(\varepsilon_1 - \varepsilon_a - \mathrm{i}\varepsilon_2)}{\varepsilon_2 + F(\varepsilon_1 - \varepsilon_a) + \mathrm{i}F\varepsilon_2}\right). \tag{8.22}$$

Hence

$$\varepsilon_1' = \frac{\varepsilon_a\{1 + q[\varepsilon_1\varepsilon_a(1-2F) + F(\varepsilon_1^2 + \varepsilon_2^2) - \varepsilon_a^2(1-F)]\}}{[\varepsilon_a + F(\varepsilon_1 - \varepsilon_a)]^2 + (F\varepsilon_2)^2} \tag{8.23}$$

$$= \frac{\varepsilon_a\{\varepsilon_a^2(1-q-2F) + F(F+q)[(\varepsilon_1-\varepsilon_a)^2 + \varepsilon_2^2] + \varepsilon_1\varepsilon_a(q+2F)\}}{[\varepsilon_a F(\varepsilon_1 - \varepsilon_a)]^2 + (F\varepsilon_2)^2} \tag{8.24}$$

and

$$\varepsilon_2' = \frac{q\varepsilon_2\varepsilon_a^2}{[\varepsilon_a + F(\varepsilon_1 - \varepsilon_a)]^2 + (F\varepsilon_2)^2} \tag{8.25}$$

where $q = NV$ = volume fraction of the film.

Most attention may be concentrated on the expression for ε_2' as this represents the optical absorption in the film, $2n'k'$. By using appropriate values for f and A, equation (8.25) can be expressed in the forms used by the various workers in this field.

Thus, setting $f = 1/3$ and, making use of the Lorentz–Lorenz interaction term, $A = N/3\varepsilon_0$, gives $AV\varepsilon_0 = q/3$. This then leads to:

$$\varepsilon_2' = \frac{9q\varepsilon_a^2\varepsilon_2}{(1-q)^2\{\varepsilon_1 + \varepsilon_a[(q+2)/(1-q)]^2 + \varepsilon_2^2\}}. \tag{8.26}$$

This is the same as equation (8.3) which was based on the Maxwell-Garnett model.

If the interaction term, A, is made zero, $F = f$ and David's equation for ellipsoids is obtained, namely

$$\varepsilon_2' = \frac{\varepsilon_2\varepsilon_a^2 q}{[\varepsilon_a + (\varepsilon_1 - \varepsilon_a)f]^2 + (f\varepsilon_2)^2}. \tag{8.27}$$

Yamaguchi *et al* (1972) assumed that the interaction between the particles could be treated on the basis of static dipoles. Substituting $f_{\parallel} - d_m/2g$ for $F_{\parallel}$ leads to

$$\varepsilon_2' = \frac{\varepsilon_2\varepsilon_a^2 q}{[\varepsilon_a + (\varepsilon_1 - \varepsilon_a)(f_{\parallel} - d_m/2g)]^2 + \varepsilon_2^2(f_{\parallel} - d_m/2g)^2}. \tag{8.28}$$

At right angles to the substrate, $F_{\perp} = f_{\perp} + d_m/g$ and, once again, $F_{\perp} + 2F_{\parallel} = f_{\perp} + 2f_{\parallel} = 1$.

A final example of the use of equation (8.25) is to apply it to the work of Polder and Van Santan (1946) on powders consisting of spherical particles. The equation they give may be transposed to produce

$$\varepsilon_2' = \frac{\varepsilon_2\varepsilon_a^2 q}{[\varepsilon_a + F(\varepsilon_1 - \varepsilon_a)]^2 + (F\varepsilon_2)^2}. \tag{8.29}$$

This is the same as the general equation, F now being equal to 1/3 as no interaction term was used.

The general nature of equation (8.25) makes it possible to derive values of F from experimental values of ε_2' and then to predict the shape factors of the film particles using the different forms of the interaction terms as suggested by the above theories. Employing this technique, Jeffesen and Barlow (1962) were able to estimate the particle sizes of colloidal gold from the absorption spectrum. Jarrett and Ward (1976) and McKenna and Ward (1981) estimated the shapes of gold and silver particles respectively from the position of the absorption maxima. They were then able to compare their shape factors with values obtained from edge-on electron micrographs of the films; it was observed that the islands became flatter, i.e. the values of F decreased as the thickness of the deposits increased.

It may be noted that the discrete island theories do not predict the percolation threshold discussed earlier in this chapter.

8.2.2 Plasma resonances

An alternative, but equivalent, explanation of the anomalous absorption peaks observable with discontinuous films is to regard them as being due to resonances in the free-electron plasma in each particle. Hampe (1958) gave the following equation for the plasma frequency, ω_s, in terms of its value in bulk material, ω_p:

$$\omega_s = \omega_p [(1 - q)/3\varepsilon_a]^{1/2}. \tag{8.30}$$

Emeric and Emeric (1967), Meesen (1972) and also Rasigni *et al* (1976) have given more detailed accounts of this theory in which Drude's equation for the dielectric constant of the island material, expressed in terms of ω_p, is substituted into a general expression for ε', leading to the prediction that maximum plasma absorption should occur at a frequency ω_s given by

$$\omega_s = \omega_p [1 + \varepsilon_0(1/f - 1)]^{-1/2}. \tag{8.31}$$

As f is increased, i.e. for thicker films and larger particles, ω_s will decrease; the absorption maximum will thus move to longer wavelengths. If the general treatment due to Jarrett and Ward is used as a basis for this theory, f in equation (8.31) would be replaced by F.

Marton and Lemon (1971) analysed the Maxwell-Garnett theory in detail. Making use of Drude's theory they showed that a resonance occurred in ε_2' when ε_1' was approximately equal to 0, similar to the bulk plasma resonance obtained when $\varepsilon_1 = 0$. However, when the energy-loss function, $-\mathrm{Im}(1/\varepsilon)$, was plotted for both ε and ε', the two peaks not only did not coincide with one another but did not coincide with that in ε_2' either. From this it was deduced that the resonance in ε_2', which occurred at a frequency ω_s', was not due to plasma in the normal sense; it was named the 'optical conduction resonance'. The true plasma resonance in ε' occurs at ω_s'' and ω_s'' increases with q, becoming equal to the bulk plasma resonance frequency, ω_p, at $q = 1$. These results are illustrated in figure 8.5. It was also shown that, approximately, $\omega_s' = \omega_p [(1 - q)/3]^{1/2}$, which is similar to both equation (8.3) produced by Doremus and equation (8.30) due to Hampe if ε_a is made equal to 1. From figure 8.5 it is seen that ω_s' decreases with increasing q and this would seem to be the resonance which has been observed in investigations on discontinuous films. When $q = 0$, $\omega_s' = \omega_s'' = \omega_p/\sqrt{3}$.

8.2.3 Dielectric constant of the ambient medium

An interesting problem which arises in all this work is the value to be ascribed to the dielectric constant of the medium in which the particles are embedded, ε_a. Some authors use the dielectric constant of the substrate material, ε_s, others the dielectric constant of vacuum, i.e. unity, whilst a third group use the average of these two quantities, $(\varepsilon_s + 1)/2$. Doremus (1966) put ε_a equal to 1 on the left-hand side of the Clausius–Mosotti equation and equal to ε_s on the right-hand side.

Yoshida *et al* (1972) used the plasma resonance concept to calculate ε_a and found that this quantity was not constant but increased with decreasing film thickness from a value of approximately 1.0 for very thick films to that

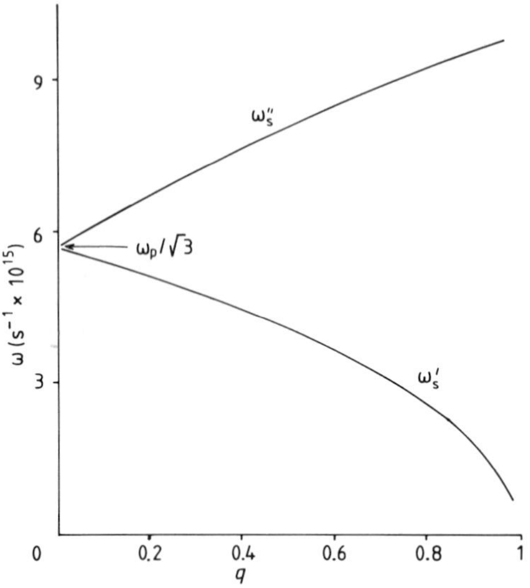

Figure 8.5 Variation of resonance frequencies ω_s' and ω_s'' of an aggregated system with q. The upper curve represents plasma resonance, ω_s'', and the lower one gives the q dependence of the conduction resonance frequency ω_s'. At small q, ω_s'' is equal to ω_s'. (From Marton J P and Lemon J R 1971 *Phys. Rev.* B **4** 275.)

of the substrate when the films were very thin. This result was regarded as reasonable because the particle size goes down as the film thickness decreases and so the substrate would be expected to have an increased effect on the optical behaviour. In a later paper, Yamaguchi *et al* (1974) were able to eliminate the use of ε_a altogether by considering the effect of the electrical images of the dipoles in the substrate. One consequence of this treatment was that the quantity $2F_{\parallel} + F_{\perp}$ was no longer exactly equal to unity. They gave the following equation:

$$2F_{\parallel} + F_{\perp} = 1 - \frac{\varepsilon_s - 1}{(\varepsilon_s + 1)\,[\frac{1}{6}(y/\beta)^2 - 0.72(2d_m/g)]} \tag{8.32}$$

where y is the axial ratio of the ellipsoids and β is the ratio of the long axis to the distance between the centres of the island and its image.

This problem is discussed in more detail when the results for individual materials are considered later in this chapter.

8.3 Effective medium theory

The second approach to the problem of the effective dielectric constant of an island film is the effective medium theory originated by Bruggemann

(1935). In this theory, the film is regarded as an array of two components as shown in figure 8.6(*a*); for example these could be the deposited material plus vacuum or two materials deposited together. Isolating a single grain of one component, of dielectric constant ε_1, the rest of the film is regarded as a homogeneous medium of dielectric constant ε' (figure 8.6(*b*)). The chosen grain is assumed to be spherical in shape so that its depolarising field may be easily evaluated in terms of a dipole whose moment is proportional to $(\varepsilon' - \varepsilon_1)/(\varepsilon' + 2\varepsilon_1)$. The calculation is then repeated with a spherical grain of the second component, of dielectric constant ε_2, embedded in the same homogeneous medium (figure 8.6(*c*)) and another dipole is obtained of moment proportional to $(\varepsilon - \varepsilon_2)/(\varepsilon' + 2\varepsilon_2)$. In order to be consistent with the postulate that the medium should appear homogeneous to a wave of sufficient length, the depolarising field must vanish. Thus

$$q_1(\varepsilon' - \varepsilon_1)/(\varepsilon' + 2\varepsilon_1) + (1 - q_1)(\varepsilon' - \varepsilon_2)/(\varepsilon' + 2\varepsilon_2) = 0 \qquad (8.33)$$

where q_1 is the volume fraction of the first component.

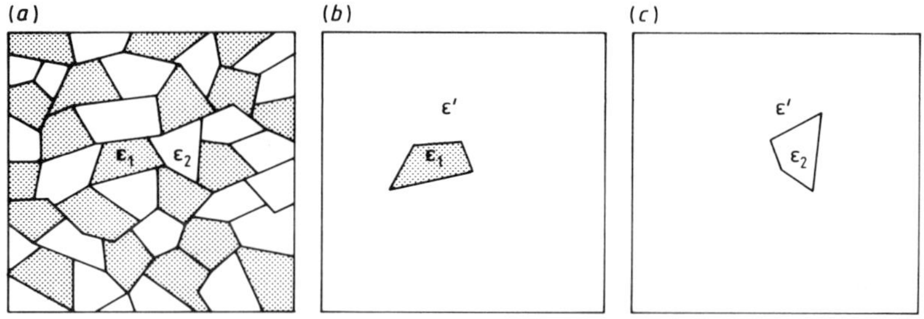

Figure 8.6 Effective medium model.

After the complex form of ε' is inserted. equation (8.33) may be split into its real and imaginary parts. If the first component is a metal and the second is vacuum, the condition for the composite to conduct electricity is that ε_1' must be negative and examination of the expression for ε_1' shows that this can only happen if $q_1 > 1/3$, this being the percolation threshold for the mixture. This implies that for $q_1 < 1/3$ there is no possibility of the metal grains forming an extended connected network and hence the DC conductivity should be zero. Although this result is in qualitative agreement with experimental results, in practice percolation thresholds lie between 0.3 and 0.55. In addition, it may be noted that the theory can give no explanation for the anomalous absorption peak observed with island films.

Bruggemann's approach is most appropriate in two cases: (i) close to, and above, the percolation threshold and (ii) composite or alloy films containing two or more components.

The effective medium theory has also been extended by Sen *et al* (1981) who considered a random microstructure of coated particles whereby the metallic grains are effectively isolated from each other.

8.3.1 Combinations of both theories

Several attempts have been made to reconcile the two theories which are both based on the Clausius-Mosotti equation. However, whereas in Bruggemann's theory the two types of particle are embedded in the same medium and are therefore symmetrical, the Maxwell-Garnett case is asymmetrical as one type of grain is embedded in a matrix of the other. Sheng's (1981) first model was based on the concept of a point cluster—a sphere, half of which was made from one of the components and half from the other—embedded in a uniform dielectric medium. A plot of the calculated conductivity against q gave almost the same results as Bruggemann's theory. In addition, peaks were predicted in the curves of T versus λ, but these were not at the same wavelengths as those for the Maxwell-Garnett theory. Further, the wavelengths of these peaks were insensitive to q, although both experimentally and in terms of the Maxwell-Garnett theory they increased with q. Sheng then went on to consider a model in which, firstly, a sphere of component 1 was coated with component 2 and then embedded in the homogeneous background (known as type 1). As the proportion of component 2 increases, the model eventually becomes transformed to type 2, in which the sphere is made of component 2 with component 1 as the coating. The probabilities of obtaining each of these types may be estimated as follows: the number of configurations of type 1, h_1, is proportional to the free volume accessible by the grain centre inside the spherical region, i.e. to $(1 - q_1^{1/3})$, q_1 being the volume fraction of component 1. Similarly, the number of configurations of type 2, h_2, is proportional to $[1 - (1 - q_1^{1/3})^3]$. Thus, for a given value of q_1, the relative probability of type 1 is $h_1/(h_1 + h_2) = p_1$ and that of type 2 is $p_2 = 1 - p_1$.

For $0 < q_1 < 0.35$, the values of p_1 are close to 1 and so type 1 will predominate; for $1 > q_1 > 0.65$, the value of p_1 tends to 0 with type 2 predominating. For $0.35 < q_1 < 0.65$ the two types of structure will occur in roughly equal proportions and, if component is metallic, the percolation threshold will lie within this region.

The condition for the average dipole moment to be zero now becomes $p_1\boldsymbol{D}_1 + (1 - p_1)\boldsymbol{D}_2 = 0$, where $\boldsymbol{D}_1$ and $\boldsymbol{D}_2$ denote the dipole moment produced by a uniform field on the specified type of configuration; it was shown that this equation will also apply to ellipsoidal units.

Aspnes (1982) has also contributed to this topic; starting with the Clausius–Mosotti equation, he was able to derive the Maxwell-Garnett and Bruggemann cases by a simple process. Consider a medium which is homogeneous to the applied field but contains two types of scattering centres, x and y, both being regarded as points. Then, equation (8.33) may written in the form:

$$\frac{\varepsilon' - \varepsilon_a}{\varepsilon' + 2\varepsilon_a} = \frac{q_x(\varepsilon_x - \varepsilon_a)}{\varepsilon_x + 2\varepsilon_a} + \frac{q_y(\varepsilon_y - \varepsilon_a)}{\varepsilon_y + 2\varepsilon_a}. \tag{8.34}$$

If y is the dilute component, $\varepsilon_x \simeq \varepsilon_a$ and

$$\frac{\varepsilon' - \varepsilon_a}{\varepsilon' + 2\varepsilon_a} = \frac{q_y(\varepsilon_y - \varepsilon_a)}{\varepsilon_y + 2\varepsilon_a} \tag{8.35}$$

which is the same form as the equation given by Maxwell-Garnett, and will thus predict an optical absorption peak. This equation describes a coated sphere microstructure with component x completely surrounded by component y and thus also represents the same situation as that proposed by Sheng.

When q_x and q_y are comparable in magnitude, it is more difficult to define the ambient medium, but let $\varepsilon_a = \varepsilon$. Then

$$0 = \frac{q_x(\varepsilon_x - \varepsilon)}{\varepsilon_x + 2\varepsilon} + \frac{q_y(\varepsilon_y - \varepsilon)}{\varepsilon_y + 2\varepsilon} \tag{8.36}$$

which is the basis of the Bruggemann model of a random mixture where x and y are considered as inserted into the medium itself; this equation will consequently predict a percolation threshold.

A further step can be taken if the microstructure does not consist of points by assuming that all the internal grain boundaries are at right angles to the plane of the film. When the electromagnetic field is applied along the surface, the boundary condition for tangential fields (equation (1.18)) implies that the electric field is uniform everywhere, and this leads to: $\varepsilon = q_x\varepsilon_x + q_y\varepsilon_y$; this is the case where no screening exists between the particles. For a field applied at right angles to the film. the boundary conditions give $1/\varepsilon = q_x/\varepsilon_x + q_y/\varepsilon_y$ and this is the case for maximum screening. These two relationships thus specify the limits to the screening effect.

Summarising the results of Aspnes' treatment:

(i) a percolation threshold is predicted lying between 0.35 and 0.65 in general agreement with experiment;

(ii) for a composite of gold in silicon dioxide, optical absorption bands are predicted which agree well with those determined experimentally; and

(iii) the disappearance of the anomalous absorption peak is predicted for $q > 0.7$, and this has also been borne out by experimental data. On the discrete island theory, this peak should persist right up to $q = 1$, which is evidently not the case.

8.4 Results for specific materials

Much of the early experimental work in this field was carried out in the 1950s and 1960s, particularly by groups of French workers, for example Philip, Payan, Rasigni, Rouard etc. In addition, Sennett and Scott (1950) reported results for eight metals, observing anomalous absorption with silver, gold, copper and aluminium but not with chromium, palladium, nickel and antimony. They used the Maxwell-Garnett theory to explain their results, showing that, in order to obtain anomalous peaks, the quantity M must be greater than zero, where

$$M = 2(k^2 - n^2)^3 - 9(k^2 - n^2)^2 + [12 + 8(nk)^2](k^2 - n^2) - 4 - 12(nk)^2. \tag{8.37}$$

Clearly, this condition is more likely to be met in materials where $k \gg n$ such as the noble metals, than in those where k and n are more nearly equal, as is the case with the transition metals.

8.4.1 Noble metals

Rasigni and Rouard (1963) studied films of gold, silver and copper. Sets of graphs for all three metals showed the characteristic absorption maximum which moved towards longer wavelengths as the mass of deposited material increased. In the cases of gold and copper these curves overlapped with the absorption bands due to interband transitions which lie in the visible spectrum. David's ellipsoidal theory was used successfully to interpret the results.

A very thorough investigation using discontinuous gold films was carried out by Norrman *et al* (1977), who measured the transmittances at normal incidence for films of mass thicknesses between 1 and 4 nm. Figure 8.7 illustrates their results showing broad minima in the transmittance which moved towards longer wavelengths as the mass thickness increased. These results could be described in terms of a Maxwell-Garnett type equation taking account of the size-dependent dielectric constant of the islands. Anderson and Granqvist (1977) had already shown that, treating the film islands as prolate ellipsoids with axes a and b, there was a most probable value for the ratio b/a, even when the film consisted of islands of different sizes. This work led to the definition of a mean value for the eccentricity, $\bar{\varepsilon}$:

$$\bar{\varepsilon} = 1 - (b/a)^2 \exp(-4\lambda n^2 \sigma) \tag{8.38}$$

where σ is the geometrical standard deviation of the diameter of the particle. Using this concept, Norrman *et al* obtained reasonable agreement between the measured and computed values of the transmittance, as long as the islands were not too irregular in shape. The effect of the substrate on the optical behaviour was shown to be small.

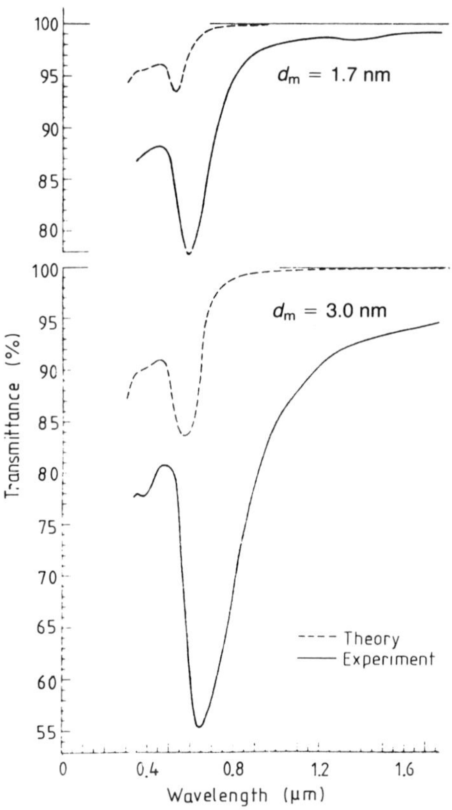

Figure 8.7 T versus λ for gold films. (From Norrman S *et al* 1977 *Solid State Commun.* **23** 263.)

Truong and Scott (1976, 1977) studied island films of silver, gold and copper by measuring the transmittance at normal incidence and the reflectance at angles of 7.5 and 75°. Their specimens contained particles of large sizes (up to 150 nm) and broadening of the absorption curves was observed; this could be accounted for by introducing a distribution function into the effective depolarisation factor of the particles. In their model they took into account the dipole–dipole interactions between the islands (Yamaguchi's model) and also the influence of the mirror images of the islands in the substrate. The dipole moment, p, induced in an island by an applied field gives rise to a mirror image, p', in the substrate. Expressions for the s- and p-components of the reflected radiation were obtained and, using Drude's free-electron model, the resonance wavelengths could be calculated.

Another study on silver and gold was made by Allen *et al* (1974) who measured the light scattered in both the forward and reverse directions from a series of silver and gold films. They too observed maxima in the light scattered backward by the film particles. These maxima shifted towards longer wavelengths as the thickness increased.

Jarrett and Ward (1976) applied their general equation to the absorption curves for discontinuous gold films in order to calculate the shape factors of the islands. These could also be estimated independently from electron micrographs of the edges of the films; the best fit between the two values was obtained by setting $\varepsilon_a = n_s^2$. It was found that, as the thickness increased, the values of F decreased, i.e. the ellipsoids became flatter.

The optical properties of very thin gold films deposited on silver have been studied by Pelletier *et al* (1984). Two methods of attenuated total reflection (ATR) were used (Otto's and Kretschmann's) and the angle of incidence giving minimum reflectance (pseudo Brewster's angle) was determined. The Maxwell-Garnett approach was employed satisfactorily to calculate the effective optical constants.

Yoshida *et al* (1971a, b) studied aggregated silver films by continuous ellipsometry during the evaporation process. They applied a modification of Schopper's thin-film model in which a logarithmic distribution of the axial ratios of the ellipsoids was used instead of the Gaussian distribution of f. The investigation indicated that the two most important factors in the modification of the optical properties by discontinuous films were the dipole interactions between particles and the optical size factor, especially when the sizes of the film particles were less than the mean free path of the electrons. They also observed that a few minutes after deposition ceased the shapes of the islands had become more globular indicating that mass transport effects were still taking place even after deposition had stopped.

McKenna and Ward (1981) applied the general equation to discontinuous silver films. Over the wavelength range used (0.5 to 1.5 μm) both ε_1 and ε_2 for bulk silver are linear with wavelength and this allowed a simplification of the calculations for F and f. For very thin films (mass thickness ~2 nm), the islands were roughly spherical but the values of f fell to about 0.20, corresponding to hemispheres, as the thickness increased to a mass thickness of 12 nm. Figure 8.8 shows the profiles of these films.

Anno and Hoskins (1984) found a change in the frequency of the plasma resonance of an island film of silver when it was exposed to an atmosphere containing hydrogen sulphide this being attributed to sulphurisation of the silver particles at the lattice defects within the islands.

Doyle (1989) investigated the reflectance of a suspension of silver spheres and was able to fit his results to the Clausius–Mossotti and Maxwell-Garnett theories. He claimed to obtain good agreement with the results obtained by Lee *et al* (1988) who studied the reflectance of porous glass media containing silver particles. Bieganski *et al* (1989) have studied the optical properties of island films of copper in the ultra-violet, visible and near infra-red regions using a modified Maxwell-Garnett formula to obtain a satisfactory explanation of the curves obtained. Yagil and Deutscher

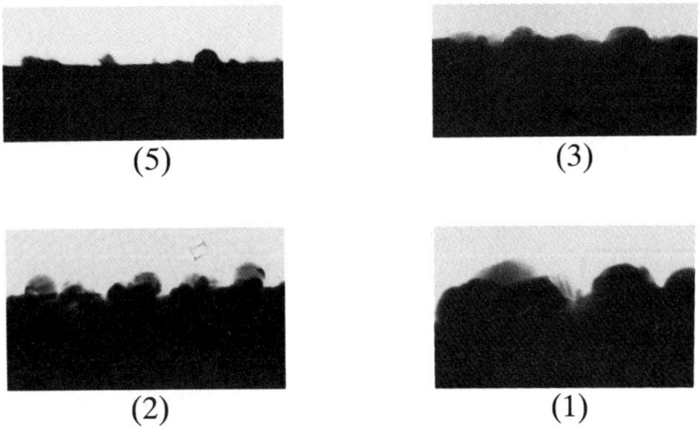

Figure 8.8 Profiles of film islands from figure 8.1. (From McKenna W T and Ward L 1981 *Phys. Status Solidi* a **68**.)

(1987) measured the transmittance of gold and aluminium films near the percolation threshold. They interpreted their results in terms of the percolation correlation length, x_c.; if $\lambda > x_c$, the sample behaved as though it were homogeneous and the Maxwell-Garnett theory was found to apply, but if $\lambda < x_c$ the sample was inhomogeneous and that theory broke down.

Results for sub-monolayer silver films on fused quartz substrates have been given by Shinozaki and Yamaguchi (1981), and these were interpreted satisfactorily in terms of a model in which ellipsoids were distributed two-dimensionally. In the free-electron region, Drude's equation and the classical size effect were used, whilst in the bound electron part (energies greater than 3 eV) the bulk values of ε had to be reduced in order to obtain a fit.

Fukui and Oda (1986) used the ATR technique with angular scan to monitor continuously the growth processes in silver films; their equipment is illustrated in figure 8.9. Yano *et al* (1990) used the same technique to measure the effective dielectric constants of discontinuous films of silver, aluminium, gold and copper up to 50 nm in thickness at 633 nm. Graphs were given of the real and imaginary parts of e against the optical thickness of the film; the one for $2nk$ showed a maximum value at one particular thickness. A good fit with the Maxwell-Garnett theory was obtained especially with the silver films.

Other contributions on discontinuous films of the noble metals have been made by Tokarsky and Marton (1974), Gasparini and Fraisse (1975) and Kazmerski and Racine (1975).

8.4.2 Aluminium

Dobierzewska-Mozrzymas *et al* (1985) studied granular aluminium films deposited on rock-salt substrates. The metallic islands were large (~40–60 nm) and could not be approximated to ellipsoids, the value of q ranging between 0.3 and 0.46. The specimens were thus close to, but still

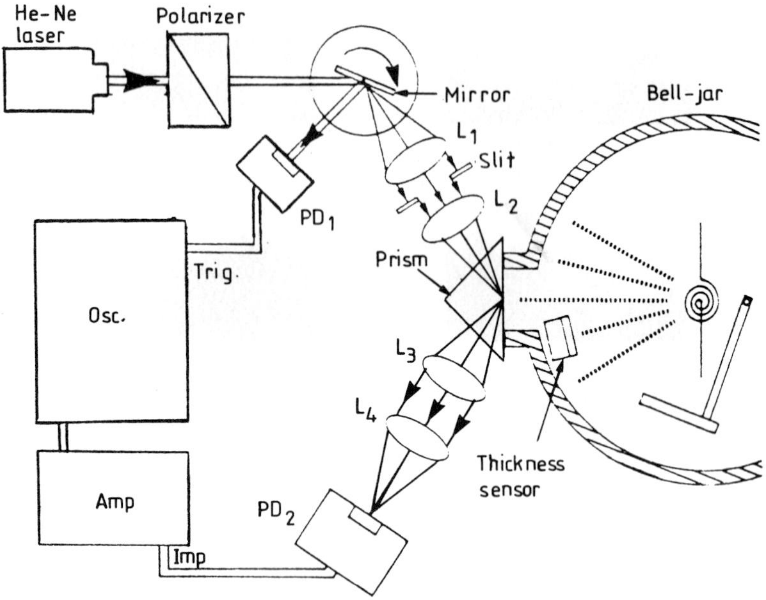

Figure 8.9 Schematic diagram of ATR instrument for study of metal film growth. (From Fukui *et al* 1986 *Appl. Surf. Sci.* **33**/34 882.)

below, the percolation threshold for aluminium films and this also meant that any of the island film models had to be used with caution. Plots of ε_2' versus λ (figure 8.10) showed absorption peaks which moved towards longer wavelengths as q was increased. Making use of the general equation of Jarrett and Ward (equation (8.25)), it was possible to express the wavelengths of the absorption maxima in terms of the shape factor, F, namely

$$\lambda_{\max} = \lambda_c [\varepsilon_0 + \varepsilon_a (1/F - 1)]^{1/2}. \tag{8.39}$$

A plot of $\lambda_{\max}^2$ versus $(1/F) - 1$ (figure 8.11) was linear and was a better fit than the Maxwell-Garnett relationship of $\lambda_{\max}^2$ versus $(2+q)/(1-q)$ indicating that, as well as the volume fraction, the shape factor of the islands was important in describing the optical behaviour.

Additional data for island films of aluminium have been reported by Dobierzewska-Mozrzymas *et al* (1987); some of these referred to films close to the percolation threshold where, surprisingly, the Maxwell-Garnett theory was still found to apply. Similar results were obtained with gold films (1988) and copper films (1993). Results for island films of silver between 2 and 18 μm have been given by Inagaki *et al* (1986).

Indium films have been studied by Rasigni *et al* (1976) who found plasma resonances near 4.8 eV, having shifted from the bulk resonance at 11.5 eV; a direct interband transition was expected at 1.4 eV but this was not

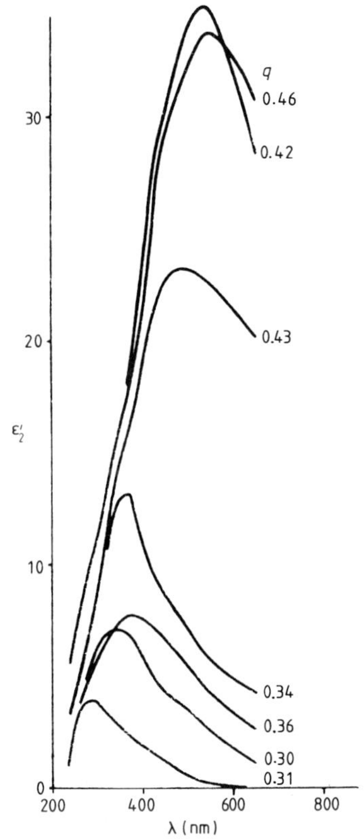

Figure 8.10 ε_2' versus λ for aluminium films, $0.31 < q < 0.46$. (From Dobierzewska-Mozrzymas E *et al* 1985 *Appl. Opt.* **24** 729.)

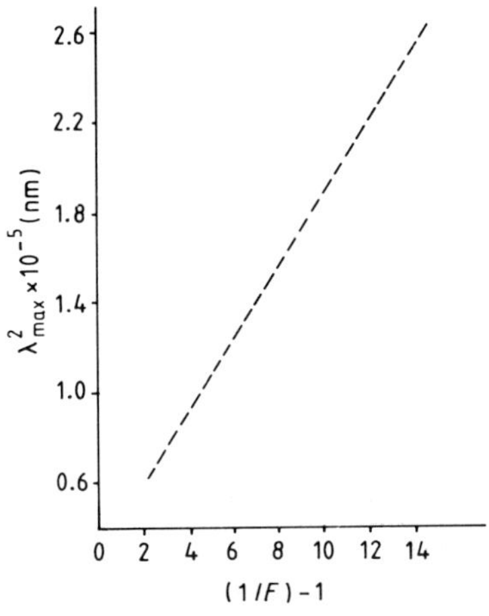

Figure 8.11 λ_{max}^2 versus $(1/F) - 1$ for aluminium films. (From Dobierzewska-Mozrzymas *et al* 1985 *Appl. Opt.* **24** 731.)

observable in the granular films. A modified Maxwell-Garnett approach produced satisfactory results.

8.4.3 Alkali metals and alkaline earths

Rasigni *et al* (1975) carried out measurements on magnesium films finding a plasma resonance in the conduction electrons at 5.12 eV. On some occasions the specimens were found to have rough surfaces and then these plasma resonances were not observed but instead a substantial fall in reflectance occurred near the plasma frequency.

Another investigation involving granular deposits of magnesium was made by Rasigni *et al* (1975) when a plasma resonance due to the islands was found at an energy of 5.1 eV.

Extensive studies on films of vapour-deposited lithium have been made by Rasigni and Rasigni. In one set of measurements (1977) they determined R, R' and T in the energy range 1–6 eV on films prepared in ultrahigh vacuum, the film structures being evaluated by electron microscopy of carbon replicas. A variety of film structures could be obtained, from spherical and hemispherical isolated particles through connected networks to microcrystals. In figure 8.12 T is plotted against U and two minima can be seen at about 1.5 and 3.5 eV. The calculated optical conductivity showed peaks in the visible and infrared due to plasma resonances, whilst, in the ultraviolet, the observed rise in conductivity was due to interband transitions. The films consisted of ellipsoidal particles and a Maxwell-Garnett theory modified for particle interactions by a Yamaguchi-type term was used to explain the results. They also obtained good agreement with the results of Davenes *et al* (1973) who investigated dispersed lithium particles in dielectrics.

Payan (1969) found two absorption bands in his work on alkali films. The band at the higher energy did not change in wavelength with q but that at the lower energies moved to longer wavelengths as q increased In addition, on passing along the alkalis from sodium to rubidium the band also moved to longer wavelengths; this latter effect had also been observed by Smith (1970) with continuous films. Payan speculated that the first peak was due to interband transition but it was also likely that both bands were caused by plasmons. The second, lower energy peak looks like the ω_s' band of Marton and Lemon (1971) and it persisted even in the thinnest films. Its maximum displacement was to the wavelength found for bulk alkalis by Mayer and Heitel (1966).

Yamaguchi and Takahashi (1978) applied multiple ATR spectroscopy to very thin deposits of alkali metals on quartz substrates, the technique giving information about the shape and the electronic properties of the islands. A film of sodium less than 0.01 Å thick could be detected. The particles were predicted to be ellipsoidal and the number of free electrons per unit volume was found to decrease with the mean thickness of the film. Their

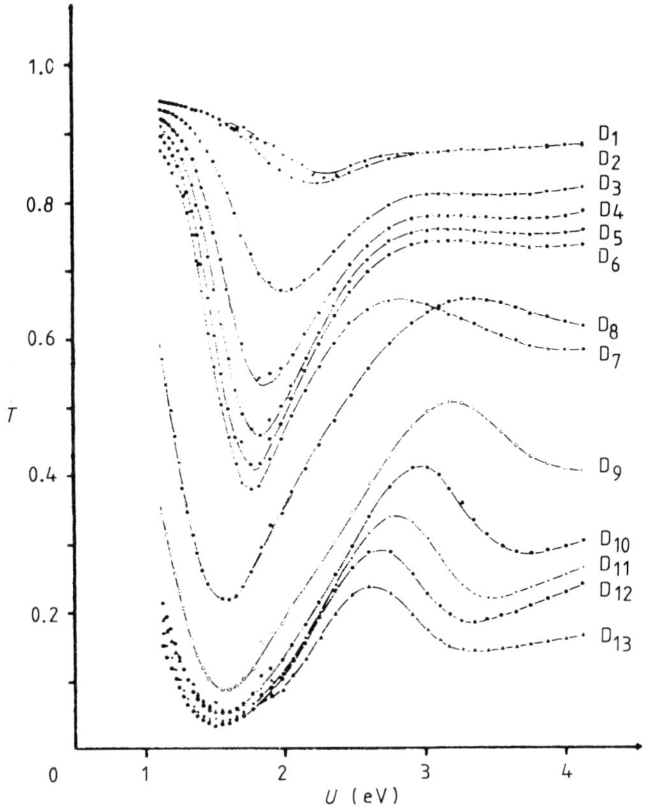

Figure 8.12 Normal incidence transmittance for 13 lithium deposits of various mass thickness (Å); D1, 13; D2, 16; D3, 31; D4, 47; D5 62; D6, 78; D7, 94; D8, 72; D9, 143; D10, 215; D11, 250; D12, 360; and D13, 430 (From Rasigni M and Rasigni G 1977 *J. Opt. Soc. Am.* **67** 511.).

calculations took into account the mirror image effect of the particles in the substrate but ignored the dipole–dipole interactions which were small.

8.4.4 Transition elements

Only a little work has been reported on the optical properties of discontinuous films of the transition elements. As already mentioned Sennett and Scott (1950) studied films of several transition metals. Palladium and chromium showed no evidence of an anomalous absorption maximum, whilst in nickel there was only a slight maximum at 500 nm observable with the thinner films.

Rasigni and Rouard (1963), working on iron, nickel and cobalt, also reported no sign of an anomalous peak. Their results for cobalt are shown in figure 8.13.

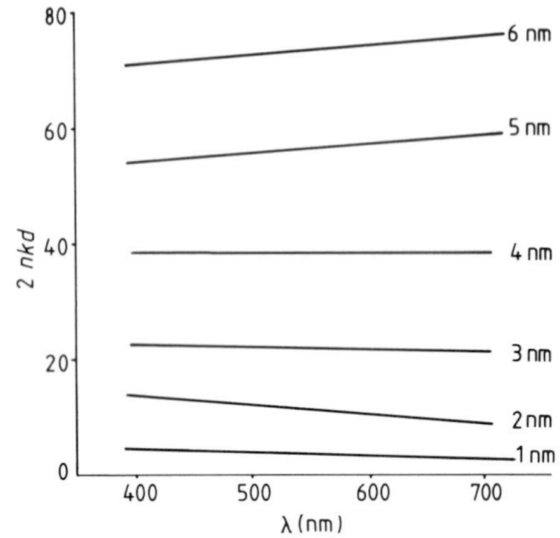

Figure 8.13 $2nkd$ versus λ for a series of six cobalt films of thicknesses 1.0 to 6.0 nm. (From Rasigni G and Rouard P 1963 *J. Opt. Soc. Am.* **53** 609.).

8.4.5 Silicon

Royer *et al* (1988), using the method of ATR, have reported measurements of the optical absorption due to oblate silicon spheroids; the particles were produced by heat treatment on silver foil. The results were analysed using a modified effective medium theory due to Carlan (1969).

8.5 Results for composites

Alloys of Cd–Pd, Sb–Pb, Cu–Fe and Bi–Sn were studied by Landauer (1952) who obtained good results by expressing Bruggemann's equation in terms of the electrical conductances as follows:

$$\frac{q_1(\sigma_1-\sigma_\mathrm{m})}{\sigma_1+2\sigma_\mathrm{m}}+\frac{q_2(\sigma_2-\sigma_\mathrm{m})}{\sigma_2+2\sigma_\mathrm{m}}=0. \tag{8.40}$$

Films of co-deposited silver and silicon dioxide of different compositions were studied by Cohen *et al* (1973) and figure 8.14 shows $-\log T$ versus λ. As the volume fraction of silver decreases, the characteristic absorption band due to SiO_2 begins to develop at about 0.5 μm. In addition, the optical behaviour of the deposits changes from metallic at large percentages of silver (T decreasing with increasing wavelength) to dielectric at small quantities of silver (T increasing with wavelength).

Niklasson and Craighead (1983) were also successful in applying Bruggemann's equation to co-deposited films of aluminium and silicon containing up to 33% of silicon. Abelès *et al* (1975) studied granular films of tungsten and aluminium oxide; the growth of tungsten grains produced by

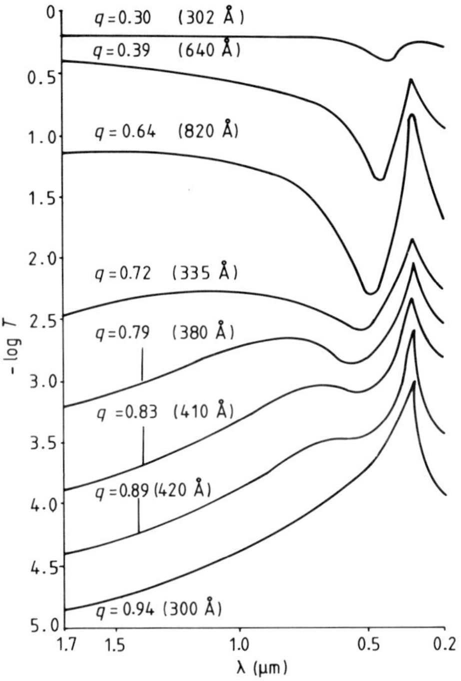

Figure 8.14 Log of optical transmission, *T*, (measured at normal incidence) versus wavelength, λ, in sputtered $AgSiO_2$ films. The volume fractions, *q*, of Ag and the film thicknesses are indicated. (From Cohen R W *et al* 1973 *Phys. Rev.* B **8** 3693.).

annealing resulted in a narrowing of the transition regions between metallic and non-metallic conductivity.

Hickey and Gibson (1985) investigated granular films of samarium sulphide and also found that Bruggemann's approach gave satisfactory results. On the other hand. Thericault and Boivin (1984), working with Cu-PbI_2 cermets, were able to use the Maxwell-Garnett theory extended by the inclusion of a shape factor and considering the metallic particles to be embedded in a dielectric matrix. Separate fits were needed for the real and imaginary parts of the effective dielectric constant.

Cummings *et al* (1984) made measurements on composites of silver particles and potassium chloride grains over the region from the ultraviolet to the far infrared; they found that the best fit with theory was the Bruggemann model. Yamaguchi *et al* (1985), working with samples consisting of alternate layers of silver and magnesium fluoride also found that the Bruggemann approach was superior to that of Maxwell-Garnett. In general, this was the case if the film was thick compared with the particle sizes; very thin films tended to follow the Maxwell-Garnett model.

The effects of oxidisation on the optical properties of discontinuous films of copper were investigated by Santucci *et al* (1983) from measurements of the transmittance between 250 and 2500 nm. Electron micrographs showed an almost continuous structure of copper and copper oxide (Cu_2O). The results agreed well with computed values using Bruggemann's theory and by considering the films as a random mixture of copper, copper oxide and air. Figure 8.15 shows T plotted against wavelength for three films of different thicknesses. In each case, at long wavelengths, oxidation produced an increase in transmittance with time because of the decrease of the metallic component.

Abelès and Gittelmann (1976) compared the Maxwell-Garnett and effective medium theories using composite films of Ag–SiO_2, Ge–Al_2O_3 and Si–SiC, but found it difficult to discriminate between the two theories.

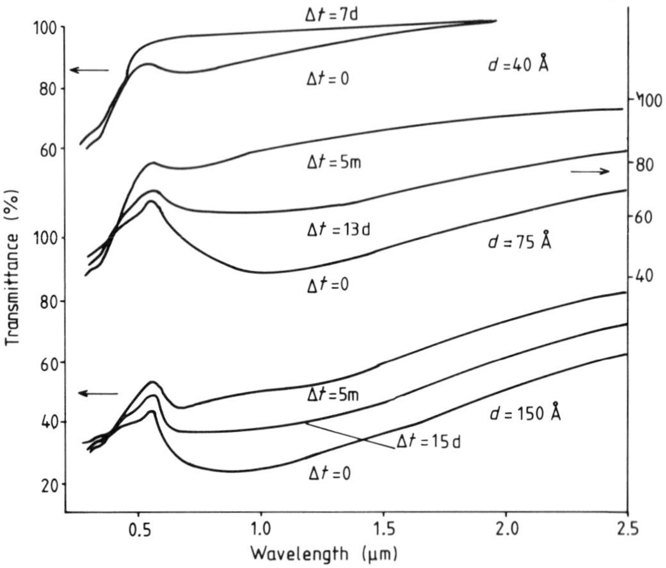

Figure 8.15 Normal transmittance at room temperature of three aggregated copper films with thickness d, at various stages of oxidation. Δt is the oxidation time after exposure of the samples to the laboratory atmosphere (d = days, m = months). (From Santucci *et al* 1983 *Appl. Opt.* **22** 3202.).

References

Abelès B and Gittleman J I 1976 *Appl. Opt.* **15** 2338
Abelès B, Pinch H L and Gittleman J I 1975 *Phys. Rev. Lett.* **35** 347

Allen E A, Scott G D, Thompson K T and Veas F 1974 *J. Opt. Soc. Am.* **64** 1190
Andersson T and Granqvist C G 1977 *J. Appl. Phys.* **48** 1673
Anno E and Hoskins R 1984 *Solid Sate Commun.* **52** 271
Aspnes D E 1982 *Thin Solid Films* **89** 249
Bieganski P, Dobierewski-Mozrzymas E and Peisert P, 1989 *Physica* **157A** 323
Brugemann D A G 1935 *Ann. Phys. Lpz.* **24** 636
Carlan A 1969 *Ann. Phys.* **4** 5
Cohen R W, Cody E D, Coutts M D and Abelès B 1973 *Phys. Rev. B* **8** 3689
Cummings K D, Garland J C and Tanner D B 1984 *Phys. Rev. B* **30** 4170
Davenes J, Perez A, Thevenard P and Dupuy C H T 1973 *Phys. Status Solidi* a **19** 679
David E 1939 *Z. Phys.* **114** 389
Dobierzewska-Mozrzymas E and Bieganski P 1987 *Opt. Acta (Pol)* **17** 231
—— 1993 *Appl. Opt.* **32** 3345
Dobierzewska-Mozrzymas E, Piesert J and Bieganski P 1988 *Appl. Opt.* **27** 181
Dobierzewska-Mozrzymas E, Radosz A and Bieganski P 1985 *Appl. Opt.* **24** 727
Doremus R H 1966 *J. Appl. Phys.* **37** 2775
Doyle W T 1989 *Phys. Rev.* B **39** 9852
Emeric N and Emeric A 1967 *Thin Solid Films* **1** 13
Fukui M and Oda K 1986 *Appl. Surf. Sci.* **33/34** 882
Gasparini J P and Fraisse R 1975 *Thin Solid Films* **30** 111
Hampe W 1958 *Z. Phys.* **152** 476
Hass G and Waylonis J E 1961 *J. Opt. Soc. Am.* **51** 719
Hickey C F and Gibson U J 1985 *J. Opt. Soc. Am.* A **2** 71
Hodkinson J R and Greenland I 1963 *J. Opt. Soc. Am.* **53** 577
Inagaki T, Goudonnet J P, Royer P and Arakawa E T 1986 *Appl. Opt.* **25** 3635
Jarrett D N and Ward L 1976 *J. Phys. D: Appl. Phys.* **9** 1515
Jeffesen M A and Barlow R B 1962 *J. Opt. Soc. Am.* **52** 99
Karlsson A V and Beckman O 1967 *Solid State Commun.* **5** 795
Kazmerski L C and Racine D M 1975 *J. Appl. Phys.* **46** 791
Kreibig U and Frastein C V 1969 *Z. Phys.* **224** 307
Landauer R 1952 *J. Appl. Phys.* **23** 779
Lee S-I, Noh T W, Gaines J R, Ko Y-H and Kreidler E R 1988 *Phys. Rev* B **37** 2918
McKenna W T and Ward L 1981 *Phys. Status Solidi* a **68** K11
Marton J P and Lemon J R 1971 *Phys. Rev.* B **4** 271
Maxwell-Garnett J C 1904 *Phil. Trans. R. Soc.* **203** 885
—— 1906 *Phil. Trans. R. Soc. A* **205** 237
Mayer H and Heitel H 1966 *Optical Properties and Electronic Structure of Metals and Alloys* (Amsterdam: North-Holland) p 47
Meesen A 1972 *J. Phys., Paris* **33** 371
Mie G 1908 *Ann. Phys., Lpz.* **25** 377
Niklasson G A and Craighead H G 1983 *Appl. Opt.* **22** 1237
Norrman S, Andersson T, Granqvist C G and Hunderi O 1977 *Solid Sate Commun.* **23** 261
Payan R 1969 *Ann. Phys., Paris* **4** 543
Pelletier D, Girouard F-E and Truong V-V 1984 *Appl. Opt.* **23** 1696

Polder D and Van Santan V A 1946 *Physica* **16** 675
Rasigni G, Palmari J P and Rasigni M 1975 *Phys. Rev. B* **12** 1121
Rasigni G, Petrakian J P, Rasigni M and Palmari J P 1976 *J Phys. C: Solid Phys.* **6** L325
Rasigni G and Rouard P 1963 *J. Opt. Soc. Am.* **53** 604
Rasigni M and Rasigni G 1977 *J. Opt. Soc. Am.* **67** 510
Royer R, Goudonnet J P, Inagaki T, Chabrier G and Arakawa E T 1988 *Phys. Status Solidi* a **105** 617
Santucci S, Picozzi P and Paoletti L 1983 *Appl. Opt.* **22** 3201
Schopper H 1952 *Z. Phys.* **131** 215
Sen P N, Scala C and Chen M H 1981 *Gephys.* **46** 781
Sennett R S and Scott G D 1950 *J. Opt. Soc. Am.* **40** 203
Sheng P 1981 *Opt. Laser Technol.* **253**
Shinozaki K and Yamaguchi T 1981 *Surf. Sci.* **103** L97
Shkylyarevskii I N and Korneeva T I 1971 *Opt. Spectrosc.* **31** 44
Shklyarevskii I N, Pakhamov I L and Korneeva T J 1973 *Opt. Spectrosc.* **34** 421
Smith N V 1970 *Phys. Rev.* B **2** 2840
Thericault J-M and Boivin G 1984 *Appl. Opt.* **23** 4494
Tokarsky R W and Marton J P 1974 *J. Appl. Phys.* **45** 305
Truong V V and Scott G D 1976 *J. Opt. Soc. Am.* **66** 124
—— 1977 *J. Opt. Soc. Am.* **67** 502
Yagil Y and Deutscher G 1987 *Thin Solid Films* **152** 465
Yano M, Fukui M, Haraguchi M and Shintani Y 1990 *Surf. Sci.* **227** 129
Yamaguchi T, Sakai M and Saito N 1985 *Phys. Rev.* B **32** 2126
Yamaguchi T and Takahshi H 1978 *Surf. Science* **78** 295
Yamaguchi T, Takakashi H and Sudoh A 1978 *J. Opt. Soc. Am.* **68** 1039
Yamaguchi T, Yoshida S and Kinbara A 1972 *Thin Solid Films* **13** 261
—— 1974 *J. Opt. Soc. Am.* **64** 1563
Yoshida S. Yamaguchi T and Kinbara A 1971a *J. Opt. Soc. Am.* **61** 62
—— 1971b *J. Opt. Soc. Am.* **61** 463
—— 1972 *J. Opt. Soc. Am.* **62** 1415

9 Applications of Optical Thin Films

9.1 Introduction

There are many devices whose operation depends on the optical properties of thin films of either metals and/or dielectrics and most of these require the accurate control of the thickness and quality of the deposited film whether it is produced by evaporation or by sputtering. Reference may be made to section 5.3 where details are given of the techniques by which film thickness may be monitored. In this chapter, we shall discuss the application of thin films in antireflection and high reflectance coatings, interference filters and beam splitters. These topics have been extensively covered in a number of books and special articles; for further details, reference may be made to the following: Macleod (1986), Heavens (1960), Berning (1963), Hass *et al* (1982), Ritter (1975), Lissberger (1970), Thalen (1969), Vossen (1977), the Handbook of Optics (1978) and Schwartz and de Sandro (1987).

9.2 Antireflectance coatings

The reflectance of the glass–air interface at normal incidence and in a region where the optical absorption is negligible is given by the formula: $((n_g - n_a)/(n_g + n_a))^2$ where n_g is the refractive index of glass and n_a that of air; this works out at approximately 4% in the visible spectrum for glass of refractive index 1.5; a transmitting optical system which has several such interfaces will thus lose by reflection a large proportion of the incident intensity. In addition, the various internally reflected beams will arrive at the final stage in different phases from that of the main beam and this will lead to problems in image formation. It is therefore important to have some mechanism by which to reduce this internally reflected radiation.

Consider a thin non-absorbing film on a non-absorbing substrate; figure

9.1(*a*) shows the ray diagram. In chapter 5, equations (5.9)–(5.14), the reflection coefficient of such an arrangement was derived and gave:

$$R = ((r_1 + r_2 \exp(-2i\delta))/(1 + r_1 r_2 \exp(-2i\delta)))^2 \quad (9.1)$$

where δ is the phase change across the film equal to $2\pi n_1 d_1/\lambda$; r_1 and r_2 are the amplitude reflection coefficients at the two interfaces of the film.

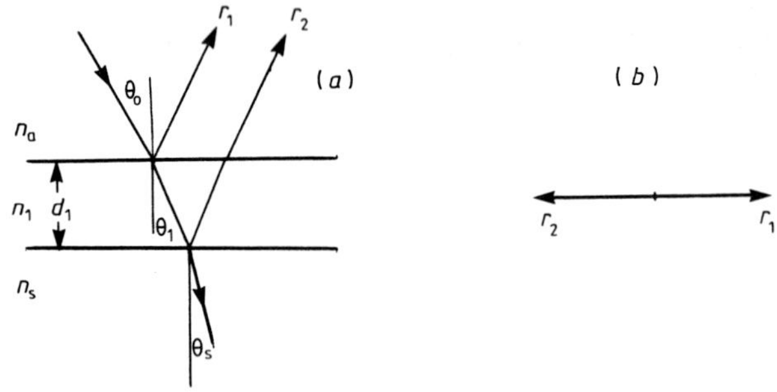

Figure 9.1 (*a*) Reflections of single film. (*b*) Amplitude-phase diagram of reflected rays.

If the optical thickness of the film, $n_1 d_1$, is made equivalent to a quarter of a wavelength, $\delta = \pi/2$ and the two exponential terms then become equal to -1. Then, $R = ((r_1 - r_2)/(1 - r_1 r_2))^2$ and this becomes equal to zero if $r_1 = r_2$. This situation may also be demonstrated by an amplitude-phase diagram as in figure 9.1(*b*).

Now considering the s state of polarisation, $r_{1s} = r_{2s}$ and $R_s = 0$ if;

$$\frac{(Y_0 \cos\theta_0 - Y_1 \cos\theta_1)}{(Y_0 \cos\theta_0 + Y_1 \cos\theta_1)} = \frac{(Y_1 \cos\theta_1 - Y_s \cos\theta_s)}{(Y_1 \cos\theta_1 + Y_s \cos\theta_s)} \quad (9.2)$$

where Y_0, Y_1, Y_s are the optical admittances at normal incidence and θ_0, θ_1, θ_s are the incidence angles in the various media.

For the p state of polarisation, $r_{1p} = r_{2p}$ and $R_p = 0$ if;

$$\frac{(Y_0/\cos\theta_0 - Y_1/\cos\theta_1)}{(Y_0/\cos\theta_0 + Y_1/\cos\theta_1)} = \frac{(Y_1/\cos\theta_1 - Y_s/\cos\theta_s)}{(Y_1/\cos\theta_1 + Y_s/\cos\theta_s)}. \quad (9.3)$$

At normal incidence, $\theta_0 = \theta_1 = \theta_s = 0$ and both conditions reduce to:

$$(n_0 - n_1)/(n_0 + n_1) = (n_1 - n_s)/(n_1 + n_s) \quad (9.4)$$

giving

$$n_0 n_s = n_1^2 \qquad \text{or} \qquad n_0/n_1 = n_1/n_s. \quad (9.5a)$$

Also:

$$R = \frac{(n_0 n_s - n_1^2)^2}{(n_0 n_s + n_1^2)}. \tag{9.5b}$$

Thus, if the substrate of refractive index n_s were to be coated with a quarter wave thickness of material of index n_1, ($n_0 < n_1 < n_s$ so that there is no phase change on reflection at either interface), there would be no reflected wave and all the incident energy would be transmitted. For glass, $n_s = 1.50$ whilst for air, $n_0 = 1.00$; hence, n_1 would need to be equal to $\sqrt{1.50} = 1.225$.

Two problems arise in the implementation of this theory. In the first place, it is difficult to find stable materials having values of refractive index as low as 1.225 in the visible spectrum; one of the closest to this value is CaF_2 ($n = 1.25$) but this material is not stable, its refractive index tending to increase with time. Another low index material, cryolite, Na_3AlF_6 which has $n = 1.33$, is soft and can easily be damaged. Secondly, refractive index is wavelength dependent and so the analysis given above only applies at a single wavelength. Nevertheless, by using MgF_2 which has $n = 1.38$ at 550 nm, the reflectance can be reduced to between 1 and 2% across the whole visible spectrum.

Single layer films can, however, be used effectively in the near intra-red with substrates such as Si, Ge or Te which have high refractive indices. For example, between 1 and 10 μm, n for Ge is about 4.0 whilst that for Te is 5.4; when coated with CdS ($n = 2.2$) the transmittance of Ge approaches 100% (Cox and Hass 1958).

9.2.1 Double film

A double layer of two different materials offers the opportunity of obtaining zero reflectance at two wavelengths; figure 9.2(*a*) illustrates the arrangement needed. In order to obtain zero reflectance and also to keep the phase angles of the reflected components correct, both films should be of thickness $\lambda/4$, the lower of refractive index n_2, and the upper of refractive index n_1; $n_0 < n_1 < n_2 < n_s$. The amplitude-phase diagram figure 9.2(*b*) shows how it is possible to obtain zero reflection at two different phase angles i.e. two different wavelengths, namely $\frac{3}{4}\lambda_0$ and $\frac{3}{2}\lambda_0$, where λ_0 is the wavelength which gives a phase difference of exactly $\pi/4$ across each film. Note that zero reflectance is not obtained at λ_0. The following relationship holds between the various refractive indices:

$$n_0/n_1 = n_1/n_2 = n_2/n_s. \tag{9.6}$$

At λ_0, $\lambda_0/2 = 2n_1 d_1 = 2n_2 d_2$, i.e. $\lambda_0 = 4n_1 d_1 = 4\, n_2 d_2$.
Also, at $\lambda = \lambda_0$,

$$R = \frac{(n_0 - (n_1/n_2)^2\, n_s)^2}{(n_0 + (n_1/n_2)^2\, n_s}. \tag{9.7a}$$

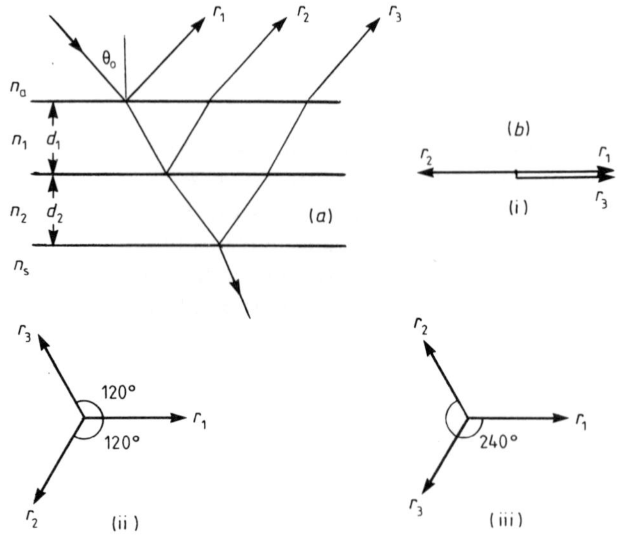

Figure 9.2 (*a*) Reflections of double film. (*b*) (i), (ii), (iii) Amplitude-phase diagrams of reflected rays.

Using equation (9.6) this may be reduced to:

$$R = \frac{1-(n_s/n_0)^{1/3})^2}{(1+(n_s/n_0)^{1/3})}. \qquad (9.7b)$$

Now, when the phase differences between the reflected rays r_1, r_2 and r_3 are 120° (figure 9.2(b)(ii)), the condition for zero reflection is:

$$(r_2 + r_3)\cos 60^\circ = r_1 \text{ i.e.: } r_2 + r_3 = 2r_1.$$

Thus, $2\pi n_1 d_1/\lambda_1 = 2\pi/3 = 2\pi n_2 d_2/\lambda_1$, i.e.

$$\lambda_1 = 3n_1 d_1 = 3n_2 d_2 = 3\lambda_0/4. \qquad (9.8a)$$

Similarly, when the phase differences are 240°, (Fig 9.2(*b*) (iii)):

$$2\pi n_1 d_1/\lambda_2 = 2\pi n_2 d_2/\lambda_2 = 4\pi/3$$

Hence

$$\lambda_2 = 6n_1 d_1 = 6n_2 d_2 = 3\lambda_0/2. \qquad (9.8b)$$

The following materials have proved to be useful for double filters in the visible region: cryolite ($n = 1.33$), LiF (1.36), MgF_2 (1.38), Al_2O_3 (1.76) and AgCl (2.06).

9.2.2 Triple and multilayer films

Continuing with the same theoretical treatment, three reflectance zeros may be obtained by using three different layers (figure 9.3(*a*)); the relation between the refractive indices now becomes $n_0/n_1 = n_1/n_2 = n_2/n_3 = n_3/n_s$. Using the same argument as before, it may be shown that zero reflection will occur at the following wavelengths; λ_0, $2\lambda_0/3$ and $2\lambda_0$. The amplitude-phase diagram figure 9.3(*b*) illustrates how this may be achieved. An example of such an arrangement is Ge coated with quarter wave layers of Si, CeO and MgF; in the intra-red the transmittance of this device increases from 50% for pure Ge to almost 100% (Cox *et al* 1961).

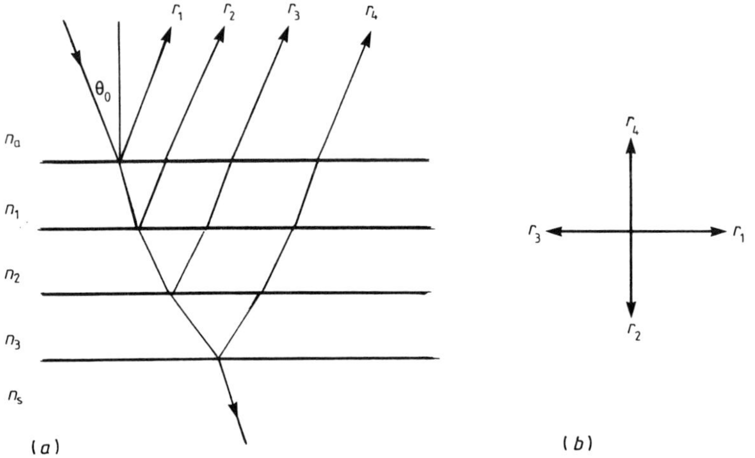

Figure 9.3 (*a*) Reflections of triple film. (*b*) Amplitude-phase diagram of reflected rays.

9.2.3 Variation with angle of incidence

In the above discussion it was assumed that the radiation was incident normally on all the interfaces. For the single layer coating the formulae are still valid up to angles of incidence of about 45°. However, in multilayer films, the tolerable angle of incidence decreases as the number of films increases. For example, with three layers, the angle will have been reduced to 25°. In addition, away from normal incidence, there will be different coefficients of reflection for the s- and p-states of polarisation; this effect, however, may be utilised in making devices for polarisation of radiation and will be discussed later (section 9.9).

9.3 High reflectance systems

There are some instances where it is desirable to produce the opposite effect to that discussed above i.e. a highly reflecting system, for example in Fabry–Perot interference etalons. Single metal surfaces, when highly polished, may have reflectances of greater than 90% in the visible and infra-red regions of the spectrum. Thus, thick films of aluminium or rhodium are often used for high reflectance systems as they keep their reflecting properties even when exposed to air for a long time. Such films, however, have high absorptances and, particularly if the films are being used as beam splitters, considerable radiation losses may occur through this cause.

Highly reflecting surfaces can also be created by using dielectric materials only and these have negligible absorptances in the visible region. The process is the inverse of that discussed above for low reflection coatings, the thicknesses and refractive indices of the film being adjusted to give the maximum reinforcement of the reflected beam. Usually the thickness of the applied film is once again $\lambda/4$ with the refractive index of the film now greater than that of the substrate so that the ray reflected from the film-substrate interface will then have suffered a total phase change of 2π (π at the interface plus π due to the double thickness of the film) and so will be back in phase with that reflected at the air–film interface which will have suffered no phase change.

The reflectance of a surface of refractive index n_2 covered by a film of index n_1 and thickness $\lambda/4$ is given by equation (9.5*a*) which indicates that, for maximum R, n_1 should be as large as possible compared with n_s. In the visible spectrum, materials such as ZnS and TiO_2 have values of n larger than 2, whilst in the infra-red, n is greater than 4 for Ge and Te. Films of these materials, deposited on glass, may be used to construct single layer reflectors having R equal to 0.4 in the visible and 0.8 in the infra-red.

9.3.1 Multiple layer reflectors

These devices employ a stack of films each of optical thickness $\lambda/4$ with alternate layers of high and low index (n_h and n_l), for example ZnS and MgF_2; an odd number of layers is always needed. The reflected rays from all the interfaces are thus in phase on emerging from the uppermost layer. It may be shown that the optical admittance of a stack of $2p+1$ layers on a substrate of index n_s is given by $Y_s = (n_h/n_l)^{2p} * n_h^2/n_s$ and the expression for the reflectance becomes $R = ((1 - Y_s)/(1 + Y_s))^2$. If $Y_s \gg 1$, R approximates to $1 - 4/Y_s^2$ and T, which is equal to $1 - R$, reduces to $4/Y_s^2$; thus, every extra double layer will reduce T by the ratio $(n_l/n_h)^2$.

Table 9.1 shows the values of the reflectances, R_1 and R_2 respectively, which may be obtained from stacks of two kinds, (*a*) ZnS + cryolite at 589 nm and $SbS_3 + CaF_2$ at 1.0 μm.

The variation with wavelength of the reflectance and transmittance of such stacks are complicated functions and are not easy to analyse; reference may be made to Frau (1952) who has derived an expression linking the transmittance with the wavelength and hence the variation of reflectance may be deduced.

However, there are problems in using the method described above. The changes in phase on reflection at the various interfaces are not constant with wavelength and this leads to high reflectance being only obtained over a limited range of wavelengths. Epstein (1955) and Heavens and Liddell (1966) have discussed this problem.

Table 9.1 Reflectance of ZnS/cryolite (R_1) and SbS_3/CaF_2 (R_2) stacks at 1 μm.

No. of layers	R_1	R_2
1	0.40	0.43
3	0.70	0.84
5	0.89	0.90
7	0.96	0.98
9	0.99	0.995
11	0.998	0.999

9.3.2 Films of metal and dielectric

Figure 9.4 shows a high reflectance device in which a metal surface is first coated with a dielectric layer and then a semi-transparent metal layer is deposited on top; the thicknesses of these layers can be adjusted so that all the reflected rays are in phase. The performance can be improved still further by depositing additional layers of dielectric and metal; for example, Al coated with layers of MgF_2 and Al can attain a reflectance of 97%. However, there is a price to pay; outside the intended wavelength region, the reflectance will be lower than for uncoated Al. In fact, by appropriate adjustment of the thicknesses of the semi-transparent layers, R can be made equal to zero at certain wavelengths.

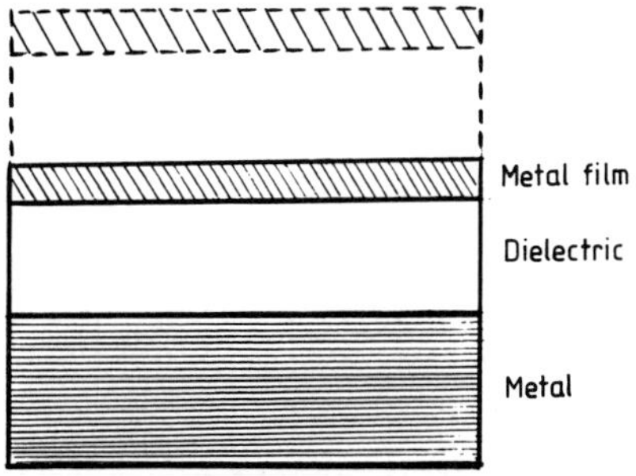

Figure 9.4 High reflectance metal–dielectric device.

9.4 Beam splitters

These devices are used to divide a beam into two parts, one reflected and one transmitted; although in many cases the beams may be divided equally in intensity, it is possible to design the device to obtain any proportions. A single thin metal foil can be used for this purpose but its disadvantage is again the large absorption in the metal. Nevertheless, films of aluminium, platinum, chromium or rhodium have been used in the visible spectrum whilst gold has proved useful in the infra-red. In the ultra-violet, aluminium may again be used provided it is evaporated very quickly (40 nm s^{-1}) to reduce oxidation and then coated with either MgF_2 or LiF to inhibit tarnishing in the atmosphere. Neutral beam splitters have constant R and T over a range of wavelengths. Beam splitters have been reviewed by Madden (1963) and by Hass and Hunter (1978).

The performance of metallic beam splitters can be improved by adding a quarter wave dielectric layer, e.g. ZnS ($n = 2.4$), between the metallic layer and the substrate. This forms an antireflection layer on the back surface of the metal thus reducing reflectance from the glass side. Shkliarevskii and Avdeenko (1959) obtained the same effect by using a layer of silver under the aluminium.

In the case of two-beam interferometers such as Michelson's, each beam passes through the splitter twice and it is therefore better to use dielectric films for this job. The effective transmission after two passes through the splitter is equal to $R \cdot T$ and the theoretical upper limit of this is 0.25; however the practical limit has proved to be closer to 0.10.

Beam splitters with 55/45 characteristics can be made by evaporating Ti and then oxidising it to the rutile form of TiO_2 whose refractive index is about 2.8, by heating in air to 420° C (Holland *et al* 1953). As in the case of antireflection coatings, better results can be obtained in beam splitters by using several dielectric layers; in particular, the effective wavelength range of the device is increased. For example, a multilayer splitter formed

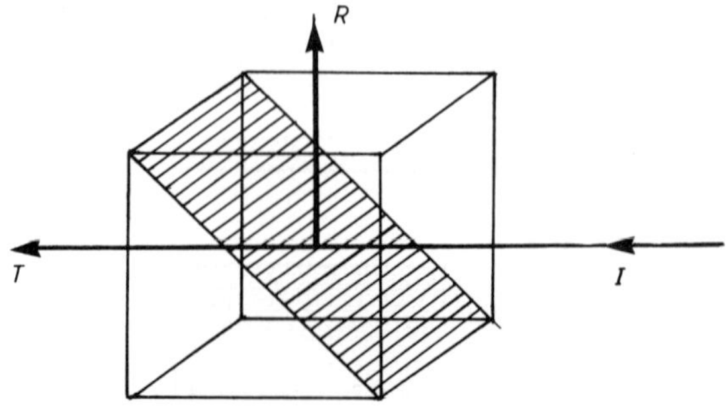

Figure 9.5 Beam splitting cube.

by air | LHLH | glass can be greatly improved by adding a double ($\lambda/2$) layer of low index material between the final H and the glass (Knittl (1976)). One of the main problems encountered with dielectric beam splitters is the low value of R_p which is obtained particularly at large angles of incidence. However, this effect can be turned to advantage in polarising devices which will tend to reflect mainly the R_s component and so transmit the T_p element.

A beam splitting cube is illustrated in figure 9.5; it is necessary to use a metallic layer in order to keep R_s and R_p roughly equal.

9.5 Polarisers using thin films

These devices utilise the principle of Brewster's angle reflections to enhance one state of polarisation whilst suppressing the other; in addition, large aperture devices may readily be produced. The method of operation is basically the same as that of the classical pile of plates (section 2.8.1) with thin films replacing the thick glass plates. The condition for the reflectance of a film to be zero can be obtained from equation (9.1). Thus, r_1 must be

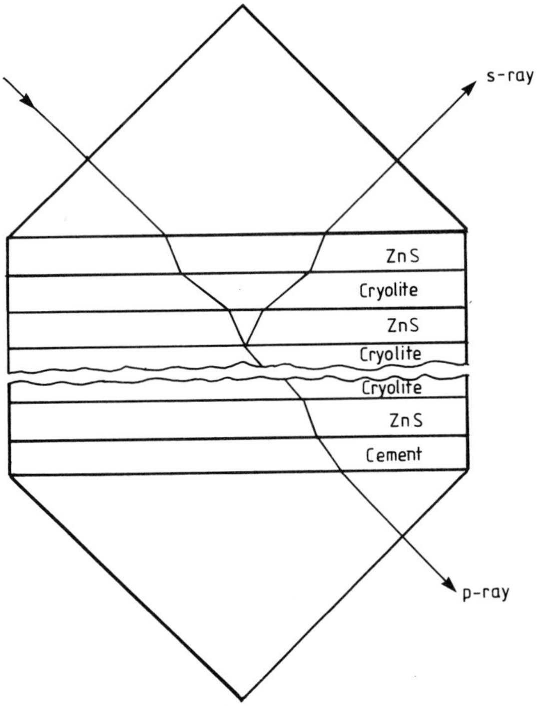

Figure 9.6 Polarising beam splitter.

equal to r_2 and $\exp(-2i\delta)$ must be equal to -1; the latter condition is obtained when $2d = (2m+1)\pi = 4\pi/\lambda nd \cos\theta$. So

$$(2m+1) = 4n_1 \cos\theta d_1/\lambda. \quad (9.9)$$

The condition for R_p to be zero can then be written in terms of d_1 and θ. At the same time, it is required that R_s should be a maximum which can be achieved by means of a multilayer stack of quarter wave layers of ZnS and cryolite.

9.5.1 Polarising beam splitters

A device using this principle to split a beam into two parts, polarised in the s and p directions, is illustrated in figure 9.6. It consists of an odd number of quarter wave layers of ZnS and cryolite deposited on the hypotenuse face of a 45° glass prism and then cemented to a similar prism. When light is incident normally on one face of the top prism it reaches the first ZnS-cryolite interface at the Brewster's angle so that the p-ray is transmitted but not reflected. The s-ray is both reflected and transmitted; however, the reflectance for the s-ray may be increased to almost 100% by the multilayer stack.

9.6 Neutral density filters

These are single metallic layers employed to reduce the intensity of a beam of radiation by a known amount. Films of nichrome (80% Ni, 20% Cr) have been used effectively for this purpose; these are easy to prepare by evaporation, are stable and do not need protection from the atmosphere.

9.7 Interference filters

A basic type of interference filter (Fabry–Perot) is shown in figure 9.7 and consists of two highly reflecting surfaces separated by a dielectric spacer. When it is illuminated by dispersed white light, multiple reflections occur in the dielectric and a set of narrow transmission (interference) bands is obtained, the separation of the bands being determined by the thickness and refractive index of the spacer. Thus, the centres of the transmission bands occur at wavelengths given by the formula;

$$2t \cos\theta = m\lambda \quad (9.10)$$

where t is the equivalent optical thickness of the spacer including any phase changes taking place at reflections and m is an integer. For normal incidence, the transmission bands will occur at the following wavelengths: $2t$, $2t/2$, $2t/3$, $2t/4$ etc; thus, a spacer of equivalent thickness 500 nm will give transmission bands at 1000, 500, 333, 250, 200 nm etc. Of these, only that at 500 nm occurs in the visible spectrum, so this filter could be used to isolate this wavelength if the other transmission bands could be suppressed or ignored.

Once again, it is possible to construct multiple layer metal and dielectric interference filters in which the pass bands are narrower, but they become increasingly less attractive because of their high absorptance.

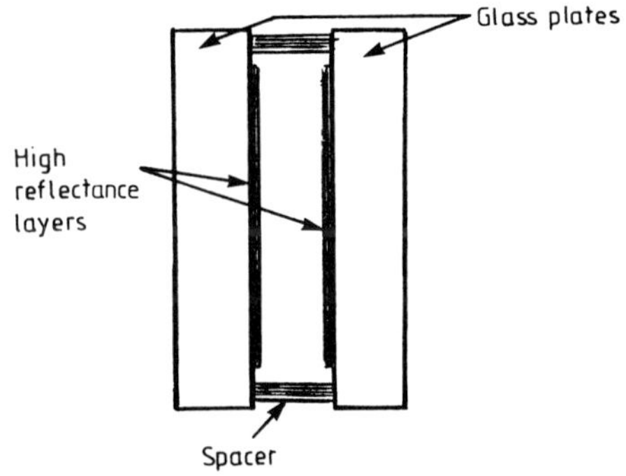

Figure 9.7 Fabry–Perot interference filter.

9.7.1 Infra-red interference filters

The high refractive indices of Ge and Te layers in the infra-red region can be combined with the low index layers of MgF_2 or cryolite to produce interference filters. Heavens (1965, p. 230) has published a table which shows the effect on reflectance and bandwidth of different numbers of layers of Te and cryolite.

9.7.2 Edge filters

These filters have an abrupt wavelength edge between their reflecting and transmitting properties. One type which is employed very frequently uses the natural absorption of the filter material and works best when the filter material has a sharp absorption edge for radiation. Examples of this form of filter are Ge and PbS which give cut off at wavelengths of 1.6 and 3.4 μm respectively.

9.7.3 Multilayer interference filters

These filters consist of a stack of quarter wave dielectric layers; a typical characteristic is illustrated in figure 9.8 and shows a number of zones of high T (low R) separated by those of low T (high R). Overall, this filter would have a long wave pass band above 6.0 μm, a short wave pass-band below 3.0 μm and a reflection (stop) band between 3.0 and 6.0 μm. The other notable features observable in figure 9.8 are the ripples within the two pass bands; clearly these are not desirable.

The width of the stop zone in this instance could be increased by combining the dielectric filter with one of the absorption type which had a cut off between 3.0 and 5.0 μm so that the short wavelength radiation would be absorbed; in fact, the absorption filter could be used as the substrate for the dielectric stack.

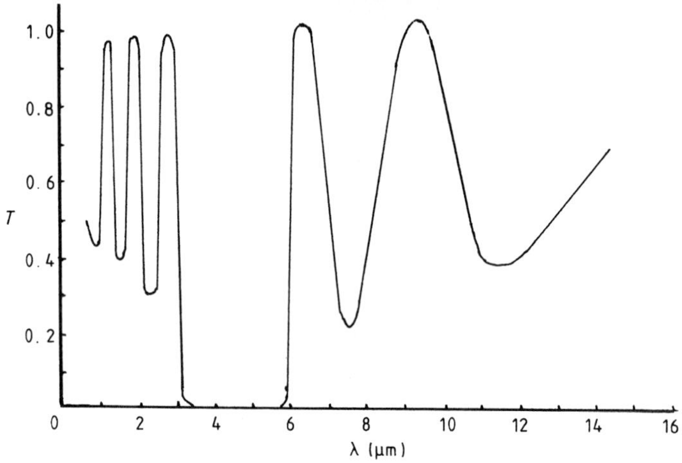

Figure 9.8 Multilayer filter characteristic.

In order to consider how the ripple might be reduced it is first necessary to analyse the operation of the filter in more detail. If we can reduce the filter to a series of identical symmetrical elements, it may be shown that the behaviour of the whole filter can be expressed in terms of the optical properties of one of these elements. This treatment was first given by Epstein (1952).

First of all, let us consider a simple three-layer filter, aba (see figure 9.9), and write down its characteristic matrix (see section 5.6,):

$$\begin{vmatrix} M_{11} & M_{12} \\ M_{21} & M_{22} \end{vmatrix} = \begin{vmatrix} \cos\delta_a & i/n_a \sin\delta_a \\ in_a \sin\delta_a & \delta_a \end{vmatrix} \begin{vmatrix} \cos\delta_b & i/n_b \sin\delta_b \\ in_b \sin\delta_b & \cos\delta_b \end{vmatrix} \begin{vmatrix} \cos\delta_a & i/n_a \sin\delta_a \\ in_a \sin\delta_a & \cos\delta_a \end{vmatrix} \qquad (9.11)$$

where n_a, n_b are the refractive indices and δ_a, δ_b are the phase differences in the various layers.

By multiplying out the matrix, expressions for M_{11}, M_{12}, M_{21} and M_{22}

may be obtained. Thus;

$$M_{11} = M_{22} = \cos 2\delta_a \cos \delta_b - \tfrac{1}{2}(A)\sin 2\delta_a \sin \delta_b \tag{9.12a}$$

$$M_{12} = [\sin 2\delta_a \cos \delta_b + \tfrac{1}{2}(A)\cos 2\delta_a \sin \delta_b + \tfrac{1}{2}(B)\sin \delta_b]\mathrm{i}/n_a \tag{9.12b}$$

$$M_{21} = \mathrm{i}n_a[\sin 2\delta_a \cos \delta_b + \tfrac{1}{2}(A)\cos 2\delta_a \sin \delta_b - \tfrac{1}{2}(B)\sin \delta_b] \tag{9.12c}$$

where $A, B = (n_a/n_b \pm n_b/n_a)$.

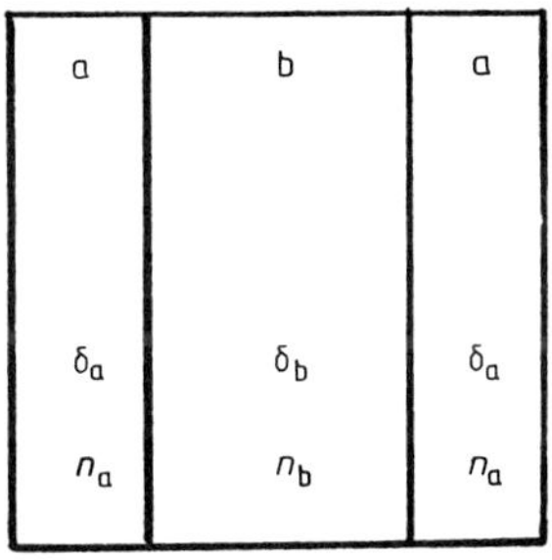

Figure 9.9 Model three-layer filter.

Adopting the same form of construction as that of the characteristic matrices, $M_{11} = M_{22}$ may be put equal to a quantity, $\cos y$, and likewise $M_{12} = \mathrm{i} \sin y/Y$ where Y is the admittance of the element and y is the phase difference across it: also, $M_{21} = \mathrm{i}Y\sin y$. Thus;

$$Y = \sqrt{M_{21}/M_{12}} \tag{9.13}$$

The matrix of the whole element now becomes:

$$\begin{vmatrix} \cos y & \mathrm{i}/Y \sin y \\ \mathrm{i}Y \sin y & \cos y \end{vmatrix}. \tag{9.14}$$

The element aba thus has the same optical properties as a single layer of phase thickness y and admittance Y; Y is sometimes called the equivalent optical admittance or the Herpin index (1947). This treatment may be applied to any symmetrical filter so that, if the stack is made up of S identical elements, the multilayer has an equivalent phase thickness of Sy and an admittance of Y. These conclusions may be demonstrated by evaluating;

$$\begin{vmatrix} \cos y & \mathrm{i}/Y \sin y \\ \mathrm{i}Y \sin y & \cos y \end{vmatrix}^S \text{ which is equal to } \begin{vmatrix} \cos Sy & \mathrm{i}/Y \sin Sy \\ \mathrm{i}Y \sin Sy & \cos Sy \end{vmatrix}. \tag{9.15}$$

When this analysis is applied to an actual filter, regions will be found where $M_{11} < -1$, i.e. $\cos y < -1$; these correspond to the stop band where both y and Y are imaginary and low transmittance, high reflectance is obtained. Outside the stop band, both y and Y become real; $\cos y$ is then greater than -1 and this corresponds to transmission. The edges between the pass and stop bands can thus be determined by the condition that: $M_{11} = \cos y = -1$.

Let us now apply these results to quarter-wave stacks. By adding a $\lambda/8$ layer at each end we can make the following two configurations;

$$\text{H/2LHLHLHLH.......HLH/2} \qquad \text{(a)}$$

$$\text{L/2HLHLHLHL.......LHL/2} \qquad \text{(b)}$$

(a) can be rewritten H/2LH/2 | H/2LH/2 | etc, i.e. | H/2LH/2 | ^{S} whilst (b) becomes L/2HL/2 | L/2HL/2 | ^{S}.

In both cases, the edges of the high reflection zone will occur when $M_{11} = -1$; in addition, δ_b is now equal to $2\delta_a$ and equation (9.11a) reduces to:

$$\cos 2\delta_b - \tfrac{1}{2}(n_b/n_a + n_a/n_b)\sin 2\delta_b = -1. \qquad (9.16)$$

This may be rearranged to give:

$$((n_b - n_a)/(n_b + n_a))^2 = \cos^2 \delta_b. \qquad (9.17)$$

The phase difference, δ_b, may also be written in terms of (*a*) the wavelength λ_0 for which the 'b' layer has an optical thickness of a quarter wavelength and (*b*) the actual wavelength, λ; thus, $\delta_b = (\lambda_0/\lambda)\pi/2$ or, in terms of frequency, $\delta_b = (f/f_0)\pi/2$.

The edges of the high reflection zone are then given by:

$$\delta_b = (f_0 \pm \Delta f)\pi/2f_0 = (1 \pm \Delta f/f_0)\pi/2. \qquad (9.18)$$

$$\text{Hence, } \cos \delta_b = \cos(\pi/2 \pm \pi/2\, \Delta f/f_0) \qquad (9.19)$$

$$= \pm\sin(\Delta f/f_0) = (n_b - n_a)/(n_b + n_a). \qquad (9.20)$$

Thus, the half-width of the high reflection zone is

$$\Delta f = (2/\pi)f_0 \sin^{-1} (n_b - n_a)/(n_b + n_a). \qquad (9.21)$$

It will be seen that this depends only on the values of n_a and n_b.

Thus, for a filter consisting of layers of ZnS ($n = 2.35$) and cryolite ($n = 1.35$), $\cos \delta_b = \pm\ 1/3.70$, i.e. δ_b is equal to either 0.413π or 0.587π. This leads to values for f (the edges of the stop band) of $0.826f_0$ and $1.174f_0$.

High reflection zones are also obtained for all wavelengths which are odd multiples of $\lambda/4$. So

$$(2p + 1)\lambda/4 = \lambda_0/4 \qquad \text{i.e.} \qquad \lambda = \lambda_0/3,\ \lambda_0/5,\ \lambda_0/7 \text{ etc.} \qquad (9.22)$$

The boundaries of the high reflection zone are: $f_0 \pm \Delta f$, $3f_0 \pm \Delta f$, $5f_0 \pm \Delta f$, etc. For the case of the same ZnS and cryolite filter, $\Delta f/f_0 = \pm\,(2/\pi)\sin^{-1}(1/3.7) = \pm\ 0.172$ which confirms the result given above.

The values of the admittance of the filter in the pass band may be calculated by computer using equation (9.13); however, when $\delta_b = 2\ \delta_a$, this equation can be treated analytically and it may readily be shown to reduce to:

$$Y = n_a \left[\frac{(1 + \frac{1}{2}(n_a/n_b + n_b/n_a) - (n_a/n_b - n_b/n_a)/2\cos\delta_b)}{(1 + \frac{1}{2}(n_a/n_b + n_b/n_a) + (n_a/n_b - n_b/n_a)/2\cos\delta_b)} \right]^{0.5} \tag{9.23}$$

Consider two extreme cases, (*a*) when $\delta_b = 0, 2\pi, 4\pi$ etc, $\cos\delta_b = +1$ and $Y = \sqrt{n_p n_q}$ and (*b*) when $\delta_b = \pi$, 3π, 5π, etc, $\cos\delta_b = -1$ and $Y = n_p(n_p/n_q)^{0.5}$. For the H/2LH/2 filter considered above, using cryolite and ZnS, for case (*a*) $f = 0$, $4f_0$, $8f_0$, etc and $Y = 1.78$ and for case (*b*) $f = 2f_0$, $6f_0$, $10f_0$, etc and $Y = 3.10$. The corresponding values for the L/2HL/2 filter are 1.78 and 1.02. The complete curves of Y versus f for both cases are shown in figure 9.10.

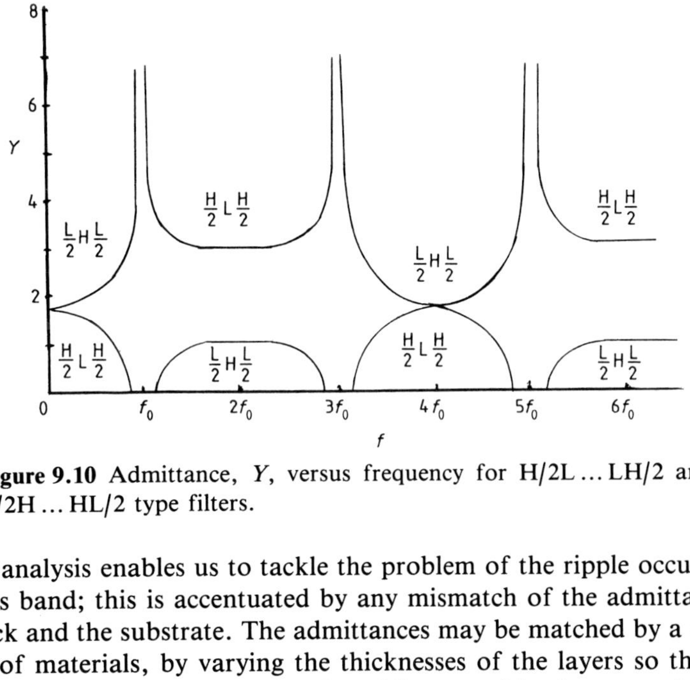

Figure 9.10 Admittance, Y, versus frequency for H/2L ... LH/2 and L/2H ... HL/2 type filters.

This analysis enables us to tackle the problem of the ripple occurring in the pass band; this is accentuated by any mismatch of the admittances of the stack and the substrate. The admittances may be matched by a suitable choice of materials, by varying the thicknesses of the layers so that their equivalent admittance is altered or by adding matching layers at either end of the filter; on the substrate side a $\lambda/4$ layer of admittance n_s may be added or a $\lambda/4$ layer of admittance n_l on the air side.

9.7.4 Bandpass filters

These types of filters may be divided into two further categories—broad and narrow band. A broad band filter with a bandwidth of 20% (i.e. $d\lambda/\lambda$ at half the height of the absorption band = 0.2) may be made by a combination of long and short wavelength filters, one of each being deposited on either side of the substrate. As above, a suitable long wavelength pass filter could comprise alternate layers of ZnS and cryolite in the $(H/2LH/2)^S$ pattern and at normal incidence would have an admittance of less than 2.0 at low frequencies. A short wave filter would be of the $(L/2HL/2)^S$ pattern; its admittance at high frequency would be about 1.0.

For narrowband filters, a typical bandwidth would be less than 15%; it could be made up of a Fabry–Perot etalon which produces narrow interference fringes together with an absorption filter. Instead of an air spacer in the etalon, a solid dielectric may be used. In order to cut out the unwanted interference fringes in the visible spectrum, an absorbing filter with a cut-off in the near infra red can be employed. A typical characteristic is shown in figure 9.11.

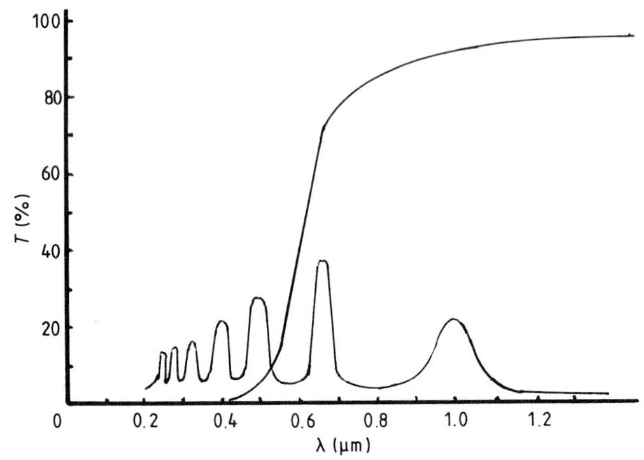

Figure 9.11 Narrowband filter characteristic.

A better performance can be obtained by replacing the metal layers in the Fabry–Perot etalon by stacks of high and low index dielectric layers. The centre layer, formed of low index material (e.g cryolite) is made half a wavelength thick and is thus double the thickness of the other layers in the stack; the high index material may be ZnS.

It is useful to consider two cases: (i) with an even number of layers(ii) with an odd number of layers, in each case built on a low index substrate.

The matrix for a $\lambda/4$ plate is given by equation (5.56), i.e.

$$\begin{vmatrix} \cos\delta & i(\sin\delta)/n \\ in\sin\delta & \cos\delta \end{vmatrix}$$

If the layers are almost $\lambda/4$ in thickness, $\delta \doteqdot \pi/2 + e$, so that $\cos\delta \doteqdot e$ and

$\sin\delta \doteqdot 1$. The matrix then becomes

$$\begin{vmatrix} -e & i/n \\ in & -e \end{vmatrix}. \tag{9.24}$$

The matrix for the low index layer, $|L|$, is obtained by substituting n_l in equation (9.24) and that for the high index layer, $|H|$, by substituting n_h.

With an even number of layers i.e $2x$, the full matrix for the filter is represented by

$$\begin{vmatrix} B \\ C \end{vmatrix} = |L||H||L| \ldots\ldots\ldots |L||H| \begin{vmatrix} 1 \\ n_s \end{vmatrix} \tag{9.25}$$

$$= \{\,|L|H|\,\}\times \begin{vmatrix} 1 \\ n_s \end{vmatrix}. \tag{9.26}$$

For an odd number of layers, i.e. $2x+1$, the matrix becomes:

$$\begin{vmatrix} D \\ E \end{vmatrix} = |H||L||H| \ldots\ldots\ldots |L||H| \begin{vmatrix} 1 \\ n_s \end{vmatrix} \tag{9.27}$$

$$= |H|\,\{\,||L||H|\,\}\times \begin{vmatrix} 1 \\ n_s \end{vmatrix}. \tag{9.28}$$

This gives the same result as that for the even number of layers but multiplied by the matrix for one high index layer.

A disadvantage of dielectric multilayer reflectors is that the high reflectance region, i.e. the rejection zone, is limited in its wavelength coverage and, far from the operating wavelength, transmission sidebands occur and must be suppressed. Those on the low wavelength side of the peak can be removed by a straightforward absorption filter if they occur in the near ultra-violet, visible or near infra-red regions. It is more difficult to remove the sidebands on the long wavelength side, but a metal–dielectric Fabry–Perot type filter can be added, designed so that its first order transmission peak coincides with that of the filter. The higher order transmission bands will only occur at shorter wavelengths (see equation (9.10)) and so will be removed by the absorption filter. As an alternative, a double cavity metal–dielectric filter may be used as described in section 9.7.6.

9.7.5 Dichroic beam splitters

The function of these devices is to split a white light beam into two or three beams each of a different colour; one use is in colour television where separate red, green and blue beams are required. The splitting can be achieved by passing the light through two filters, one after the other as illustrated in figure 9.12 which uses two long-wavelength pass filters each inclined at 45°. At the first filter, the short wavelength blue light is rejected (i.e. reflected) and the complementary colour then passes to the second filter where the

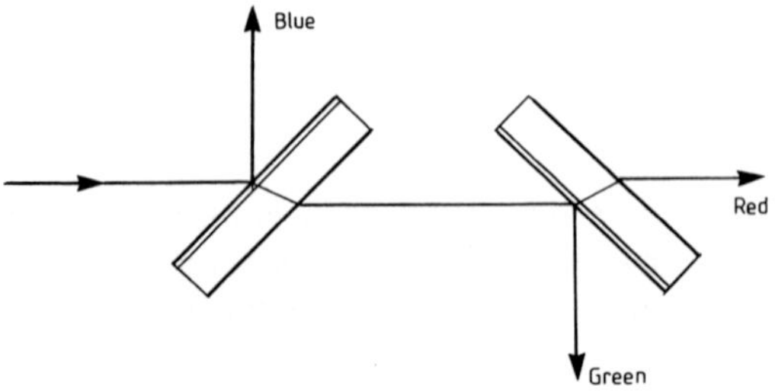

Figure 9.12 Dichroic beam splitter using long wavelength pass filters.

shorter wavelength green light is reflected and the long wavelength red light transmitted. Dichroic splitting can also be effected by means of two short wavelength pass filters, the first reflecting the red and the second the green; the blue light then passes directly through the whole system.

9.7.6 Multiple cavity filters

These devices consist basically of two Fabry–Perot etalons in series with a $\lambda/4$ coupling layer between them as illustrated in figure 9.13. The theory of these filters has been given by Smith (1958). It emerges that such filters behave similarly to two tuned electrical circuits in series and give a much sharper and squarer transmission peak than those devices discussed earlier. This is clearly a desirable feature for an optical filter.

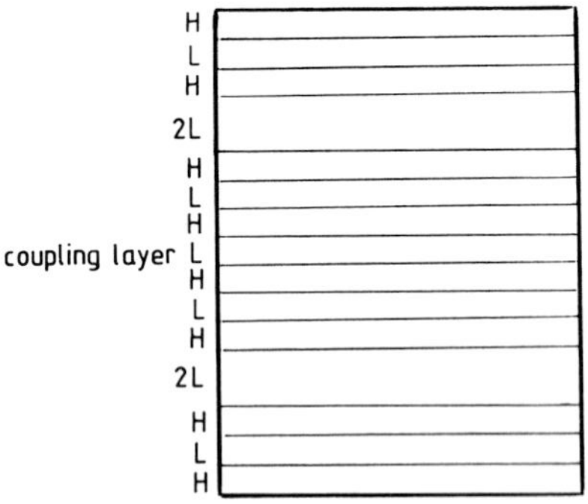

Figure 9.13 Multicavity filter.

9.8 Rugate filters

In this type of thin film filter, the layers are made up of materials whose refractive index, n, varies continuously in the direction at right angles to the layer surface. Baumeister (1986) has described how a sinusoidal index profile may be obtained by the evaporation of many sub-layers each containing a mixture of two dielectrics, one of high and one of low index. Provided the thickness of each sub-layer is much less than the wavelength of the radiation, the device can be treated theoretically by the method of characteristic matrices using macroscopic optical parameters (Jacobsson 1975, Bovard 1988). For certain wavelengths, Bragg-type reflections occur from structures of this kind and the device behaves as a narrow band reflection filter.

Calculating the optical characteristics by the matrix method gives satisfactory results but is very tedious as there may be 100 layers in each filter. Southwell (1988) has proposed using complex wave theory for this purpose whilst Dobrowolski and Lowe (1978) and Bovard (1990) used a Fourier transform method; both methods are simpler to use than the matrix technique.

Now that techniques are available for the precise deposition of very thin epitaxial layers (see section 5.4) the practical realisation of rugate filters can readily be achieved. The filter is constructed as a train of many layers in each of which the refractive index is made to vary with a sinusoidal profile by using appropriate mixtures of two components of refractive indices n_h and n_l; a typical thickness of each complete layer is 0 25 μm for a filter with a stop band centred at 1.0 μm. The bandwidth of the stop filter produced is given by the relation: $\Delta\lambda/\lambda = (n_h - n_l)/(n_h + n_l)$.

One of the problems associated with rugate filters, as with all the filters dealt with earlier in this chapter, is the generation of multiple reflecting side bands on each side of the main reflection zone. Southwell (1989) has given details of how these may be suppressed in rugate filters by (*a*) apodisation, i.e. producing an amplitude modulation (linear, Gaussian or quintic) of the sinusoidal index profile and (*b*) adding gradient index matching regions to each side of the filter.

In practical terms, rugate filters may be employed for many purposes, for example as antireflection coatings when non-absorbing materials are used and for sunglasses which make use of absorbing materials.

References

Baumeister P 1986 *Appl. Opt.* **25** 2644
Berning P H 1963 *Phys. Thin Films* **1** 69
Bovard B G 1988 *Appl. Opt.* **27** 1998
—— 1990 *Appl. Opt.* **29** 24
Cox J T and Hass G 1958 *J. Opt. Soc. Am.* **48** 677
Cox J T, Hass G and Jacobus G F 1961 *J. Opt. Soc. Am.* **51** 714
Dobrowolski J A and Lowe D 1978 *Appl. Opt.* **17** 3039
Jacobsson R 1975 *Phys. Thin Films* **51** 8
Epstein L I 1952 *J. Opt. Soc. Am.* **42** 806
—— 1955 *J. Opt. Soc. Am.* **45** 360

Frau D C 1952 *Rev. Opt.* **31** 161
Handbook of Optics 1978 Optical Society of America (New York: McGraw Hill)
Hass G, Heaney J B and Hunter W R 1982 *Phys. Thin Films* **12** 1
Hass G and Hunter W R 1978 *Phys. Thin Films* **10** 71
Heavens O S 1960 *Rep. Prog. Phys.* **23** 1
—— 1965 *Optical Properties of Thin Films* (New York: Dover)
Heavens O S and Liddell H M 1966 *Appl. Opt.* **5** 373
Herpin A 1947 *C. R. Acad. Sci. Paris* **225** 182
Holland L, Hacking K and Putner T 1953 *Vacuum* **3** 159
Knittl Z 1976 *Optics of Thin Films* (London: Wiley)
Lissberger P H 1970 *Rep. Prog. Phys.* **33** 197
Macleod H A 1986 *Thin Film Optical Filters*. 2nd edn (Bristol: Hilger)
Madden R P 1963 *Phys. Thin Films* **1** 123
Ritter E 1975 *Phys. Thin Films* **1** 8
Schwartz C and de Sandro L F 1987 *Appl. Opt.* **26** 3140
Shklyerewskii I N and Avdeenko A A 1959 *Opt. Spectrosc.* **6** 439
Smith S D 1958 *J. Opt. Soc. Am.* **48** 43
Southwell W H 1988 *J. Opt. Soc. Am.* A **5** 1558
—— 1989 *Appl. Opt.* **28** 5091
Thalen A 1969 *Phys. Thin Films* **5** 47
Vossen J L 1977 *Phys. Thin Films* **9** 74

Appendix. Values of optical constants of various materials at two laser wavelengths (0.488 and 0.633 μm)

The values of n, k, e_1 and e_2 given in this table have been obtained from many sources. Particularly useful sets of data are contained in;

(*a*) Weaver J H, Krafka C, Lynch D W and Koch E E 1981 *Physik Daten* (Karlsruhe: Fachsinformationszentrum)

(*b*) *Handbook of Optical Constants of Solids I* 1985 ed E D Palik (Boston: Academic)

(*c*) *Handbook of Optical Constants of Solids II* 1991 ed E D Palik (Boston: Academic)

(*d*) Landolt-Bornstein 1985 *New Series III* vol 15b (Berlin: Springer)

Subscripts: o denotes ordinary ray, e extraordinary ray.

	$\lambda_1 = 0.488$ μm,		$E = 2.55$ eV		$\lambda_2 = 0.633$ μm,		$E = 1.95$ eV		
Material	n	k	e_1	e_2	n	k	e_1	e_2	Ref
Alkali metals									
Cs	0.315	0.68	−0.36	0.43	0.242	1.18	−1.33	0.57	70
K	0.045	1.21	−1.46	0.11	0.051	1.73	−2.99	1.18	1
Li	0.217	2.11	−4.41	0.92	0.221	2.94	−8.59	1.30	2
Na	0.065	1.97	−3.88	0.26	0.051	2.62	−6.86	0.27	1
Rb	0.107	1.01	−1.00	0.22	0.080	1.53	−2.32	0.24	70

(*continued*)

Table continued

	$\lambda_1 = 0.488\ \mu$m,	$E = 2.55$ eV			$\lambda_2 = 0.633\ \mu$m,	$E = 1.95$ eV			
Material	n	k	e_1	e_2	n	k	e_1	e_2	Ref
Alkaline earths									
Ba	0.97	2.00	−3.06	3.88	1.09	2.48	−4.96	5.41	4
Be	3.22	3.17	0.32	20.41	3.47	3.23	1.61	22.42	3
Ca	0.28	1.97	−3.80	1.11	0.30	2.84	−8.00	1.69	34
Mg	0.28	4.70	−22.01	2.63	0.38	5.74	−33.00	4.41	5
Sr	0.60	2.52	−10.00	4.00	0.62	3.22	−6.00	3.00	28
Transition elements									
Ag	0.24	3.01	−9.07	1.43	0.27	4.18	−17.40	2.26	5
Ag	0.13	2.80	−7.82	0.73	0.14	4.02	−16.14	1.09	6
Au	1.03	1.51	−1.22	3.13	0.12	3.32	−11.06	0.80	7
Au	1.08	1.82	−2.15	3.93	0.18	3.05	−9.27	1.10	8
Co_e	1.85	3.36	−7.87	12.43	2.26	4.09	−11.62	18.47	9
Co_o	1.88	3.30	−7.36	12.41	2.34	4.18	−12.02	19.52	9
Cr	2.48	4.42	−13.39	21.22	3.56	4.36	−6.26	31.13	9
Cr	2.75	3.30	−3.28	18.15	3.20	3.30	−0.65	21.12	25
Cu	1.14	2.53	−5.09	5.78	0.24	3.46	−11.90	1.67	5
Fe	2.87	3.37	−3.12	19.34	2.40	3.31	−5.20	15.89	10
Ir	1.95	3.93	−11.64	15.33	2.54	4.63	−14.98	23.52	14
Hf_e	2.91	3.13	−1.36	18.24	3.64	2.63	6.37	19.14	18
Hf_o	2.60	3.23	−3.72	16.79	3.58	3.20	2.55	22.92	18
Mn	2.53	3.60	−6.56	18.22	2.23	3.11	−4.70	13.87	25
Mo	3.29	3.67	−2.64	24.15	3.71	3.55	1.16	26.34	19
Nb	2.83	2.92	−0.52	16.53	2.82	2.86	−0.23	16.13	7
Nb	2.02	3.04	−5.16	12.30	2.27	3.30	−5.74	15.00	68
Ni	1.67	2.87	−5.45	9.59	1.98	3.74	−10.07	14.81	11
Ni	1.50	3.20	−7.99	9.60	1.75	3.80	−11.38	13.30	69
Os	5.32	2.60	21.54	27.66	3.88	1.67	12.27	12.96	16
Pd	1.50	3.49	−9.93	10.47	1.72	4.32	−15.70	14.86	13
Pt	1.94	3.36	−7.53	13.04	2.34	4.17	−11.91	19.52	12
Pt	2.75	3.30	−3.33	18.15	2.20	3.30	−6.05	14.52	25
Re_e	2.99	2.74	1.18	16.68	2.94	2.96	−0.11	17.37	21
Re_o	3.96	2.78	8.01	22.02	3.59	2.46	6.82	17.68	21
Rh	1.86	4.60	−17.70	17.11	2.16	5.64	−27.14	24.36	15
Ru_e	2.69	4.33	−11.51	23.30	4.17	4.71	−4.80	39.28	21
Ru_o	3.21	4.72	−11.97	30.30	4.79	4.92	−1.26	47.13	21
Sc	1.12	2.69	−5.98	6.03	1.19	3.38	−10.01	8.04	43
Ta	2.77	1.99	3.71	11.02	1.70	2.12	−1.60	7.21	19
Ti	1.77	2.38	−2.53	8.43	2.16	2.93	−3.42	12.68	18
Ti	2.36	3.19	−4.61	15.06	2.67	3.72	−6.71	19.86	25

(*continued*)

Table continued

	$\lambda_1 = 0.488\ \mu m$,		$E = 2.55$ eV		$\lambda_2 = 0.633\ \mu m$,		$E = 1.95$ eV		
Material	n	k	e_1	e_2	n	k	e_1	e_2	Ref
Transition elements									
V	3.45	3.30	1.01	22.77	3.57	2.96	3.98	21.13	17
V	1.96	2.92	−4.67	11.43	2.35	2.83	−2.49	13.30	19
W	3.35	2.64	4.25	17.69	3.65	2.92	4.80	21.32	20
Zr_o	1.58	2.78	−5.23	8.81	2.18	3.42	−6.87	14.83	18
Other elements									
Hg	1.28	4.25	−16.42	10.88	1.93	5.20	−23.43	20.07	22
Zn	1.18	4.85	−22.13	11.45	3.54	5.70	−20.00	40.40	23
Cd_e	0.52	2.88	−8.02	3.00	1.44	3.47	−9.98	10.00	24
Cd_o	0.62	3.22	−9.98	3.99	1.34	4.10	−15.01	11.00	24
Eu	1.02	2.46	−5.01	5.02	1.15	3.05	−7.98	7.02	28
Gd	1.65	2.70	−4.57	8.91	2.15	3.12	−5.11	13.42	39
Tb	1.35	1.88	−1.71	5.08	1.72	2.46	−3.09	8.46	39
Yb	1.12	2.69	−5.98	6.03	1.19	3.38	−10.00	8.04	28
Bi	1.5	2.6	−4.5	7.8	1.6	3.0	−6.4	9.6	26
Sb	2.5	4.0	−10.0	20.0	2.75	4.0	−8.0	22.0	27
Se_o	2.98	0.55	8.58	3.19	3.36	0.60	10.93	4.03	31
Se_e	4.37	1.29	17.43	11.27	4.71	0.65	21.76	6.12	31
Te_o	3.76	3.76	0.00	28.28	6.05	2.55	30.10	30.86	30
Te_e	3.46	4.85	−11.55	33.56	6.35	4.45	20.52	56.52	30
Al	0.73	5.92	−34.48	8.61	1.39	7.65	−56.60	21.39	32,33
Al	0.77	5.80	−33.05	8.93	1.51	7.37	−52.03	22.26	34
In	1.00	2.00	−3.00	4.00	1.03	2.23	−3.90	4.60	35
Ga_o	1.15	4.28	−17.00	9.84	2.40	4.85	−17.76	23.28	29
Ga_e	0.84	3.96	−14.98	6.65	2.25	4.52	−15.37	20.34	29
Tl	1.08	2.32	−4.20	5.00	1.14	2.41	−4.50	5.50	35
Th	1.28	3.52	−10.76	8.99	1.71	3.42	−12.45	13.40	36
U	2.40	3.50	−6.51	16.85	2.95	4.30	−9.86	26.41	37
Pb	1.67	3.27	−7.90	10.92	1.95	3.48	−8.31	13.57	38
Semiconductors									
$Graphite_o$	2.65	1.31	5.31	6.94	2.88	1.75	5.23	10.08	40
$Graphite_e$	2.03	0.64	3.71	2.60	2.01	0.72	3.52	2.89	40
Ge	4.24	2.33	12.55	19.76	5.51	0.84	29.65	9.26	41
Si	4.37	0.08	19.09	0.70	3.88	0.20	15.01	1.55	42
$CdSe_e$	2.59	0.58	6.37	3.00	2.40	0.16	5.60	0.77	44,45
CdS_o	2.79	0.38	7.64	2.12	2.46	0.01	6.05	0.05	44,45

(continued)

Table continued

	$\lambda_1 = 0.488\ \mu m$,		$E = 2.55$ eV		$\lambda_2 = 0.633\ \mu m$,		$E = 1.95$ eV		
Material	n	k	e_1	e_2	n	k	e_1	e_2	Ref
Semiconductors									
$CdSe_e$	2.705	0.52	7.05	2.81	2.836	0.29	7.96	1.645	46
$CdSe_o$	2.793	0.57	7.48	3.18	2.740	0.275	7.43	1.51	46
CdTe	3.16	0.53	9.71	3.32	2.87	0.24	8.18	1.38	47
GaAs	4.41	0.486	19.21	4.29	3.85	0.192	14.79	1.48	48
GaP	3.64	0.09	13.31	0.65	3.31	small	10.96	—	41
GaSb	4.31	2.29	13.37	19.71	5.24	1.38	25.55	14.44	41
InAs	4.17	1.83	14.04	15.38	3.96	0.61	15.31	4.83	41
InSb	3.51	2.19	7.52	15.37	4.31	1.80	15.34	15.52	41
InP	3.85	0.54	14.53	4.16	3.53	0.31	12.39	2.16	41
PbS	4.35	2.00	14.92	17.40	4.28	1.46	16.19	12.50	49
PbSe	2.36	3.64	−7.68	17.68	3.65	2.90	4.91	21.17	50
PbTe	2.14	5.00	−20.42	21.40	6.30	4.00	23.69	50.40	51
ZnS_{cub}	2.43	—	5.90	—	2.35	—	5.52	—	52,53
ZnS_{hexo}	2.43	—	5.90	—	2.35	—	5.52	—	53
ZnS_{hexe}	2.43	—	5.90	—	2.36	—	5.57	—	53
ZnSe	2.44	0.50	5.93	2.44	2.45	0.02	6.00	0.10	54
ZnTe	3.25	0.09	10.55	0.59	2.98	0.07	8.88	0.42	55
Dielectrics									
Diamond	2.44	—	5.95	—	2.41	—	5.81	—	67
CaF_2	1.437	0.00	2.06	0.00	1.433	0.00	2.05	0.00	56
CsI	1.810	0.00	3.28	0.00	1.781	0.00	3.17	0.00	57
LiF	1.39	—	1.93	—	1.39	—	1.93	—	56
KCl	1.499	—	2.25	—	1.488	—	2.214	—	57
KBr	1.571		2.468		1.556		2.421		57
MgO	1.747		3.052		1.735		3.010		60
MgF_{2o}	1.382	0.00	1.909	0.00	1.377	0.00	1.896	0.00	59
MgF_{2e}	1.394	0.00	1.943	0.00	1.389	0.00	1.929	0.00	59
NaCl	1.553	—	2.41	—	1.542	—	1.542	—	57
NaF	1.328		1.764		1.325		1.776		58
SiO	2.04	0.05	4.16	0.20	1.96	0.01	3.84	0.04	61
SiO_{2o}	1.550	—	2.40	—	1.542	—	2.378	—	62,63
SiO_{2e}	1.559	—	2.43	—	1.551	—	2.406	—	62,63
SiO_{2glass}	1.463	—	2.14	—	1.457	—	2.129	—	64
TiO_{2e}	3.08	—	9.49	—	2.87	—	8.24	—	65
TiO_{2o}	2.75	—	7.56	—	2.58	—	6.66	—	65
Polythene	1.492				1.480				66

References

1. Na, K: Smith N V 1969 *Phys Rev.* **183** 634
2. Li: Rasigni M and Rasigni G 1977 *J. Opt. Soc. Am.* **67** 54
3. Be: Serynac A and Robin S 1976 *Solid State Commun.* **19** 343
4. Ba: Marfaing J 1980 *J. Physique* **41** 971
5. Ag, Cu, Mg: Hagemann, H J Gudet W and Kunz C 1975 *J. Opt. Soc. Am.* **65** 742
6. Ag: Winsemius, P van Kampen F F, Lengkeek H P and van Went C S 1976 *J. Phys. F: Met. Phys.* **6** 1583
7. Au, Nb: Weaver J H, Lynch D W and Olson C G 1973 *Phys. Rev.* **B** 7 4311
8. Au: Theye M L 1970 *Phys. Rev.* B **2** 3060
9. Co, Cr: Bos L W and Lynch D W 1970 *Phys. Rev.* B **2** 4567
10. Fe: Weaver J H, Colavita E, Lynch D W and Rosei R 1979 *Phys. Rev.* B**19** 3850 (also Co)
11. Ni: Lynch D W, Rosei R and Weaver J H 1973 *Solid State Commun.* **9** 2195
12. Pt: Weaver J H 1975 *Phys. Rev.* B 11 1416
13. Pd: Weaver J H and Benbow R L 1975 *Phys. Rev.* **B 12** 3509
14. Ir: Weaver J H, Olson C G and Lynch D W 1977 *Phys. Rev.* B **15** 4115
15. Rh: Weaver J H and Olson C G 1976 *Phys. Rev.* B **14** 3252
16. Os: Lynch D W, Olson C G and Weaver J H 1981 *Physik Daten* p 253
17. V: Nestell J E and Christy R W 1980 *Phys. Rev.* B **21** 3173
18. Ti, Zr, Hf: Lynch D W, Olson C G and Weaver J H 1975 *Phys. Rev.* B **11** 3617
19. Mo, Ta, V: Weaver J H, Lynch D W and Olson C G 1974 *Phys. Rev.* B **10** 501
20. W: Weaver J H, Olson C G and Lynch D W 1975 *Phys. Rev.* B **12** 1293
21. Ru, Re: Weaver J H, Krafka C, Olson C G and Koch E E 1981 *Physik Daten* p 153
22. Hg: Inagaki T, Arakawa E T and Williams M W 1981 *Phys. Rev.* B **23** 5246
23. Zn: Yaravaya R G, Shklyarevskii J N and El-Shazly A S A 1974 *Sov Phys.* **38** 331
24. Cd: Graves R H W and Lenham A P 1968 *J. Opt. Soc. Am.* **58** 126
25. Mn, Ti: Johnson P B and Christy R W 1974 *Phys. Rev.* B **9** 5052
26. Bi: Hodgson J N 1954 *Proc. Phys. Soc.* **B 67** 269
27. Sb: Shklyarevskii I N, Adveenko A A and Padulka V G 1959 *Opt. Spectrosc.* **6** 326
28. Eu, Sr, Yb: Endriz J C and Spicer W E 1970 *Phys. Rev.* B **2** 1466
29. Ga: Lenham A P 1963 *Proc. Phys. Soc.* **82** 933
30. Te: Grosse P and Richter W 1983 *Landolt–Bornstein Numerical Data and Functional Relationships in Science and Technology* vol 17. ed K-H Hellwege (Berlin: Springer)
31. Se: Titikasi S and Chen I 1967 *Phys. Rev.* **158** 623
32. Al: Shiles E, Sasoki T, Inokuti M and Smith D Y 1980 *Phys. Rev.* B **22** 1612
33. Al: Ditchburn R W and Freeman G H 1966 *Proc. R. Soc.* **294** 20
34. Al, Ca: Matthewson A G and Myers H P 1971 *Phys. Scr.* **4** 291
35. In, Tl: Bor J and Bartholemew C 1967 *Proc. Phys. Soc.* **90** 1153
36. Th: Weaver J H and Olson C G 1977 *Phys. Rev.* B **15** 4602
37. U: Faldt A and Nilsson 1980 *J. Phys. D: Appl. Phys.* **10** 2573

38. Pb: Golovashkin A I and Motulevich G P 1968 *Sov. Phys.–JETP* **26** 881
39. Gd, Tb: Miller R F, Julien L S and Taylor A J 1974 *J. Phys. F: Met Phys.* **4** 2338
40. Graphite: Bassini F and Tossati E 1968 *Phys. Lett.* **27A** 446
41. Ge, InAs, InSb, InP, GaP, GaAs: Aspnes D E and Studna A A 1983 *Phys. Rev.* B **27** 985
42. Si: Aspnes D E and Theeten J B 1980 *J. Electrochem. Soc.* **127** 1359
43. Sc: Sigrist M, Chassaing D, Francois J C, Antonangeli F, Dema N and Piacentrini M 1987 *Phys. Rev.* B **35** 3760
44. CdS: Debenham F and Tossati E 1968 *Phys. Lett.* **27A** 446
45. CdS: Cardona M and Harbeke G 1965 *Phys. Rev.* A **137** 1467
46. CdSe: Logothatides S, Cardona M, Leutenschlager P and Gerrigan M 1986 *Phys. Rev.* B **34** 2458
47. CdTe: Marple D T F and Ehrenreich H 1962 *Phys. Rev. Lett.* **8** 87
48. GaAs: Theeten J B, Aspnes D E and Chang R P H 1978 *J. Appl. Phys.* **49** 6097
49. PbS: Rossi C E and Paul W, 1967 *J. Appl. Phys.* **38** 1803
50. PbSe: Cardona M and Greenaway D L 1964 *Phys. Rev.* A **133** 1685
51. PbTe: Korn D M and Braunstein R 1972 *Phys. Rev.* B **5** 4837
52. ZnS: Pikhtin A N and Yas'hov A D 1978 *Sov. Phys. Semiconduct.* **12** 622
53. Zns: Bieniewski T M and Czyzek S J 1963 *J. Opt. Soc Am.* **53** 496
54. ZnSe: Li H H 1976 *J. Phys. Chem. Ref. Data* **5** 329
55. ZnTe, CaF2: Malitson I H 1961 *Appl. Opt.* **2** 1103
56. CaF_2, LiF: Laporte P and Subtil J L 1982 *J. Opt. Soc. Am.* **72** 1558
57. KCl, NaCl, KBr, CsI: Li H H 1976 *J. Phys. Chem. Ref. Data* **5** 329
58. NaF: Harting H 1948 *Sitzungsber Deutsch Akad. Wiss. Berlin* **4** 1
59. MgF_2: Dodge M J 1984 *Appl. Opt.* **23** 1980
60. MgO: Dodge M J 1986 *CRC Handbook of Laser Sci* vol iv, ed M J Weber (Roca Baton, FL: CRC Press Inc.)
61. SiO: Philipp H R 1971 *J. Phys. Chem. Solids* **32** 1935
62. SiO_2: Sosman R B 1927 *The Properties of Silica* (New York: Chem Catalog Co. (Tudor) p 591
63. SiO_2: 1972 *American Institute of Physics Handbook* 3rd ed., ch 6 (New York: McGraw Hill) p 27
64. Glass: Brixner B 1967 *J. Opt. Soc. Am.* **57** 674
65. TiO_2: Devore J R 1951 *J. Opt. Soc. Am.* **41** 416
66. Polythene: Panter L R, Arakawa E T, Williams M W and Ashley J C 1980 *Rad. Res.* **83** 1
67. Diamond: von Rosch S 1965 *Opt. Acta* **12** 253
68. Nb: Golovashkin A I, Laksin I E, Motulevich G P and Shukin A A 1969 *Sov. Phys.–JETP* **29** 27
69. Ni: Gushchin V S, Shvarew K M, Baum B A and Geled P V 1977 *Sov. Phys. Solid State* **20** 948
70. Cs, Rb: Monin J and Boutry G A 1974 *Phys. Rev.* B **9** 1309

Index